VIRTUAL LEGO®
The Official LDraw.org Guide to LDraw Tools for Windows

VIRTUAL LEGO®

The Official LDraw.org Guide to LDraw Tools for Windows

by Tim Courtney, Steve Bliss, and Ahui Herrera

NO STARCH PRESS

San Francisco

Printed on recycled paper in the United States of America

1 2 3 4 5 6 7 8 9 10 – 06 05 04 03

Publisher: William Pollock
Managing Editor: Karol Jurado
Cover and Interior Design: Octopod Studios
Copyeditor: Kenyon Brown
Compositor: Wedobooks
Proofreader: Riley Hoffman
Indexer: Broccoli Information Management

Distributed to the book trade in the United States by Publishers Group West, 1700 Fourth Street, Berkeley, CA 94710; phone: 800-788-3123; fax: 510-658-1834.

Distributed to the book trade in Canada by Jacqueline Gross & Associates, Inc., One Atlantic Avenue, Suite 105, Toronto, Ontario M6K 3E7 Canada; phone: 416-531-6737; fax 416-531- 4259.

For information on translations or book distributors outside the United States, please contact No Starch Press, Inc. directly:

No Starch Press, Inc.
555 De Haro Street, Suite 250, San Francisco, CA 94107
phone: 415-863-9900; fax: 415-863-9950; info@nostarch.com; http://www.nostarch.com

Library of Congress Cataloguing-in-Publication Data

Courtney, Tim, 1982-
  Virtual LEGO : the official LDraw.Org guide to LDraw tools for Windows
/ Tim Courtney, Steve Bliss, and Ahui Herrera.
        p. cm.
  ISBN 1-886411-94-8
  1. LEGO toys. I. Bliss, Steve, 1965- II. Herrera, Ahui. III. Title.
  TS2301.T7C69 2003
  688.7'25--dc21
                                        2003000472

FOREWORD

As parents of James Jessiman, the creator of the original LDraw, we are proud and privileged to have been asked to write the foreword to this book on LEGO drawing. It is the culmination of years of work by a dedicated group of LEGO building enthusiasts and computer experts, a few of whom we were delighted to meet at Brickfest 2001 in Washington, D.C.

LDraw began as James's endeavour to illustrate a LEGO parts listing simply for his own use. After many attempts, including changing programming languages a number of times, James found the combination that worked best. The move from a collection of static illustrations to the use of the drawings as elements in model constructions seemed so natural to James. Another very natural thought was that something as enjoyable as LDraw should be shared with other LEGO fans — hence the use of the "Net."

The enthusiastic reception to this new toy, virtual LEGO, was wide and varied. James received countless emails daily, most requesting that new parts be drawn. These requests were immediately filled as James devoted most of his night hours to the task. What he next planned was for fans to write their own parts with his encouragement.

All came to an end, or so we thought, when on July 25, 1997, James died very suddenly. He was only 26 years of age. However, not only was there great compassion and sadness from his "Net" friends for us, but they also took up the task of continuing the development of LDraw and other allied systems; their enthusiasm was only matched by their technical expertise.

So it is with pleasure that we see this work come to fruition, whereby many people, young and old, may join the worldwide community of building, playing and computing LEGO fans.

With grateful thanks and all good wishes for the success of this venture —

Don and Robyn Jessiman
Wagga Wagga, New South Wales
Australia

BRIEF CONTENTS

CONTENTS IN DETAIL

1
WHAT IS LDRAW?

2
INSTALLING THE SOFTWARE AND USING THIS BOOK

3
DIVING IN: CREATING YOUR FIRST MODEL

4
THE LDRAW PARTS LIBRARY

5
PLACING, MOVING, AND ROTATING PARTS

6
EXPLORING MLCAD

7
CAPTURING A SIMPLE MODEL: THE CAR

8
COMPLEX MODELS: SUB-MODELING AND SNOT

9
MINIFIGS, SPRINGS, RUBBER BELTS, AND MORE

10

CREATING AND USING FLEXIBLE ELEMENTS

11

MANAGING YOUR MODEL FILES

12

INTRODUCTION TO BUILDING INSTRUCTIONS

13

INTRODUCTION TO RAYTRACING SOFTWARE

14

L3P AND L3P ADD-ON

15
LPUB: AUTOMATE BUILDING INSTRUCTION RENDERINGS

16
POV-RAY

17
MEGAPOV

18

POST-PROCESSING YOUR BUILDING INSTRUCTIONS

19

LDRAW AND THE WEB:
VIEWING AND PUBLISHING MODELS ONLINE

20

BUILDING INSTRUCTIONS: MOBILE CRANE

21
CREATING YOUR OWN LDRAW PARTS

22
LDRAW AND LEGO BRICKS: TAKING THE HOBBY FURTHER

A
GLOSSARY OF TERMS
383

B
WEB LINKS

C
LDRAW COMMUNITY HISTORY
395

D
LDRAW FILE FORMAT SPECIFICATIONS

E

EXTENDED AND DITHERED COLOR INFORMATION

F

LDRAW PRIMITIVES REFERENCE

INDEX
441

ACKNOWLEDGMENTS

The authors would like to collectively thank:

Jaco van der Molen – Jaco provided the most help to us of anyone here. He tirelessly edited our material and even took it upon himself to jump-start a few chapters by giving it a go himself. We are deeply in debt to him for his help in this project.

Jake McKee – Jake was with us from the beginning on this project. He provided an honest, in-depth critique of many chapters, which was invaluable in our efforts to make them understandable for the reader. He also worked to provide material for Chapter 12.

Erik Olson – Erik assisted by compiling much of the CD-ROM gallery, taking a complex task off of our hands and allowing us to focus on book material. Thanks, Eric!

Steve Barile – Steve was both an encouragement and a help with material; he provided information for Chapters 12 and 18.

Larry Pieniazek – Larry provided some critical support in business matters surrounding the book, and also provided material for Chapter 18.

Michael Lachmann – Michael provided support for his MLCad software, and also ensured we had access to a new version before its release, for the purposes of writing the material.

Kevin Clague – Kevin was an invaluable resource when it came to understanding his programs, LPub and LSynth. Kevin also co-authored the first LDraw-related book, LEGO Software Powertools, which all of us are anxious to read after wrapping up this book.

Chris Dee – Chris gave us permission to re-print his excellent (and lengthy!) Primitives Reference in Appendix F for the benefit of our readers.

Brandon Grifford – Brandon designed the amazing cover model and was gracious enough to allow us to use it.

Thanks to *Jennifer Clark, Lars Hassing, Dan Jassim, Bram Lambrecht, Cale Leiphart, Chris Maddison, Jon Palmer, Jin Sato, and Willy Tschager* – for contributing images, models, and artwork to this book.

Thanks to everyone who has contributed to LDraw.org – software developers, parts authors, web page writers. Anyone who has discussed Lego CAD online, or asked questions about the tools, or requested special part files has contributed, in some way, to this amazing online community.

And especially, thanks to you for buying this book!

Tim Courtney would like to thank:

First and foremost, I must publicly give praise to Jesus Christ, my Lord and Savior, for carrying me through the difficult times through the course of this project. I could not have finished my work on this book without His divine help.

I would also like to thank:

Don & Robyn Jessiman – I would like to thank James Jessiman's parents, for their blessing of this project and encouragement throughout the process.

Steve and Ahui – My coauthors. Thanks for tackling the things that are beyond the scope of my knowledge, for keeping me in check when I needed it, and for following this through to the end.

Jake McKee – Jake also wrote a book for No Starch Press at the same time we wrote *Virtual LEGO*, so we were able to share in this experience of being published. Jake was a big encouragement to me throughout the long and lonely journey.

Matthew Gerber – My good friend Matt was an encouragement and a support the entire way through the writing process.

OnDrew Hartigan – One of my few local friends in this endeavor (as many of those mentioned here are spread out over the globe) was yet another sounding board and source of moral support throughout the project.

Ralph Hempel – Ralph spent considerable time as an experienced author, giving me advice before I even signed a contract.

Anton Raves – Anton is regarded by many as a master of POV-Ray, and rightfully so. Anton sits on the POV-Team (the group responsible for POV-Ray) and spent quite a bit of time with me, helping me learn the nuances of the POV-Ray scene language. He also offered his computers to share the load of rendering. Anton also served as a liaison to the POV-Team when it came time to seek permission to re-distribute the software in the book's CD-ROM.

I would also like to thank these additional individuals for their help reviewing material: *Rob Dreschel, Kyle Keppler, Frank McKeever, Molly Sumner, and Rebecca Swartz.*

I also thank my parents, David and Dotty Courtney, for putting up with me and putting me up through the course of this project. Many thanks to my grandparents for their love and support, as well as to Kristin, Jamie, and David Bliss (Steve's wife and kids). Also, Tamy Teed, Eric "Legomaster" Sophie, Tim Saupé, Lars Hassing, and my best friend, Brandon Grifford.

Finally, to everyone not mentioned here whose excitement about this project kept me going throughout the many long months of writing and editing.

Steve Bliss would like to thank:

My beautiful and wonderful wife, Kristin – Thank you for putting up with my LEGO habit, and for being a wonderful hostess when "LEGO friends" come for the weekend.

Jamie and David, my amazing sons – Thanks for your support and hard work. Next time, you guys can write the book!

Tim Courtney – Thanks for taking on this monster project, and wrangling it through to completion. Everyone should know that without you, this book never would have happened. And thanks for not giving up on me, when maybe you should have.

Ahui Herrera – Your enthusiasm and good spirit are amazing!

The LEGO Company – Thanks for making all those bricks!

Dan and Jennifer Boger – Thank you for reviewing material.

Don and Robyn Jessiman – Thank you for continuing to share LDraw with everyone.

Ahui Herrera would like to thank:

My lovely wife, Marisela – You put up with my many hours of sitting, standing, dancing and laying next to my laptop as I worked on this book, all the while understanding how important this was for me. Thank you SO, SO MUCH! I love you forever (minus 1 day).

Fidel & Judith Herrera – Mom & Dad, you supported me for 10 months while I was unemployed and working full-time on this book. Without your support, my chapters would not have been possible. *Hay que les pague Dios lo que les debo!*

Iriana Hererra (N.B.A. player "Namps") - Thanks, little sis, for reading all those chapters. I know you thought they were boring since they did not deal with your "cool things," but you helped me out a lot by looking at the book through the eyes of a non-LEGO fan.

Adolfo & Maria Herrera – Abuelitos gracias por todo el apollo que me an dado turante toda mi vida. Thanks for letting me run my laptop hours on end, typing this book while you kept my cookies warm for me.

Daniela & Rebecca Sosa – "Pulga" and Becky, thank you for reminding me time and time again while you girls played with your DUPLO, making those tall, skinny towers that always fell which was always my fault, that playing with LEGO is fun but writing this book at times was not.

"Little Boy" Tim Courtney – Had you not "suckered" me into becoming the LDraw Help Desk Coordinator, I don't think we would have ever done this project together.

Jamie & David Bliss – Thank you guys for your all your hard work in helping me with the installer. You guys are the BEST for sticking it out with me as I attempted to make a decision.

Steven Barile – Thanks for teaching me the "art of building instructions." Your knowledge provided me with a solid foundation that helped me while writing this book and my many personal BIs, which I have posted.

Kevin Clague –Thanks for putting up with my many emails on bugs and upgrade requests for LPub. Without LPub, I would still be making BIs for this book now! Have you blacklisted me from your email inbox yet? =)

Jeff Boen – Without L3PAO, I don't think I would have ever become a serious LDraw user. L3PAO opened the world of LDraw raytracing to me and showed me how easy it was to build in that world.

Michael Lachmann – Your application is the one that got me started and hooked on LDraw. I remember first downloading the original DOS LDraw, and deleting it after a few minutes because it was in DOS! Then MLCad came around, and everything became so much easier. Keep up the good work.

Other LDraw Application Developers – You guys have put up with me sometimes on a daily basis as I "bugged" you about "bugs" in the software, requested and re-requested permission to redistribute your applications, and asked the general "I need help" questions. Thanks for NOT blocking my email address from your inbox. I hope you continue to make wonderful LDraw applications.

No Starch Press – Thank you for giving us the opportunity to show the world what LDraw is all about, and never once complaining about the massive amounts of CC emails you got from us while working on the book.

The LEGO Company – Thanks for helping me imagine . . .

Don & Robyn Jessiman – I never had the pleasure of meeting or talking to James, but he provided all of us "LEGO fanatics" with the best gift ever: UNLIMITED PINK LEGOS! They may be virtual bricks, but it's the closest I'm ever going to get in owning PINK bricks, as long as LEGO doesn't make them in larger quantities.

1

WHAT IS LDRAW?

Chances are, you bought this book because you are already an avid LEGO fan who wants to enhance your building experience by learning about the amazing free CAD and rendering tools we describe. That's great! Since 1996, the worldwide community of LEGO fans has used the LDraw parts library and file format to do amazing things with virtual LEGO bricks. The LDraw community has come a long way from the small group of users it once was. Now tens of thousands of LEGO fans worldwide use the software as a part of their building experience!

Before we dive in, let it be said that the entire community of LEGO enthusiasts owes a great deal of gratitude to one man, James Jessiman. James is the author of the original LDraw and LEdit programs. He set the standard for everything you see here with his work. Sadly, James Jessiman passed away in July 1997, leaving behind countless friends, a loving family, and a dedicated group of fans from all parts of the world. Today James' memory lives on in the LEGO community through the work of the countless people who now use the LDraw system.

The Scope of LDraw

Individuals in the LEGO community have raised the bar of creativity in the realm of virtual LEGO building, and have amazed us all with their incredible models. Using the available LDraw tools requires some technical skill; however, these tools enable a wide range of individuals to easily achieve stunning results. Perhaps you will be one of the next people to push the envelope of what can be done with virtual LEGO models and LDraw.

The original LDraw package, released by James Jessiman in 1996, set a standard for 3D LEGO building on the Web. Today, the word "LDraw" is used to refer to a system of tools, instead of a single program. Many programs are designed to interact with LDraw model and part files. We like to say these programs make up the "LDraw system of tools." Such programs include the ones we will discuss in this book: MLCad, L3P, L3PAO, LPub, and many more.

To help introduce you to this diverse system, we have broken it into five major categories:

1. Editors
2. Viewers
3. Utilities
4. File Format Converters
5. Raytracers

You perform different tasks with each tool as you digitize your LEGO model. This book walks you through how to use the most essential and most popular tools in each of these categories. When you are finished with this book, you should have the confidence and ability to document your models and create building instructions to share with the world.

The Possibilities

The official LDraw website, LDraw.org (www.ldraw.org), features Model of the Month and Scene of the Month competitions. These competitions showcase the best in LDraw model building and high-quality rendered scenes (www.ldraw.org/community/contests/). Anyone is allowed to enter an image of their favorite model or scene for the honor of being featured on the site's main page during the following month. Figure 1-1 is an example of a Scene of the Month winner.

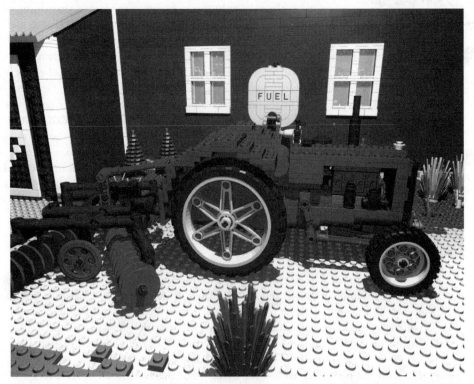

Figure 1-1: *Ready for Work*, by Cale Leiphart: LDraw.org Scene of the Month contest winner for March 2002.

There are numerous notable achievements outside of these contests as well. From a technical as well as an aesthetic perspective, Jin Sato's MIBO robotic dog, documented in LDraw (Figure 1-2), is an achievement worth mentioning in itself. If you Technic-heads aren't satisfied with that, the inner workings of Jennifer Clark's Demag AC50-1 all-terrain crane (Figure 1-3 on the following page) are documented in LDraw, and are freely available on her site at www.telepresence.strath.ac.uk/jen/lego/demag_crane.htm.

NOTE   *On Jennifer Clark's site, you can scroll down past the photos of real cranes to get to details about the amazing LEGO model, and download the LDraw files at the bottom.*

Figure 1-2: MIBO, by Jin Sato.

Figure 1-3: Demag AC50-1 All-Terrain Crane, by Jennifer Clark.

Bram Lambrecht's assortment of high-quality LEGO models, coupled with his firm grasp on POV-Ray, make for stunning scenes, many of which are available on his website at http://lego.bldesign.org. Figure 1-4 shows a rendering of his Pentapterigoid (pen'-ta-ter-a-goid) Fighter.

This is just a taste of what can be done with the programs you'll learn about in this book. As with physical LEGO bricks, the possibilities are limitless when you use LDraw. This book enables you to build your own LEGO creations by teaching you the skills necessary to generate 3D models, render 3D models, and create building instructions based on your 3D models. The rest is entirely up to you and your imagination.

Defining "LDraw"

If you are at all familiar with LDraw, you are probably familiar with the names MLCad, LPub, POV-Ray, L3P, LDLite, and others. How do these programs relate to LDraw? LDraw.org defines "LDraw" as the foundation for a "system of tools." All of the above mentioned software titles (with the exception of POV-Ray) use the LDraw file format and parts library as their native file format. Each type of program interacts with an LDraw model file in a different way. Editors create and edit LDraw models, other programs convert the file to different formats, and still other programs generate complex construction directions like a spring or a rubber belt. Viewers let you look at a model and manipulate it in 3D. Together, the programs we discuss in this book compose a cross-section of the total LDraw experience, just as many distinct compatible LEGO elements assemble to become a complete model in real life.

Figure 1-4: Pentapterigoid Fighter, by Bram Lambrecht.

Definitions We Use

The word "LDraw" can be used to refer to the LDraw system of tools; it can also reference the original DOS software package from 1995, the LDraw File Format, and the LDraw Parts Library. This can get a little confusing, regardless of your experience level with the software.

For the purposes of this book, we will refer to each of the above possible definitions for LDraw as follows. Hopefully, this will avoid confusion later.

LDraw / LDraw system

The entire system of tools, which uses the LDraw File Format and LDraw Parts Library.

Original LDraw Package

The DOS software package, written by James Jessiman, first distributed in 1996, containing ldraw.exe and ledit.exe.

LDraw Parts Library

The official LDraw parts library as distributed on www.ldraw.org/library/.

LDraw File Format

The official LDraw file format as described at www.ldraw.org/reference/specs/.

LDraw File or LDraw Model

An LDraw-formatted virtual LEGO model file.

Virtual LEGO Toolbox

In essence, [the] LDraw [system] is a virtual LEGO toolbox. Its components, when used in conjunction with one another, allow you to create wonderful 3D "virtual" LEGO models on your computer. Your raw materials are your imagination and the assortment of LEGO parts. Just like a physical toolbox — with its saw, hammer, tape measure, level, screwdriver, and other necessary tools — LDraw allows you to combine raw materials and craftsmanship to create beautiful furnishings.

Describing LDraw as a toolbox is a great analogy. Just like in physical construction, be it with wood or be it with LEGO, the job is not completed in one step. Instead, you often use several tools together across multiple phases to complete a project. When learning to use LDraw tools, you will understand and appreciate the multi-step process involved when you are documenting your creations on your computer.

Community Driven

LDraw developments are driven by a community of volunteers who have agreed (most of the time) to follow the same standard. Each component tool has its own software author. Each author develops his or her own application at their own pace during their spare time. Developers usually make their programs available for download through their personal websites. LDraw.org, the official LDraw site, links to each program through its own central download index at www.ldraw.org/download/.

LDraw.org provides the official distribution of the LDraw Parts Library, maintains the specification documents for the LDraw File Format, and serves as a central launch-point for people inquiring about the LDraw system. The site also hosts tutorial documents and provides the latest news relevant to the LDraw community. The website is maintained by individuals who volunteer their time and technical and organizational skills to keep things running smoothly in the LDraw community.

Individuals in the LDraw community communicate with each other via the Internet. Currently, the most popular place online to find LDraw discussion is LUGNET™, the (unofficial) LEGO Users Group Network, located at www.lugnet.com. You can join in discussion on LUGNET by signing up to post (free) at www.lugnet.com/news/.

Getting Oriented with LDraw

Because the LDraw system of tools contains many programs that all interact with the same file format, getting oriented can be rather intimidating to new users. We will take some time in this chapter to introduce you to the major categories of tools. Throughout this book, we will introduce you to popular programs used by the LDraw community. In the appendices, we will provide you with resources so you can find out more information about these and other tools.

What We Will Teach

This book will take you through a handpicked group of software programs that allow you to create your own 3D LEGO models, render high-quality images of your models, and create your own building instructions.

It is beyond the scope (and publishing budget) to cover all available programs in-depth. However, we will provide in-depth discussion of some of the more popular programs, which will give you a firm understanding of how the system works. From there, you can take your knowledge and use it to investigate other LDraw-based programs that may interest you. You can view a list of available LDraw tools online at http://www.ldraw.org/download/.

This book will show you how to create models using MLCad, prepare them for rendering with L3P and L3PAO, generate building instructions with LPub, and hand-edit the POV-Ray scene language in POV-Ray and MegaPOV to create high-quality renderings. You'll be introduced to concepts that help you post-process the building instruction images you generate. When you finish this book, you will be able to create a model in 3D, render a high-quality scene of the model, create building instructions for it, and publish your instructions online for others to see. Our goal is to enable you to document your creativity and show it to the world. At the end of this book, we talk about how you can create your own LDraw parts from scratch.

The Tools

Here is a brief introduction to the general LDraw-based tool categories and the functions they perform.

Editors

Editors are programs that allow you to create and edit LDraw model files in a graphical or a text-based environment. An editor is the starting point of all model or part files. Graphical editors are programs that let you place parts on the screen and arrange them to build your model. Text LDraw editors offer more than standard text editors do. They include special functions that help you manipulate elements of LDraw files. To use one of these editors requires an advanced knowledge of the file format, something we will discuss in depth in Chapter 21, "Creating Your Own LDraw Parts."

NOTE *When we say "graphical environment," we mean that you can see the model being assembled on-screen as you place the parts. We do not mean to imply that a graphical user interface (or GUI) surrounds your model. The original LDraw editor, LEdit, is a DOS program that uses keyboard commands to select and position parts. However, we still consider this a graphical environment because it displays the model as it is being assembled.*

Throughout the bulk of this book, we will discuss the use of MLCad, the most popular LDraw-based editor for Windows (see Figure 1-5). The other major Windows editor is LeoCAD. For the sake of brevity and clarity, we have limited our discussion to the more popular MLCad.

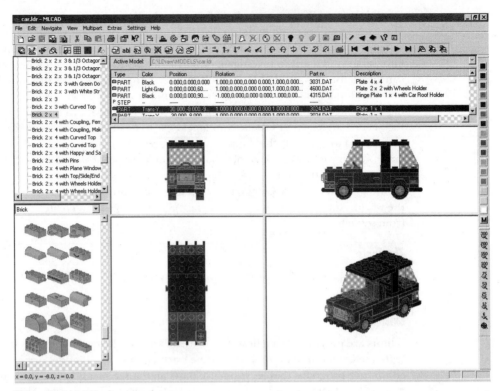

Figure 1-5: MLCad screenshot.

Viewers

Some programs are designed only to display an LDraw file or generate images. These programs have no editing capability, and therefore are referred to as "Viewers." Viewers, such as LDView, LDLite, and L3Lab, often allow you to spin a model with your mouse, and set various options that determine how the parts of the model are rendered (drawn) on the screen. These are useful, "lite" programs that provide an easy way to see a model or a part without editing it.

Utilities

Some LDraw-related programs don't fall either of the above categories; they're simply "utilities." They are usually geared towards advanced users and perform specific functions, such as creating a spring, rubber belt, or flexible element, or mapping an image to the surface of a part like a sticker.

File Format Converters

File Format Converters, or "converters" for short, take LDraw model files and convert them to non-LDraw formats. These programs make possible the high-quality renderings you saw in the examples earlier in this chapter. We will teach you to use L3PAO and LPub. Neither of these programs are converters in and of themselves; they merely provide a graphical Windows interface that generates DOS commands for L3P, the most popular DOS LDraw to POV-Ray converter. However, both L3PAO and LPub are very powerful interfaces that bring many of their own features to the table, beyond the capabilities of L3P itself.

Raytracers

These programs lie outside of the LDraw system of tools; but because they are an important part of the LDraw community, they are given a category so they can be easily referred to. Raytracers are programs used to generate high-quality, photo-realistic images of LDraw models. These programs are able to render objects far more complex than the LEGO shapes generated by LDraw. Raytracers are what make possible the stunning scenes on display at LDraw.org's Scene of the Month competition.

Developing Area: Animation Tools

Animating LDraw files is a relatively new capability, one that has not yet been fully developed. At the moment, there are two major methods of animating LDraw files: by rendering in POV-Ray format, or by using the program LD4DModeler. The POV-Ray technique is not for the faint at heart, and as of this printing, LD4DModeler is still in the process of being developed. Despite the field's primitive nature, animation promises to be an exciting new area of development in the LDraw community — just imagine the possibilities! However, in this book, we don't go into depth on animation techniques.

Summary

As you turn the page, you will be on your way to enhancing your LEGO building experience by learning how to document your models using the LDraw tools. The LDraw system has grown over the last several years, thanks to countless hours of work by devoted LEGO fans. This tool system is based on James Jessiman's original file format and parts library, which now, years after his death, continues to be maintained by the community of users. We will teach you the necessary tools to document your own models, create your own building instructions, and generate photo-realistic images of your models. We will also guide you in publishing your creations so others can enjoy your work. By the end of this book, we hope you will enjoy creating virtual LEGO models as much as we do! Happy building!

2

INSTALLING THE SOFTWARE
AND USING THIS BOOK

Before diving into LDraw, we'll walk you through the installation process and teach you some of the conventions we'll use throughout this book.

As we said in Chapter 1, the LDraw system is made up of many tools, written by many different people. If you were downloading these tools individually through the Internet, you would need to install and configure each separately. We've assembled an installation program that installs and configures most of the software discussed in this book, to minimize the effort you need to put into configuring the tools up front.

Installing the Software

To install the software, insert the CD-ROM into your computer. Most new systems should automatically run the installation program. If it automatically runs, skip the following section and proceed to the sub-heading "Step 1: Installing the LDraw Tools." *Be sure you read and follow all three major steps to installing the software!*

Starting the Installation Program Manually

If the installation program doesn't launch automatically, follow these instructions. On your Windows desktop, select **Start** > **Run**, then type **D:\vlego-install.exe** and click **OK** (we assume your CD-ROM is assigned drive letter D; if it is not, change it accordingly). The installation program should start. Now proceed to Step 1.

Step 1: Installing the LDraw Tools

We have compiled this installer to extract and install the various LDraw-based tools we discuss in this book. Setup for LDraw tools is rather complicated because it has evolved from a mostly informal, open-source online community. We were able to greatly simplify the installation process, but we couldn't totally eliminate the need for you to manually configure a couple of things. Here we will walk you through the process of running the installation program.

The first two screens are the Information and License Agreement steps, typical for installing any program.

POV-Ray and MegaPOV Notice

Figure 2-1 shows an important message about POV-Ray and MegaPOV, two programs that our installer loads onto your machine. The message informs you that POV-Ray will run as a separate install program after the LDraw install has finished, and outlines related issues.

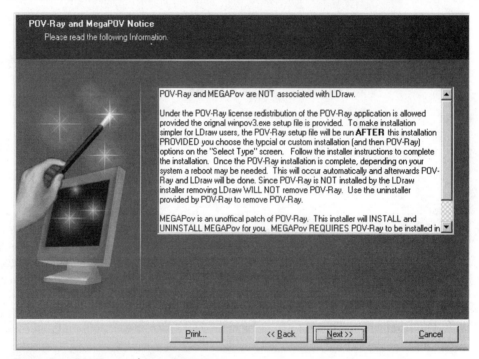

Figure 2-1: POV-Ray and MegaPOV Notice.

Select Destination Directory

After you read the POV-Ray and MegaPOV notice, the installer will prompt you for a directory to install the files (see Figure 2-2). In this book, we assume you are installing your software into the directory **C:\LDraw**. All references to the locations of files and programs throughout the text make this assumption. Keep this in mind as you use the book.

Be aware that per the notice at the bottom of this screen, you must not install the programs into a location on your hard drive where there are spaces in the path. This would break functionality on many of the LDraw-based programs you are installing. For example:

CORRECT	**INCORRECT!**
C:\LEGO\LDraw\	C:\LEGO Stuff\LDraw\

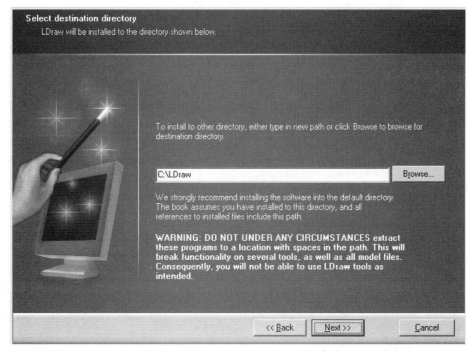

Figure 2-2: Select Destination Directory screen.

Setup Type

After selecting your Destination Directory and Program Folder, you can select the Setup Type (Figure 2-3). For most users, we recommend Typical Installation, or Custom Installation with everything selected from the Options screen.

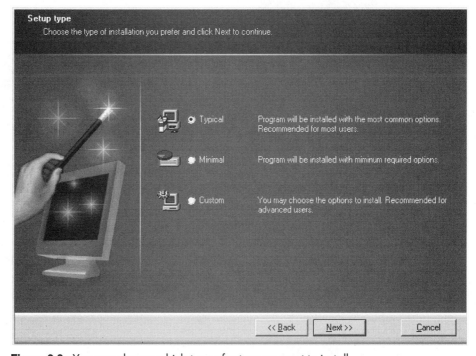

Figure 2-3: You can choose which type of setup you want to install.

Typical Installation

We have configured the "Typical" installation to give you all of the programs and book files you need to learn the material in this book. *Use this option unless you want to place all of the files included on the CD-ROM on your hard drive.*

Minimal Installation

Minimal Installation only installs the original LDraw software. It does NOT install MLCad or any other LDraw-based program. Therefore, by choosing this option, you won't have all of the software needed to work your way through this book. *We DO NOT recommend this option unless you want to manually configure the additional tools. This requires that you generally are familiar with configuring computer software!*

Custom Installation

The Custom Installation Options (Figure 2-4) allow you to select exactly which programs and groups of files to install. *Use this option if you want to decide exactly what to install, or if you want to add the POV-Ray Add-Ons and Gallery files to your hard drive as well.* See Table 2-1 and later sections in this chapter for a description of those two items.

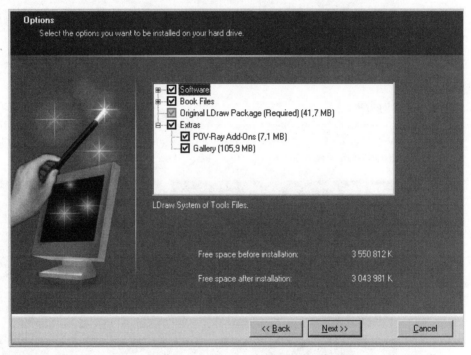

Figure 2-4: Using Custom installation, you can select exactly what you want to install.

In Figure 2-4, you can see a tree structure in this window, which contains different categories of files to be installed. You can expand and collapse each category via the plus sign to the left of each category. By default, Custom installation will install everything on the CD-ROM to your hard drive. We recommend you install all of the Software and all of the Book Files. The Gallery uses 105 MB of hard drive space; therefore, it is optional.

The POV-Ray Add-Ons and Gallery files install to the same subdirectory as the Book Files, which is C:\LDraw\VLEGO\ (where C:\LDraw\ is your base install directory).

Table 2-1: Custom Installation Options Categories

Name	Description
Software	Allows you to select manually which specific software titles you want installed on your machine.
Book Files	These are the working files for the various chapters in this book. We reference these files throughout the text, so we recommend you install these on your hard drive. See the "Book Files" section later in this chapter for more details.
Extras: Gallery	The Gallery is a compilation of LDraw files and scenes from various LEGO builders like you. We have provided this information for your enjoyment. You can either install the Gallery to your hard drive, or access the files directly from the CD-ROM.
Extras: POV-Ray Add-Ons	These files and programs enhance your POV-Ray experience. Chapter 16 references some of these files. You can either install these files to your hard drive, or access them directly from the CD-ROM.

Installation Summary

After clicking **Next** from either the Setup Type screen or the Options screen, the installer will present you with an Installation Summary. Select the **Install** button to begin copying the files to your hard drive.

Step 2: Installing POV-Ray

After the Virtual LEGO installer is finished installing the LDraw tools, it will automatically launch POV-Ray's installer. This is a separate installation program, and none of your options (such as what you chose for your installation directory) will carry over from the Virtual LEGO installer to the POV-Ray installer. Follow the POV-Ray installation process and use the options it recommends.

IMPORTANT: POV-Ray Installation Directory

When you arrive at the Set Destination Directory screen (Figure 2-5), we *STRONGLY RECOMMEND* that you install POV-Ray to the path **C:\LDraw\Programs\POV-Ray**.

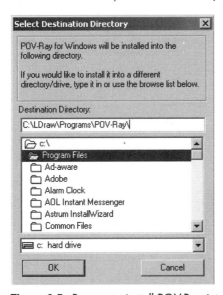

Figure 2-5: Be sure to install POV-Ray into the same path as your other LDraw programs.

After setting your POV-Ray install directory, follow the POV-Ray installation the rest of the way through until it is complete. Close the Virtual LEGO setup window that may still be open in the background.

Step 3: Moving Installed MegaPOV Files

After the installation process, you must do a critical step. LPub (covered in detail in Chapter 15), a program that automatically runs POV-Ray and MegaPOV (covered in detail in Chapters 16 and 17), depends on the MegaPOV executable (program file) being located in a specific path. Because the POV-Ray installation was a separate program, we couldn't do this for you automatically. You need to move these files yourself.

MegaPOV originally installed to C:\LDraw\Programs\MegaPOV\; you NEED to move all the contents of that folder to **C:\LDraw\Programs\POV-Ray\bin** in order for MegaPOV to function properly via LPub. To do this, open Windows Explorer, and move the files.

All Done!

After you have finished these three steps, you are finished with the installation process. Congratulations!

Running the Software

The installer created a Start Menu folder for the programs, under Start > Programs > LDraw. Inside that folder are icons so you can quickly launch the programs.

Creating Desktop Icons

You may notice that the installer didn't automatically create desktop icons for you. We decided it would be best to let you choose which icons you want to have, rather than give you a dozen or so icons automatically. The easiest way to create desktop icons for the LDraw software is to copy and paste them from the Start Menu folder. To do this:

1. Open the LDraw Start Menu folder and move your mouse over the icon you want to copy.
2. Right-click the icon and a menu will appear (Figure 2-6).
3. Select **Copy** from the menu.
4. Right-click on the desktop, and select **Paste**. You should now see the desired icon on the desktop.

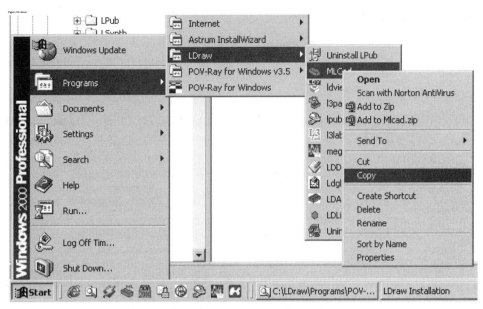

Figure 2-6: Copying a shortcut from the Start menu to the Desktop.

Repeat this procedure for all programs for which you want to create shortcut icons. We recommend you create icons for MLCad, LSynth, LDDesignPad, LPub, L3PAO, LPub, POV-Ray, and MegaPOV.

Using This Book

Throughout this book, we assume you know a few basic things about how we have organized the files that support the chapter material, as well as a few conventions about the way the LDraw system is configured. Pay careful attention to this section as we spell out these basic conventions the book will follow.

Basic Assumptions

If you picked up this book and started to read it, we assume you have a basic knowledge of how to use a computer, specifically how to use Windows. You should be comfortable using menus, the cut/copy/paste functions, the Start menu, and the Run dialog.

Book Files

Throughout the text, we refer to files that are examples of the material we are discussing. We assume you will open, examine, and use these files to help you better understand what we are teaching. Some concepts in this book are rather difficult for first-time users, so we have included the files to give you a practical example of what we are talking about.

We call these files the "Book Files." They are located on the CD-ROM included with this book. The installation program also added these to your hard drive. They are subdivided by chapter in the format C:\LDraw\VLEGO\CHXX\. Whenever you see a reference to a file in this path, it is a Book File that applies to the topic we are discussing. These references are usually coupled with figures to point out specific things we want you to pay attention to.

Default LDraw Folders

The original LDraw package contains a few default folders for models, parts, and primitives. LDraw tools today still use this folder structure. It is critical for you to understand this structure when you interact with the tools.

Models

LDraw models — files you create with LDraw tools — belong in the folder **C:\LDraw\ MODELS**. You can create subfolders underneath this path, but as a rule, keep all of your LDraw models here.

Parts/Subparts

LDraw parts — the individual files in the parts library that represent LEGO parts — are stored in **C:\LDraw\PARTS**. Sub-parts, which are files that part files reference, are found in **C:\LDraw\PARTS\S**.

Under normal circumstances, you won't have to access files in these folders, but at times you may need to edit the contents of a part file, or add unofficial parts (parts that aren't distributed via LDraw.org's Official Parts Updates) to the Parts folder. Keep this in mind in case you need to do something with the contents in the future.

Primitives

Part Primitives — components that parts are made out of — belong in **C:\LDraw\P**. Unless you are a parts author, you will not need to access the contents of this folder. Learn more about Primitives in Chapter 21, "Creating Your Own LDraw Parts," and Appendix F, "LDraw Primitives Reference."

LDraw File Conventions

LDraw files have certain important conventions that deal with naming issues.

File Names

Ideally, names of LDraw files should be no longer than eight characters, to ensure they are compatible with the original tools. If you don't plan to open your file in the original LDraw package, or programs that date back to 1997, you can probably safely ignore this. However, *be sure you never include spaces in your file name*, since the LDraw language does not support this.

File Extensions

LDraw files use three common extensions: .LDR, .MPD, and .DAT. LDR is the standard extension for models, MPD is the standard extension for multi-part models (explained in Chapter 8, "Complex Models: Sub-modeling and SNOT"), and DAT is the default extension for part files and primitives.

Summary

In this chapter, we walked you through installing the software and opening the programs you just installed, and discussed topics relating to using this book. You are now prepared to dive in and start learning how to create your own virtual LEGO models using LDraw tools! Congratulations!

3

DIVING IN: CREATING YOUR FIRST MODEL

Now that you have installed everything you need to get started with LDraw, let's start having some fun! We're going to skip ahead briefly and dive right into MLCad, the editor we will focus on throughout the book. This is a bit of a detour — the LDraw system of tools is pretty complex, and there are some important lessons we need to teach you before you can fully comprehend everything about it. For now though, let's dive in!

The sample model we will walk you through is synonymous with the original LDraw package, released in 1996. The infamous Pyramid model is one of two examples James Jessiman included with the DOS programs LDraw and LEdit. These two models were widely known among the early community of users. Since then, the number of LDraw users has grown exponentially. Third-party applications like MLCad have taken the spotlight, and today these two original LDraw models are not so widely known.

At first you may find it difficult to document, or "LDraw," (yes, we like to use it as a verb, too) these models. It's OK, though — we will take you every step of the way. You will master the techniques needed to build LEGO models on your computer sooner than you think. All it takes is a little bit of patience and the diligence to stick with it, and soon your friends will envy your new skill.

Launching MLCad for the First Time

When you installed the software from the CD-ROM in the last chapter, the installation program placed an icon to Virtual LEGO Launcher on your desktop. Virtual LEGO Launcher is a small application designed especially for use with this book, containing quick shortcuts to all of the software we will discuss. It also allows you to easily browse the other book materials installed by the CD-ROM, but we will get to that part later.

The MLCad Interface

To get started, go ahead and click on the Virtual LEGO Launcher icon on your desktop. Next, click the MLCad icon, to load MLCad.

The first window you will see is the LDraw Base Path window (Figure 3-1). MLCad is looking for the LDraw parts library on your computer. Remember that MLCad is an LDraw editor and it needs the LDraw parts to function properly. Note that initially, it cannot find them. MLCad is only looking in its own directory, as indicated in the Base Path window. The error is shown in the Status window.

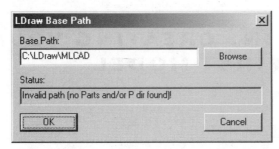

Figure 3-1: MLCad looks for the LDraw parts library upon first launch.

You will need to tell MLCad where to look for the LDraw parts manually. Follow these steps:

1. Click the **Browse** button.
2. A browse window displaying your computer's directory tree will appear. Select the **C:** drive and then the **LDraw** folder. Click **OK**.

Your LDraw Base Path window should now look like Figure 3-2.

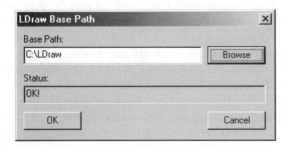

Figure 3-2: LDraw Base Path window checks out OK!

Once you click **OK** on the LDraw Base Path window, this dialog will appear, asking you if you want to register MLCad's file types (Figure 3-3). Click **No**; the installer has already properly registered the LDraw file types.

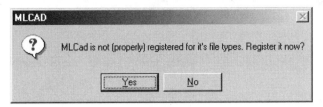

Figure 3-3: Register file types dialog.

Clicking No will bring you into the main MLCad window (Figure 3-4).

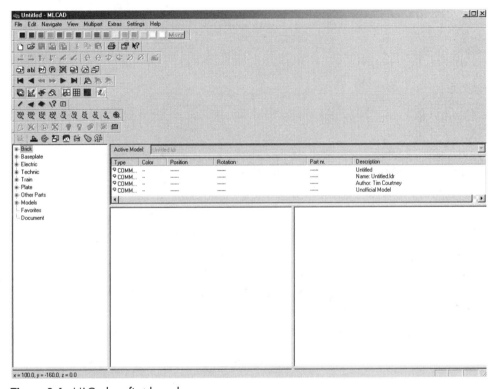

Figure 3-4: MLCad on first launch.

Right now it looks pretty messy. The default locations of the toolbars crowd the rest of the workspace. Click and drag each toolbar to rearrange the workspace so it resembles Figure 3-5 on the following page.

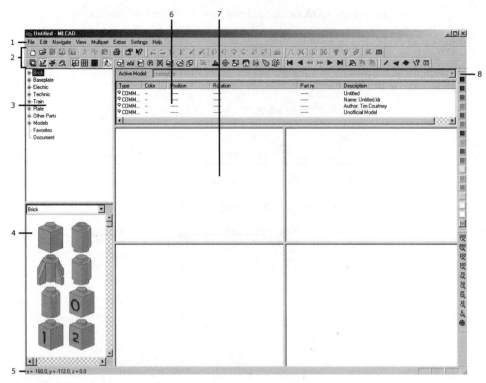

Figure 3-5: The MLCad interface.

Here is a quick orientation of the main MLCad window:

Menus (1): These are standard on all Windows applications, and MLCad is no different.

Toolbars (2): These are the buttons you will learn to use when manipulating your model files.

Parts List (3): This is an expandable/collapsible menu for browsing the LDraw parts library. Parts are broken down into categories and sub-categories. This functions like the directory tree in Windows Explorer.

Parts Preview Window (4): This window shows you which part you are selecting. When you browse through the Parts List, the part that is currently selected displays as the first part in this window. The parts that follow fill the rest of the area.

Status Bar (5): This passes useful information to you, such as where your cursor is located in the 3D geometry plane.

Model File Window (6): This window shows you the raw LDraw file (cleaned up for user-friendliness). Here you will be able to see the actual part lines written as you place pieces on the screen.

Workspace (7): These four views simultaneously display different angles of the current model. You can interact with the parts files here by using your mouse.

Color Palette (8): Here you can select a color for your parts. See more colors by pressing the **M** button on the bottom of this bar.

Configuring MLCad

Before you begin to build your model, you will need to change a few of the configuration settings in MLCad.

1. Select the **Settings** menu, the **General** tab, then **Change** . . . (**Settings** > **General** > **Change** . . .).

Figure 3-6 shows you the MLCad Options window. Note that the options have been divided up into groups called General, Rendering, Printing, Document, "Step, Grid, Snap," and Viewing. These are set apart by the folder tabs near the top of the window. We will not be discussing every pane right now; we only need you to change a couple options for our quick start building session.

2. Change the **Author Name** field to your name.
3. Uncheck **Show Warnings**. This feature can get rather annoying when dealing with multiple versions of parts, or parts that change part numbers. Most of the error messages you will experience are alerts, and mean little to what is on the screen. It is generally best to leave this option off.
4. Uncheck **Register File Types**. We don't want MLCad taking over the file associations the installer set up.

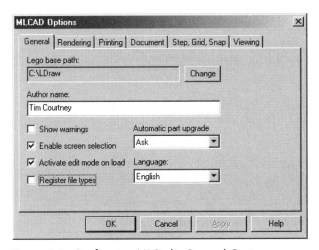

Figure 3-6: Configuring MLCad's General Options.

5. Click the **Rendering** tab near the top of the window (Figure 3-7 on the next page).
6. Change Optimisations [sic] to **Maximum**.
7. Change Stud-Mode to **Normal**.
8. Check the **Draw to selected part only** box. This option, when checked, only renders the active file on the screen up until the current selected part. If you have selected a part in the middle of the file sequentially, only a portion of your model will be drawn. This is useful when you are creating building instructions, but it can be inconvenient when you are simply documenting a model.

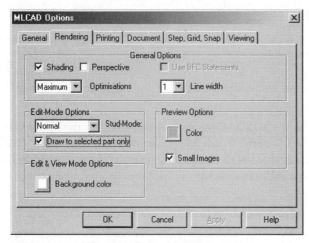

Figure 3-7: Configuring the Rendering Options.

9. In the Preview Options area, click the **Color** box (Figure 3-8). Change the color to **Light Gray**, the color in the top right of the palette. Click **OK** to exit the color palette window.

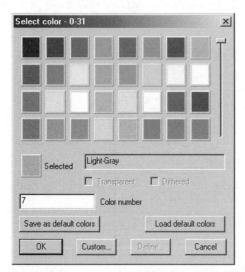

Figure 3-8: Selecting a preview color in the Color Palette window.

10. Finally, click **OK** to exit out of the Options dialog. You are finished there for now.

Whew! Now MLCad is configured and you are ready to dive in and build the **sample model**. Let the fun begin!

The Pyramid

The pyramid is one of the most basic possible creations. This makes it an ideal first model for new users. To assemble this model, all you have to do is move parts on the three axes, X, Y, and Z, and learn how to rotate a part 90 degrees.

Placing the First Part

Let's start building the pyramid. Here's how to place the first part on the screen.

MLCad's Modes

*Before you place the first part, note that MLCad has several "modes" of operation. You need to make sure you are in Place Mode before you can actually place the first part on the screen. Check to be sure you are in Place Mode by selecting the **Settings** menu and clicking **Place Mode** from the list. Now you are free to place the first part. We will talk more about MLCad's modes in Chapter 6, "Exploring MLCad."*

1. In the Parts List window, click the **plus sign** next to **Brick**. This will expand the Brick category, displaying for you every part that LDraw refers to as a Brick.

   We will explain more about parts and the reasons behind their names in a later chapter. For now, just follow along and don't worry about all that stuff; remember this is the quick start chapter!

2. In the Parts List window, scroll down until you find **Brick 2×4** on the list. This is the most basic LEGO part, one of the original parts molded when plastic LEGO bricks hit Europe in the late 1940s.

   Notice that the 2×4 brick also appears as the first brick in the Parts Preview pane, directly below the Parts List.

3. Select the **Brick 2×4** entry in the parts window. Now drag and drop it over onto the top-left **Workspace** pane. You can accomplish the same task by dragging the 2×4 brick from the **Parts Preview** pane onto the **Workspace**, if you prefer.

NOTE *A good habit to develop is dragging the first part of each model into the bottom right pane of the Workspace. This is the 3D view window. Dragging the first part into this window places it at the origin (0, 0, 0) by default.*

4. On the Color Palette (far right hand side), click the **Blue** box, second from the top. This changes the brick's color to blue.

Now you have one part placed on the screen. Good job! Let's pause our pyramid-building project for a moment so you can get used to moving a part around on the screen. Be grateful you don't have to mix your own straw to make these bricks!

Moving Parts in the Workspace

There are three main techniques to move parts around in the Workspace area:

- Using the mouse
- Keyboard shortcuts
- Element Bar buttons

Most likely, you will find it easiest to manipulate parts with the mouse. Some find it even more convenient to use a combination of mouse and keyboard when assembling more advanced constructions.

Using the Mouse

This is the most straightforward method of moving parts in MLCad. It literally is as easy as it sounds. Click on a part, drag it to the desired location and release. Go ahead and give it a try —click and drag the 2×4 brick in any pane but the 3D view, and move it around a bit.

3D View

You are probably wondering why you can't drag the part in the 3D view. Or, you may have tried it (even though we told you not to, shame on you!) and noticed that it actually rotated your part. By doing this, you only changed the VIEW — the angle from which you are looking at the part. You did NOT change the actual ANGLE of the part in the file. This is a very useful feature MLCad offers; it allows you to look around the model quickly while building.

*If you want to go back to the original 3D view, **right-click** in the 3D pane, and select **View Angle** > **3D** from the menu that pops up.*

Wasn't that fun? You should have noticed the parts didn't glide smoothly; instead, they moved across the screen incrementally. This is due to MLCad's built-in Grid system. We will talk about the details of the Grid system a little later. All you need to know right now is that each increment is one-half of a stud, the most basic LEGO unit (a *stud* is LEGO jargon for the round peg on top of a brick or plate; an *antistud* is LEGO jargon for the socket underneath a brick or a plate).

Using the Keyboard

The keyboard can be used to accomplish the same task. The four arrow keys are used to move along the X- and Z-axes. Home and End (the middle two keys in the group of six keys located above your arrow keys) move the parts up and down respectively along the Y-axis. Left and Right arrows move along the X-axis, while Up and Down arrows move along the Z-axis. Go ahead, give this a try; move the part on each axis using the keyboard.

While this method may seem clunky to you at first, it can be useful in certain cases. The mouse is probably easier for you right now though, so file this away and we will revisit it later.

Using the Element Bar Buttons

MLCad's Element Bar accomplishes the same task the mouse and keyboard do. As you can see in Figure 3-9, the arrows on the left-most portion of the bar are for moving parts along the three axes. The two horizontal arrows represent the X-axis, the verticals represent the Y-axis, and the angled arrows represent the Z-axis. Give these a try as well.

Figure 3-9: The Element Bar.

It is likely you won't be using the move arrows on the Element Bar. Mouse and Keyboard are so much easier.

We don't have to worry about the rest of the bar yet, but to satisfy your curiosity, the right-most button is for manually entering position and orientation. This is complex, and unless you have an advanced understanding of 3D coordinate planes and 3D matrix math, it is likely you won't use this feature. The group of circular arrows with axes represented is used for part rotation. Please leave these alone for now; we will discuss them later when we have time to go more in-depth.

Placing Additional Parts

Now that you understand how to place a part on the screen and manipulate it within the Workspace, it is time to continue building the pyramid.

1. To add another 2×4 to the screen, select **Edit** > **Duplicate**. Note that the keyboard shortcut for this function is **CTRL+D**. Remember this, as it will come in handy editing models in the future.

2. Move the part so that the 2-stud end butts up next to the 2-stud end on the previous part.

3. Duplicate the current 2×4 by pressing **CTRL+D**. For this part, you will want to rotate it on the Y-axis so it turns 90 degrees. Press the **A** key on your keyboard once. The A key is MLCad's keyboard shortcut for rotating a part along the Y-axis. You can also use the ⟲ ⟳ buttons on the Element Bar to achieve the same result.

4. Finish assembling the base level of the pyramid by dragging and dropping 2×4 bricks appropriately. Your model should now look like the model in Figure 3-10.

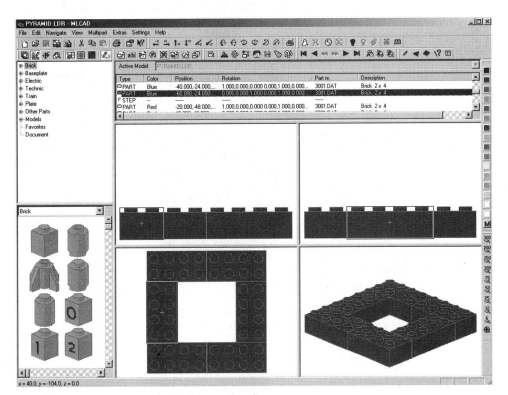

Figure 3-10: The pyramid's first level completed.

CAUTION *Parts Don't Click*

Unlike physical LEGO parts, the LDraw file format does not allow for parts to "snap" together. When you are placing parts, be careful to make sure they are aligned exactly where you want them. You can do this by checking the different view windows against each other. Take notice that even if a part appears to be in place in one view, it may not be where you think it is. Make sure you check all of the views to ensure your selected part is shown in the proper location from all angles. The default views in MLCad are Top, Right, and Front.

NOTE *View Panes and Axes*

Each 2-dimensional view pane in MLCad allows you to view two axes at a time. Top view allows you to view and manipulate along the X- and Z-axes, Front view allows for X- and Y-axes, and Right view allows manipulation in the Y- and Z-axes.

Completing the Pyramid

Now that you know how to place the parts where they need to go for the first level, additional levels should be easy. Go ahead and complete the pyramid. Figure 3-11on the following page shows you the final pyramid model.

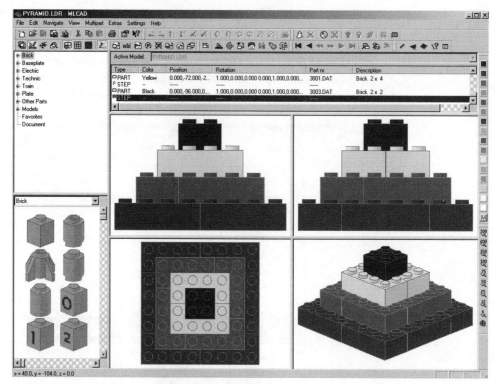

Figure 3-11: The completed pyramid.

Congratulations! You just completed your first LDraw model with MLCad. If you wish, you can save your work by choosing **File > Save**. Save your file in the **C:\LDraw\Models** directory. You will want to keep all of your model files stored here, because several programs use this directory when searching for LDraw model files. It is a smart idea to develop several subdirectories to organize your models in; we'll discuss that topic in depth later in Chapter 11, "Managing Your Model Files."

Summary

In this chapter, you learned the most basic movement techniques for placing parts in MLCad. The model you completed is just a small taste of what you will be able to accomplish using this book. Remember the amazing Technic models we showed you in the first chapter? Believe it or not, we will teach you all of the skills required to build models with the same intricate level of detail!

4

THE LDRAW PARTS LIBRARY

Before we teach you how to model with MLCad, let's introduce you to the LDraw parts library. It is important for you to understand how the library works, so you will be able to find the parts you need when you build your own models. The LDraw parts library includes nearly 2,000 different parts. This library is constantly growing, as individual parts authors are always creating new parts. New LDraw parts go through a certification process to ensure quality, and certified parts are released periodically in Official LDraw.org Parts Updates. You can download these at http://www.ldraw.org/library/updates/.

Whether you are new to the LDraw system or are an experienced user, it is important to be familiar with the parts library. If you know your way around the library, you should have little problem finding the parts you need — even if you don't know the name of the part you are looking for. However, if you are not familiar with how the library is set up, it will be difficult to find parts. In this chapter, we introduce you to the library conventions, teach you several methods of searching for parts, discuss library terminology, give examples of common part categories and their attributes, and provide notes on the locations of unique or odd parts.

A Wide Variety of Elements to Choose From

The parts modeled in the LDraw library accurately represent a wide cross-section of the amazing variety of parts LEGO has produced. Some of the parts in the library date back to the 1970s, while others are new in sets this year. There are many LEGO System parts (from the most common, mini-figure-scale LEGO sets) and Technic pieces, as well as parts from Belville, Freestyle, Scala, and DUPLO (renamed LEGO Explore for 2003), as shown in Figure 4-1 on the next page. There are even a few LEGO Baby parts. LEGO people, called *minifigs* (short for mini-figure), are well represented. The parts library contains a large number of printed faces, torso designs, and accessories. There are roughly 600 decorated (printed or stickered) parts, specialized parts for building vehicles, doors and windows, robots, and more. There are even a few parts for animals!

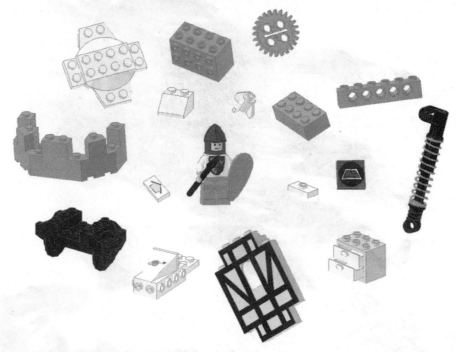

Figure 4-1: There are many different types of LDraw parts. All of them represent parts from the LEGO System.

Only LEGO® Parts Are Official Parts

The LDraw.org Parts Library only includes parts that represent actual official LEGO® brand elements. Since parts go through a certification process to be included, LDraw.org ensures only these parts get added. Why do we do this?

James Jessiman, LDraw's creator, began writing the software in order to catalog the various LEGO parts he owned. Soon thereafter, he discovered his invention was also useful for creating CAD drawings of his models. The early fan community, around the time of James' death, was composed of individuals incredibly loyal to LEGO products, feeling that imitation brands, or "clones," don't achieve the same quality as LEGO. Today, much of that spirit remains.

The LDraw community has generally decided it is best to maintain James' original goal of creating a CAD system for LEGO parts. Some LDraw users do in fact model parts for other (and even competing) building systems. There exist Mega Bloks® LDraw libraries, as well as K'Nex™. Now, fans of the Hirst Arts Castlemolds™, a rubber mold system for plaster-casting building blocks to create painted models, are even developing an LDraw library! None of these non-LEGO libraries are distributed officially via LDraw.org. They have their own websites and distribute their own libraries independent of LDraw.org (see the section "Non-LEGO LDraw Parts" section in Chapter 22, "LDraw and LEGO Bricks: Taking the Hobby Further" for more information).

Having Many Parts Poses an Organizational Challenge

You can create models freely with the many parts in the LDraw library. Unfortunately, when you have many parts at your disposal, it can be difficult to find the piece you need. Being new to the LDraw Parts Library is like having a huge pile of pieces in the middle of the floor; it can be hard to find the one piece you need, hidden in the pile! The more you know about the LDraw library, the easier it will be to find those elusive elements. With a good working knowledge of the library and the part naming conventions we will describe, you will eventually find that you can instinctively find parts you have never used before, based on the part's properties.

Library Conventions

Two things are consistent for all the parts in the LDraw library: part numbers and part titles.

Every Part Has a Unique Number

Each part has its own *part number*, a unique identifier that is used to catalog the parts. Most often, the part number matches the number inscribed on the real brick. This isn't always the case. In some instances, the LDraw number is entirely made up. LDraw numbers are usually fabricated because the actual LEGO number is not known (it may not be printed on the brick). In other cases, the LDraw number is based on the LEGO number, but includes extra information. Patterned parts, for example, have a "pattern code" suffix added to the basic part number.

NOTE *These part "numbers" can contain both letters and numbers.*

Parts Are Individual Files

Each part is a separate file, named according to the part's number. Parts carry an extension of .DAT, a carry-over from the original LDraw package that used the .DAT extension for model and part files.

Every Part Has a Descriptive Title

Every LDraw part also has a descriptive title or name. This title gives basic information about the part: first the general category of the part, followed by its dimensions (in "studs," the bumps on top of LEGO parts), and then any special attributes of the part. Some parts have more than one special attribute, so part titles can get quite long! However, no title is allowed to be longer than 64 characters. The title for a part is usually unique, so if you know the title, then you can be pretty sure you've got the right part. Here is an example of a part title:

```
Technic Brick  1 x 16 With Holes
```

This title represents the part seen in Figure 4-2.

Figure 4-2: The LDraw part entitled "Technic Brick 1x16 With Holes."

Every Part Belongs to a Category

Every part in the LDraw library belongs to a category. Categories can be based on the shape of the part (Brick, Plate, Slope, and so on) or on the function or feature of the part (Wheel, Electric, Minifig, and so on). There is more information on parts and part categories later in this chapter.

Some Parts Contain Keywords

Some parts contain keywords. These are additional words defined to help users search for parts by words that aren't used in the title. For example, a part might be named and categorized a certain way. This may make perfect sense within the LDraw part naming conventions, but it might make little sense to a builder who knows the part by a different name. The common name or function of a part could be a keyword, to help people find it.

Finding Parts in the Library

There are a number of different ways you can track down parts in the LDraw library. While it isn't practical to document every utility or every website that you can use for this purpose, we will teach you a few common methods. This section will show you how to access and navigate the library via MLCad, show you how to use the LDList utility program to search the library, and also teach you how to access the library data directly from your hard drive.

Finding Parts in MLCad

MLCad gives you several ways of adding parts to your models. You can search for parts by name and category, or by number.

Finding Parts by Name

In Chapter 3, we introduced you to the Parts List and the Parts Preview panes. Remember that you can use the collapsible menus of the Parts List to find a part by name, and you can preview the part in the Parts Preview pane. To place a part on the screen, you can click and drag either the part name or the preview image of the part onto the Workspace.

NOTE *If a part's name is too long to be displayed in the parts list pane, you can hover your mouse over it for a few seconds and a little popup window will appear, showing the complete name.*

Using the Preview Pane's Drop-Down List and Scrollbars

MLCad offers two other methods of browsing for parts graphically.

Using the Drop-Down List

If you don't want to use the folder view the Parts List pane offers, you can also browse parts groups to display in the Preview pane using the drop-down list directly above the preview images. When you use this box, the group you select will be displayed in the Preview pane. If you like working this way, you can resize the two parts panes and hide the Parts List folders entirely by dragging the bar that separates the two panes (see Figure 4-3).

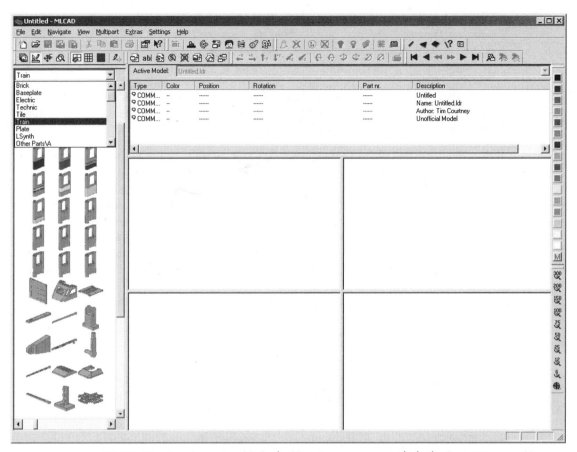

Figure 4-3: You can resize MLCad's Parts Preview pane to hide the Parts List pane. You can navigate between part groups by using the drop-down list box at the top of the pane.

Using the Scrollbars

Another way to browse the parts is by using the horizontal scrollbar at the bottom of the Preview pane. Moving the scrollbar to the right advances the Preview pane through the groups alphabetically, starting with the MLCad top-level groups, then through the "Other Parts" folder letter by letter. This is not a very efficient way to browse the parts, but is available nonetheless. We prefer using the tree in the Parts List pane to browse parts by group and name, and then preview them in the bottom pane before adding them to a model.

Special Folders in the Parts List Pane

In addition to the parts groups themselves, MLCad's Parts List pane contains four special folders: "Other Parts," "Models," "Favorites," and "Document."

Other Parts: Alphabetically lists the parts not covered by the top-level groups in the Parts List tree.

Models: Contains all of the models in your C:\LDraw\MODELS\ folder.

Favorites: Gives you quick access to your favorite parts. You can drag-and-drop your favorite parts into this category.

Document: Displays all of the sub-models in your current document.

NOTE   *The Models and Document folders both deal with the concept of sub-modeling, or placing a model within another model. This is an advanced concept, so we won't discuss it in depth right now. We will introduce you to sub-modeling in Chapter 8, "Complex Models: Sub-Modeling and SNOT."*

Using the Favorites Folder

The Favorites folder is designed to allow you to keep a quick-access list of your favorite parts. You can drag parts into this folder, and they'll be there later when you need to find them again.

TIP   *To add a part from the Parts List pane, select the part name, right-click it, and select **Add to favorites**.*

Managing MLCad's Part Groups

By default, MLCad gives you just a few common part groups: Brick, Baseplate, Electric, Technic, Train, and Plate. All other parts are listed alphabetically in the folders within "Other Parts." After you use MLCad for a while, you might want to organize parts your own way, especially to get some of your frequently-used part categories out of the "Other Parts" folder. Fortunately, MLCad offers a powerful configuration tool so that you can set up the groups however you want.

To launch the Parts List Configuration dialog, open the **Settings** menu, then select **Groups > Group Configuration**. This will give you the Parts List Configuration dialog window, as shown in Figure 4-4.

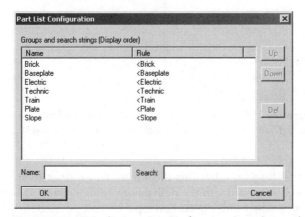

Figure 4-4: MLCad's Parts List Configuration window with the default groups.

When you first open this dialog, you will see the default groups listed.

Adding, Editing and Deleting Groups

You can add, edit, or delete the part groups used in the Parts List tree. Here we will teach you how to do each of these tasks.

Adding Groups

To add a group, enter new data into the Name and Search fields at the bottom of the dialog. As an example, let's add a new group for "Tiles," since they are common parts.

1. In the Name field, type **Tile**. This field determines what MLCad displays on the Parts List tree. While we have used the LDraw category name, "Tile," you could just as easily enter "Smooth Plates" or another name of your preference.

2. In the Search field, type **<Tile**. This tells MLCad to match all parts whose name begins with "Tile." Unlike the Name field, you must tell MLCad the exact LDraw part term you are searching for, or it will not find the desired parts.

Adding your new entry to the list requires a slightly odd interaction. If you simply click "OK" here, MLCad will update the Parts List, but it will also close the Parts List Configuration window. If you want to make several entries at once, you would have to re-open the window for each entry. To prevent this, you can save your new entry and start a new one by clicking anywhere within the large whitespace area where it lists the groups.

NOTE *If you click an existing entry, your new entry will be saved, but the Field and Search boxes will be filled with the entry you clicked. That won't work; if you type over the entry in the box, you will replace the existing entry. Simply click on the whitespace around the list to clear the Field and Search boxes and start a new entry.*

Later in this chapter, we'll discuss the categories of parts built into the LDraw Parts Library. Take a look at the "Important Categories and Their Attributes" section for ideas on part groups you might want to create.

Editing Groups

To edit any existing group, just click the group in the list to send it to the Field and Search boxes. Type your changes, then click the whitespace (or another group entry) to save them.

Deleting Groups

If you want to delete a group, select the group from the list and click the DEL button.

Changing the Order of Groups

After you've made several entries, you'll probably notice that MLCad does not keep the groups in alphabetical order. If you exit the Parts List Configuration, you will see that the Parts List window displays the groups in the same order as the Parts List Configuration window does. You can easily rearrange the groups to your liking by selecting an entry and using the Up and Down buttons on the left-hand side.

Using Advanced Lookup Functions

In the Parts List Configuration, you can do more than simply look up parts whose names start with a specific word. You can also look for words anywhere in the part name, or even restrict the search to parts ending with a specific word. You can exclude a specific word from matching, and you can combine multiple words in a single search. Table 4-1 lists the possibilities. As you grow accustomed to MLCad and want to customize further, experiment with these functions to create a more robust Parts List, tailored to your needs.

Table 4-1: MLCad Parts List Configuration Commands

Symbol	Meaning	Example
<	Match at the beginning of the part name (must go in front of the search word).	<WORD (where WORD is the word you are searching for)
>	Match only at the end of the part name (must go at the end of the search word).	WORD>
&	Search with two words, both of which must match.	$WORD_1$ & $WORD_2$
|	Search with two words, either of which may match.	WORD | WORD
!	Exclude. The search word must not match.	!WORD
( )	Grouping. Use with & and | to control complex searches.	(A & B) | C is different than A & (B | C)

NOTE *Information in the above table was derived from a LUGNET post at http://news.lugnet.com/cad/mlcad/?n=943.*

An Advanced Configuration Example

A common request from MLCad users is to separate out patterned parts from non-patterned parts. Here's an easy way to accomplish this for the Brick group. You can use the same technique for other part groups.

Go into the Parts List Configuration, and make a new entry for "Patterned Bricks." In the Search field, enter **<Brick & Pattern**. This will match all parts whose titles include both "Brick" and "Pattern." Then click the existing **Brick** group, and change its Search entry to **<Brick & !Pattern**. This will exclude any parts that have the word "Pattern" in their title. Close the Parts List Configuration, and your new group should be ready to use.

Finding Parts by Number

If you know the number of the part you want to add, you have yet another option for inserting the part into your model. Pressing the **I** key will open the Select Part dialog, as shown in Figure 4-5. You can scroll through the list of parts by number or description, or if you know the part's number, you can key it in the field at the bottom. When you've found the desired part in this window, click **OK** or press ENTER to insert it into your model.

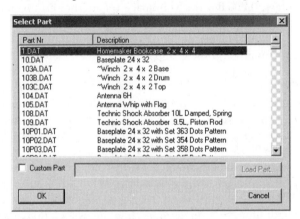

Figure 4-5: MLCad's Select Part dialog.

NOTE *In MLCad 3.00 (the version we are using in this book), it currently does not allow you to use a "hotkey," or type the first letter to jump to that part listing within the Select Part dialog. Using these hotkeys will jump to that letter in the Number column, rather than the Description column.*

Searching for Parts in MLCad

To access MLCad's Search Part dialog, right-click in the **Parts List** pane, and select **Find**. You can search by the part name or part number. We entered **danger** to search for parts that have "danger stripes" patterns (see Figure 4-6). Clicking **OK** will bring you to the first part that has the search term in its name. To find the next entry, right-click the **Parts List** and select **Find Again**.

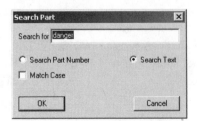

Figure 4-6: You can search for parts in MLCad via its Search Part dialog.

Using LDList to Find Parts

MLCad has good part organization tools, but its search utility is rather clunky. LDList is a small program that was created specifically for part searches. If you are having trouble tracking down a part in MLCad, LDList is often a good tool to use. Figure 4-7 shows a screenshot of LDList.

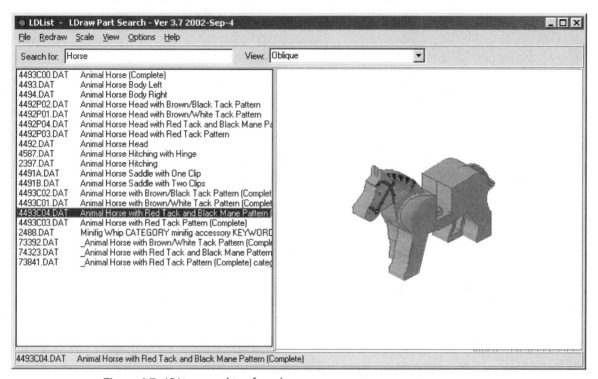

Figure 4-7: LDList: searching for a horse.

You can open LDList from its icon on the desktop or in the LDraw group on the Start Menu.

Running LDList for the First Time

When you run LDList for the first time, the left pane will read, "No LDRAWLIST.TXT file found. Press SCAN to generate it." To scan for parts, select **File > Scan**. LDList will work for a minute or so. When it is finished, you should see the list pane filled with part titles and numbers. To view any part, simply select it from the parts list. (LDList creates its own parts list file that includes category and keyword data. LDraw's PARTS.LST only includes part number and title).

NOTE   *Set the view drop-down to* **Oblique** *to see the default 3D view.*

Searching for Parts in LDList

To search for parts, type your search term in the **Search for** box. LDList will automatically narrow down the list on the left as you type. Try typing "Horse" to see what you get. Besides the various parts that make up LEGO horses, you should also see saddles, traces (called "Hitching"), and some other parts. You get all of these results because LDList is searching not only the part titles (names), but it is also searching for keywords embedded in the part files. Because LDList searches keywords as well as names, it can be an incredibly useful search tool. Many times a keyword is more intuitive than the part names, as part names are designed to fit into a rigid classification system; we'll talk more about this later.

If you enter more than one word at a time, LDList will give you results that match all of the words you entered. Try searching for "animal." Besides all the horse parts, now you'll get all the animal parts as well — the parrot, pieces of the dragon, and other parts. If you want to find only exact matches for your search, you can put quotation marks (" ") around your search words. Try **"Animal Horse"**. Now you should only see the horse parts again.

NOTE   *If you search using quotes around your terms, be careful what you type. Dimensions in part titles are often double-spaced. This is done to help align alphabetical lists of part names. If you search for "Brick 1×1," you won't find anything. You have to put spaces in ("Brick·· 1·×· 1") to find your parts.*

Once you've found the part you want, look it up by name in the MLCad parts list and insert it into your model.

Peeking at the Parts List File

You can also access the parts data by opening up the master PARTS.LST file. This file contains all of the part numbers and descriptions for the parts you currently have in your library. You can find this file under **C:\LDRAW\PARTS.LST**. Open this file in your favorite text editor (like Notepad or WordPad), then use the Find function to look for any word in the part titles. Note that keywords and categories are not included in this file, and once you've found your part in PARTS.LST, you will still have to track it down again in MLCad.

PARTS.LST is very useful for utility programs or scripts, because it puts all the part numbers and titles into a simple text file.

Websites with Part Information

There are several websites that contain information on LEGO parts. They don't all use the exact same part numbers and titles that LDraw uses, but most of them cross-reference with LDraw data.

Partsref: http://guide.lugnet.com/partsref/

Peeron Inventories: http://www.peeron.com/inv/

Technica: http://w3.one.net/~hughesj/technica/technica.html

BrickLink: http://www.bricklink.com

Part Categories

In this section, we will introduce you to the various categories of parts in the LDraw Parts Library. Before you continue, it's important to note that categorizing parts can sometimes be challenging. Some parts don't fit neatly into existing categories, and some could fit into multiple categories. If you can't find a part where you think it belongs, try thinking of other places it might be located. We've included some notes for popular part categories later on, highlighting were certain common parts can be found.

NOTE   *Categories in MLCad?*

MLCad, as we showed you earlier, uses part "groups." You can configure these groups to display one category or multiple categories. These groups are editable, while LDraw part categories have been determined by the Parts Library administrators. All types of parts that aren't referenced by a top-level group can be found alphabetically under the group "Other Parts."

Category Confusion
Most parts have been carefully assigned to categories, but there are few "loose ends" in the library; some parts are located in categories with few parts (and thus are harder to find), and yet others are in categories with names that might seem odd to new users. The parts library administrators are working at cleaning up these issues over time.
The organization of part categories is evolving with every LDraw.org Parts Update. Most updates re-issue a few old parts with new names that reflect their new category placement. Why is the LDraw library evolving in this way? In the earlier days of the LDraw community, some parts were placed in a category for logical and legitimate reasons. However, even if a placement follows convention, it isn't always intuitive for the user. Some parts are difficult for the average enthusiast to find in the LDraw library, so the Parts Library Administrators often move parts to more sensible locations.

Category List

Table 4-2 is a list of all the categories in the LDraw parts library, current as of the 2003-01 Parts Update. We will describe the important categories in detail in a later section.

Table 4-2: Categories in the LDraw Parts Library

Animal	Forklift	Round
Antenna	Freestyle	Scala
Arch	Garage	Slope
Arm	Gate	Space
Bar	Glass	Staircase
Baseplate	Grab	Sticker
Belville	Hinge	Streetlight
Boat	Homemaker	Stretcher
Bracket	Hose	Support
Brick	Jack	Tail
Brush	Ladder	Tap
Car	Lamppost	Technic
Castle	Lever	Tile
Cockpit	Magnet	Tipper
Cone	Maxifig	Town
Container	Minifig	Tractor
Conveyor	Monorail	Trailer
Crane	Panel	Train
Cylinder	Plane	Turntable
Door	Plant	Tyre
Duplo	Plate	Wedge
Electric	Platform	Wheel
Excavator	Propellor	Winch
Exhaust	Rack	Window
Fabuland	Roadsign	Windscreen
Fence	Rock	Wing
Flag		

Terminology

Table 4-3 lists some common words used in the library to describe parts, along with pictures.

Table 4-3: Terminology

Term	Definition	Image
Bar	Cylindrical attribute of a part, which minifigs can grasp with their hands.	
Clip	Clips are attachments on parts (Bricks, Plates, etc.) that can grasp a bar-sized cylinder. A "Clip Vertical" holds a bar vertically (relative to the studs on top of the part), and a "Clip Horizontal" holds the bar horizontally.	
Hole	Hole through a part, which pins can fasten into, most commonly found in Technic parts.	
Hinge	Describes an attribute of a part that allows it to connect to a corresponding part and pivot.	
-L	This abbreviation is used to indicate length. For example, "Bar 4L" is a Bar that measures 4 studs in length.	

Table 4-3: Terminology

Term	Definition	Image
Pattern	Suffix that denotes features of a printed element.	
Pin	Protrusion that fastens into a "hole," most commonly found in Technic parts.	
Stud	"Bumps" that are found on LEGO parts, which enable them to stack and interlock (a stud fits into an "antistud"; some people call it the "socket").	
Towball/ Towball Socket	Small ball protruding from a part, which connects to a Towball Socket. This feature allows the two connected parts to pivot. These parts are often used to connect cars and trailers in official LEGO models.	
Turntable	Part that pivots freely on one axis.	
"with"	Element of a part title that separates a part's broader description from a special attribute of the part.	N/A
"without"	Element of a part title that describes a common element the specific part does not contain.	N/A

Important Categories and Their Attributes

Here are some important categories, with descriptions and images of the types of parts they contain. The parts library contains too many categories to fully describe them all. Some categories we do not describe here are self-explanatory, while others may make little sense to you. If a category name confuses you, it's a good idea to familiarize yourself with the parts it contains.

Category: Bar

The elements in this category are based on "bar" cylinders, solid rods that are the right size to be grasped by minifig hands. Some of the parts look like fences, others are like ladders, but all of them have the bar cylinder as their basis (Figure 4-8).

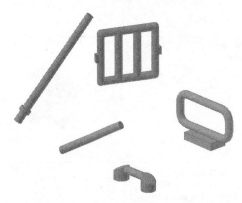

Figure 4-8: Bars are parts that are composed mostly of bar cylinders.

NOTE *The Star Wars™ Lightsaber™ blade is in this category (Bar 4L). Some other elements contain "bar" cylinders, and they can be found in other categories. For example, the classic antenna, a Bar with an antistud on one end, and a rounded top, is in the Antenna category.*

Category: Baseplate

Baseplates are the large, flat, somewhat flexible plates used as ground or roads in LEGO creations. You can't attach parts to the bottom of a baseplate. The most common baseplate size is 32×32 studs, or 10"×10". Baseplates come in sizes ranging from 8×8 studs up to 48×48 studs. Most baseplates are completely flat, but some have raised portions to make hills or ramps or pits (Figure 4-9).

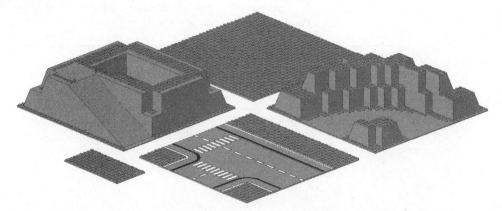

Figure 4-9: Baseplates are large building bases. They can be flat or raised, and can also have some smooth decorated areas, such as the road plate.

NOTE *Large, flat parts that allow you to connect parts underneath them are called Plates, not Baseplates. If you are looking for the raised piers found in seaport sets, look in the Platform category.*

Category: Bracket

Brackets are plates with a 90 degree bend (Figure 4-10). Usually, brackets have studs facing different directions, so you can mount bricks sideways on your creations. Some brackets can have multiple bends.

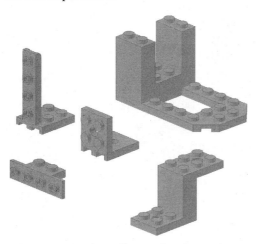

Figure 4-10: Brackets offset studs either in the same plane, or in a different one.

Category: Brick

Bricks are the most basic and most common LEGO parts. Dimensions for bricks are usually given as width and length (X × Z). Bricks that are taller than the standard height also have their height specified, measured in brick-heights (X × Z × Y). The Brick category includes standard rectangular bricks (like everyone's favorite 2×4 brick), but it also includes bricks with extra features, such as pins, wheel holders, and studs on their sides. Sizes range from the diminutive 1×1 brick (part number 3005) up to the large 10×20 brick (Figure 4-11).

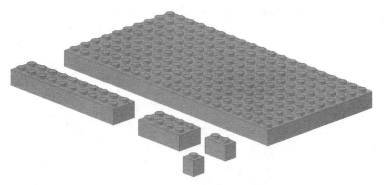

Figure 4-11: Bricks are standard, rectangular LEGO parts.

NOTE *Rounded bricks carry the word "Round" at the end of their name.*

Categories: Cylinder, Cone, Round

Cylinder and Cone are two small categories and are based entirely on part geometry (Figure 4-12). These categories include parts that have these specific geometric properties. Because these categories are relatively new, not all parts that fit this description are found here yet. Cylinders are cylindrical parts that are not bricks.

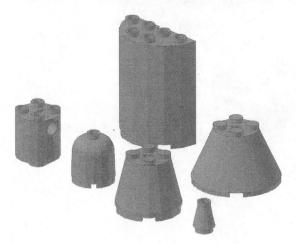

Figure 4-12: Cylinders and Cones.

NOTE *Many round parts may be found in the "standard part" categories like Brick, Plate, and Tile. Therefore, if you're trying to track down a part that looks like a brick with one curved side, you should check both the Round and the Brick categories. Again, rounded brick names will end with the word "Round."*

Category: Duplo

The Duplo category covers parts released under the LEGO Duplo system. Note that all DUPLO parts are twice the scale of standard LEGO parts — a 2×4 DUPLO brick is as big as a regular 4×8×2 brick (if there were such a part). Because of the difference in scale, and because there are not very many DUPLO parts modeled in LDraw (yet!), all DUPLO parts are placed in a single category (Figure 4-13).

Figure 4-13: Duplo parts are twice the size of LEGO parts.

Category: Electric

Electric parts are parts that work with the LEGO Electric System. They contain wires or metal pieces to conduct electricity, or are motors, lights, or sound devices. Nearly every electrified part is put in the Electric category. This includes wires, plates with electric contacts, motors, Light & Sound elements, battery boxes, fiber optics, and LEGO MINDSTORMS® sensors and computer bricks (Figure 4-14).

Figure 4-14: Almost all electric elements are found in the Electric category.

NOTE *Train motors are included in Electric, but the tracks are in the Train category. Some ancillary parts (such as part number 2871, the decorative siding for the train motor) are also included in Electric. Because of the wide variety of parts in this category, their names are not very standardized.*

Category: Glass

The Glass category contains the "glass" (usually clear or translucent) inserts for Windows, Panels, and Doors (Figure 4-15).

Figure 4-15: The Glass category contains the clear insert parts that fit into frames, often called "Windows."

NOTE *This category doesn't include the frames the parts are inserted into — look in categories like Window, Panel, or Door for those.*

Category: Hinge

Hinges are parts that have hinge attachments on them (Figure 4-16). Table 4-3 describes the hinge attribute.

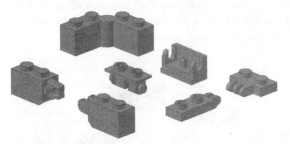

Figure 4-16: Hinges are bricks, plates, or other parts that can pivot around a central point.

NOTE *Hinge is a relatively new category; many of these parts once existed in different categories, such as Plate, Brick, and Panel, based on their other properties.*

Category: Minifig

The Minifig category contains parts that are used to build minifigs: heads (with different patterns or decorations on them), torsos, arms, hands, legs, hips, and prosthetic devices such as the hook and wooden leg (Figure 4-17).

Figure 4-17: The Minifig category contains parts that make up minifigs, including the complete minifig shortcut part, pictured above.

Category: Minifig Accessory

Minifig Accessory includes items that minifigs wear or hold in their hands (Figure 4-18).

NOTE *Not all programs recognize the Minifig Accessory category. In these programs, the accessory parts are included in the general Minifig category. MLCad is one of the programs that currently don't recognize this category, so if you are looking for parts here, check Minifig.*

Figure 4-18: Minifig Accessories are parts that minifigs wear, hold, or interact with.

Category: Panel

Panels can best be described as wall sections that do not take up a full brick-width in area. Some people call these parts *thin walls*, because of this very attribute. Panels can be flat, faceted or curved (Figure 4-19).

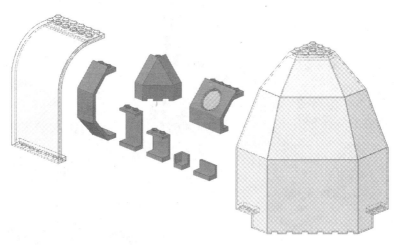

Figure 4-19: Panels are wall sections of all sizes and shapes.

NOTE *The Panel category includes odd parts such as the clip-on panel found in Exploriens sets, the Space corridor wall system, and the large quarter- and half-dome parts.*

Category: Plate

Plates are fundamental to the LEGO System in the same way as Bricks. Plates are basic shapes with one-third of the height of a brick (Figure 4-20).

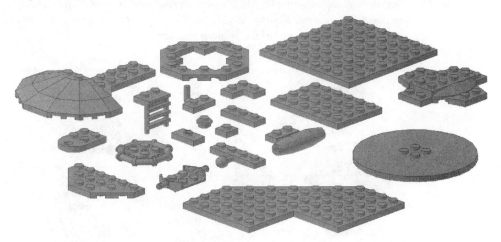

Figure 4-20: Plates are one-third of the height of Bricks. There are a wide variety of plates, categorized as such because the plate is the principal connection area of the part.

Slope (aka Slope Brick)

Slope is yet another category filled with commonly used LEGO parts (Figure 4-21). Undecorated slopes can be used for creating roofs, while decorated slopes feature anything from racecar markings to computer consoles.

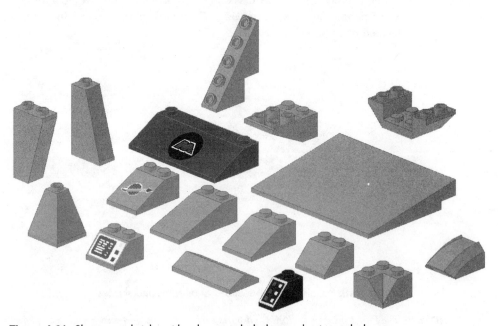

Figure 4-21: Slopes are bricks with edges angled along a horizontal plane.

Category: Sticker

Sticker is a new category that includes stickers found in common LEGO sets. As of the 2003-01 parts update, the only sticker in the library was the Technic Supercar logo (Figure 4-22). More sticker files are in the process of being certified and released.

Figure 4-22: The first sticker in the parts library, the Technic Supercar logo.

NOTE *Stickers differ from decorated parts in that decorated parts are printed, while stickers come in LEGO sets on a sticker sheet and must be applied manually.*

Category: Support

Supports are parts that commonly function as support braces. They're designed to look like they are for that purpose as well (Figure 4-23).

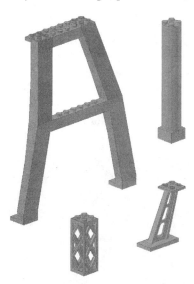

Figure 4-23: Supports are tall columns of different types.

NOTE *Supports include Lattice Pillars, the common Space angled support (Stanchion Inclined), and the large "A" frame commonly used to build cranes in official LEGO sets (Crane Stand Double).*

Category: Technic

Parts in the Technic category were introduced in LEGO's Technic system. Technic parts include bricks with holes in them, pegs, axles, pulleys, gears, universal joints, pneumatic cylinders, and more (Figure 4-24).

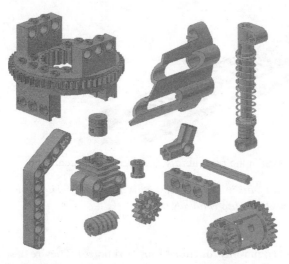

Figure 4-24: Parts in the Technic category are from the Technic System.

Category: Tile

Tiles are fundamentally Plates without studs on the top of them — they're either totally smooth on top, or have only a few studs (Figure 4-25). Some Tiles have attachments other than studs, such as clips, handles, and pins. In the LDraw library, a specialty part that's one-third of a brick-height is a plate if more than half its surface has studs, and a tile if less than half the surface is studded. Also, many tiles are decorated with logos, signage, computer panels, food items (mmm, pizza!), common objects such as life jackets, and other patterns.

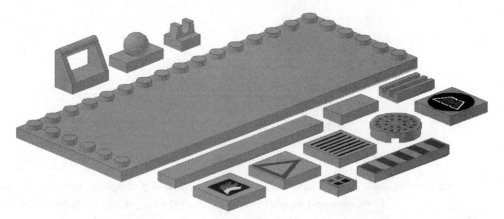

Figure 4-25: Tiles are plate-height, but don't have studs. Many tiles are decorated, and some have attachments similar to plates.

Category: Train

Train parts are specifically a part of the Train system. They include wheels, bogie plates, crossing gate elements, coupler holders, buffers, parts of train bodies, sliding cargo doors, and more (Figure 4-26).

Figure 4-26: Train-specific parts belong in the Train category.

NOTE *The train coupler magnet can be found under "Magnet." The clear inserts for Train doors and other parts can be found under "Glass."*

Category: Turntable

Turntable is an important yet small category. It contains only two complete parts: the 2×2 Turntable Plate and the 4×4 Turntable (Figure 4-27).

Figure 4-27: There are only two parts in the Turntable category.

NOTE *The large Technic turntable can be found in Technic.*

Category: Tyre

Tyre (the Australian spelling of Tire) is a self-explanatory category. It contains all of the tires for various types of LEGO wheels (Figure 4-28).

Figure 4-28: Tyres are rubber LEGO parts that fit over wheels.

NOTE *Because James Jessiman created the original LDraw parts, he used the proper spelling in his native Australia. This has remained unchanged since his death. There are other part names and categories in the library that retain the Australian spelling as well.*

Category: Wedge

Wedges are bricks, plates, or slopes with an angled side face (Figure 4-29). They're common in airplanes and Space models. Some wedges can be considered complex slopes, while others are bricks or plates with one or both sides shaved down.

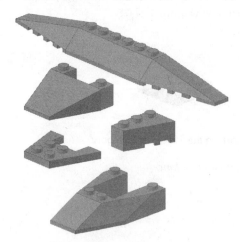

Figure 4-29: Wedges are like slopes, but they are cut along a vertical plane.

NOTE *For more wedge-like parts, see "Wing."*

Category: Wheel

Wheels are the plastic counterparts to Tyres (Figure 4-30). Most wheels allow rubber LEGO tires to fit over them, much like car wheels in real life. However, some wheels, like the ones from Space sets, do not have a rubber tire that fits over them.

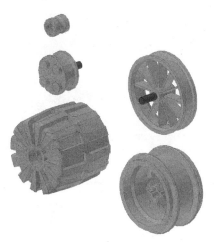

Figure 4-30: Wheels are plastic parts you can mount to special wheel pegs (found in the Plate category) or Technic Pegs or Technic Axles. Some wheels allow you to slip rubber tires over them.

Category: Window

Window is a relatively self-explanatory category. Many windows provide the frames for Glass parts, while others have their own translucent plastic panes pre-inserted. This category holds many "classic" windows that are no longer made, as well as windows that have been released in recent years (Figure 4-31).

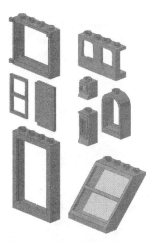

Figure 4-31: The Window category includes "classic" windows as well as new ones with preinserted "Glass."

NOTE *The "classic" window frame that holds swinging panes and shutters can be found in this category, as well as the large bay window that has been recently used in XTreme Team, Alpha Team, and Studios sets. Windshields are not in this category (see "Windscreen").*

Category: Windscreen

"Windscreens" are what North Americans call "windshields." These are the front and rear windows for cars, airplanes, helicopters, spaceships, construction vehicles, and more (Figure 4-32).

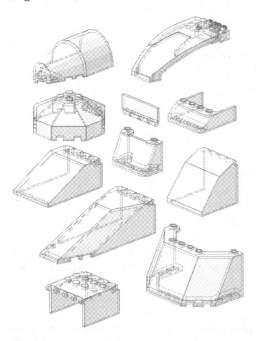

Figure 4-32: Windscreens come in many translucent colors, and sometimes are molded in opaque colors as well.

Category: Wing

Wings are basically Plates with unusual shapes. One or more sides are angled to give the wing its swept shape. A few Wings break away from the flat-plate mold and have a sloped surface (Figure 4-33).

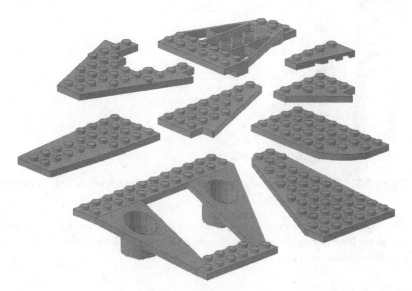

Figure 4-33: The Wing category contains parts used to build wings on airplanes and spaceships.

NOTE *You can find some plate-versions of Wedges here in the Wing category. You can also find some airplane-related parts in the Tail category.*

Parts That Are Hard to Find

Some parts are difficult to find because of an unusual name or categorization. Here is a list of our common hard-to-find parts. You can use this section as a guide, to help teach you more about the LDraw part naming system, or you can refer to this when you're having trouble finding one of the parts we picture.

Lever Control

The Control Lever is a common part (Figure 4-34). It has a hinge on one end, and a hollow stud on the other. Despite these properties, it is stored in the library under "Lever."

Figure 4-34: Lever Control

Lever Small

This common two-piece part is often used in the cockpits of helicopters and airplanes, or as an antenna on the top of vehicles (Figure 4-35).

Figure 4-35: Lever Small

Boat 2x2 Stud

This piece was originally used to attach to the bottom of boats in the Nautica series (Figure 4-36). They're smooth on the bottom and studded on the top. Since then, these pieces have appeared in other official LEGO sets as well.

Figure 4-36: Boat 2x2 Stud

Panel 3x5 Solar/Clip-On/Deltoid

This is a difficult part to categorize (Figure 4-37). It contains a clip, but is not a hinge, and its primary function in LEGO sets is as a solar panel.

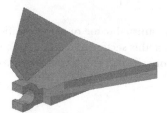

Figure 4-37: Panel 3x5 Solar/Clip-On/Deltoid.

Downloading Official Parts Updates

As we mentioned earlier in this chapter, LDraw.org periodically releases Official Parts Updates. These updates add to the parts in the official LDraw.org Parts Library. LDraw.org offers two methods of download; you can either download the entire library in one file, or you can download the library update-by-update. Once you've downloaded the entire library once, at each update you only need to obtain the latest update file.

LDraw.org parts updates are hosted at http://www.ldraw.org/library/updates/. They are made available in two formats: Zip and self-extracting .exe (run from the DOS command line). If you are familiar with Zip files, we suggest you download the zipped updates. If you do not have Zip software (such as WinZip or PKZip), you can either download WinZip at www.winzip.com or get your LDraw parts updates using the .exe files.

Downloading Zipped Updates

When you download zipped updates, save these to your root directory (**C:**). Extract them, and they should place the parts in the appropriate directories on your hard drive (provided your LDraw directory is C:\LDraw\; if you have chosen another directory for LDraw, place the file in the folder directly above your LDraw folder).

Downloading .EXE Updates

When you download an .exe file of the latest parts update, there is a required process for extracting the contents of the file into your C:\LDraw\PARTS\ folder. These files must be run from the DOS command line to work properly and to properly overwrite previous parts. Here are the steps for extracting a parts update:

1. Download the file and save it on your hard drive at the root (**C:** — not in a subfolder). Typical file names for parts updates are complete.exe (the complete archive) or LCAD200302.exe (the second update in 2003).

2. In Windows, select **Start > Run** to open the Run dialog.

3. Type in **C:\filename.exe -y** (the -y is *very important*, it tells the program you are running to overwrite parts in your \LDraw\PARTS\ folder, an important part of the process for updating your library).

4. Click **OK** or press **ENTER**. A DOS command window will open, and you should see lines of text scrolling across the screen. When the program is finished, the window will close, or it will return you to a prompt (C:\>). If you get a prompt, type **exit** and hit **ENTER** to close the window.

Updating the Parts List

Once you have downloaded and extracted a parts update, you need to update the PARTS.LST file using the MKList (make list) program, so that MLCad and other programs will recognize the new parts you just added. Here are the steps:

1. In Windows, select **Start > Run** to open the Run dialog.
2. Type in **C:\LDraw\mklist.exe**.
3. Click **OK** or press ENTER to launch the program.
4. A DOS window will open, and the program will prompt you to enter **N** to sort by part number, or **D** to sort by part description. Since most people are more responsive to names than numbers, we recommend sorting by description, should you need to create (or recreate) the PARTS.LST file. Type **D** to continue, and MKList will create the PARTS.LST file for you.

Summary

In this chapter, we introduced you to the LDraw Parts Library and the part naming conventions. We taught you how to find parts in MLCad using LDList and by accessing the PARTS.LST data directly. We also listed common terminology to help you when you are browsing the library, and taught you how to use the official LDraw.org parts updates that are downloadable via the website. You should now be well-equipped to use the parts library in the exercises throughout the rest of this book, and to find parts when you are building your own models.

5

PLACING, MOVING, AND ROTATING PARTS

Now we transition from looking at parts and part types back to building again. MLCad's commands for placing, moving, and rotating parts are easy to grasp with a little practice. Soon you will be able to handle parts connections with confidence in this WYSIWYG (What You See is What You Get) environment.

In Chapter 3, we discussed the steps you need to take when launching MLCad for the first time. This chapter assumes you have already configured MLCad and gone through the contents of Chapter 3. If you have not, please do so now.

If you haven't already done so, please start MLCad by clicking its desktop icon, or by accessing it from the Virtual LEGO Launcher.

Placing Parts

Placing parts in MLCad is very straightforward. We define parts "placing" as adding a part to the workspace area. Manipulating the part into its desired location is covered later under the topics "moving" and "rotating."

You should recall the basic drag-and-drop placement technique we taught you in Chapter 3. There you used the mouse to drag a part from the parts list window onto the workspace. There are a total of three ways to place parts in MLCad. They are:

Drag from the Parts List window: This is the first technique you learned. Find a part by name within the Parts List, and drag that entry onto the screen. An outline of the part will appear and you can place it.

Drag from the Parts Preview window: When you select a part in the Parts List window, it shows up as the first part below in the Parts Preview window. You can drag parts from that preview window onto the workspace as well. This is advantageous, especially if you are working with multiple parts in the same category: Pick the part you want off the screen, drag, drop, and go back for another. (Note that if your selected part is listed towards the end of a category, or if the category you are browsing has fewer parts than will fill the Parts Preview window, MLCad may not display the selected part first.)

Select Part Dialog: This dialog window gives a list of parts and allows you to pick from the list alphabetically (see Figure 5-1). It does not, however, give you a preview of the part you are selecting. This is recommended for users who are already familiar with the desired part name or number. Also, this window allows you to insert a "sub-model" by clicking the Browse button. Sub-models are complete LDraw models which when referenced in a separate model file act as one part — they can be moved and oriented as one single unit. We will introduce you to sub-models in Chapter 8, "Complex Models: Sub-modeling and SNOT."

The Select Part dialog can be accessed via the Add Part button on the toolbar 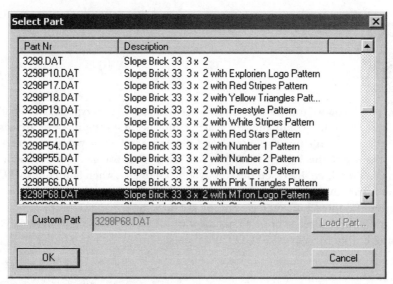, **Edit > Add > Part**, or by right-clicking on the workspace and selecting **Add > Part**. It is also accessible using the Modify command (**Edit > Modify** or CTRL+M). Note that this command uses the dialog for the purpose of changing the current part, not adding a new part.

Figure 5-1: The Select Part dialog box allows you to select parts by number or description, or load a separate LDraw file and represent it as a "sub-model" of your current file.

Moving Parts

When we say "moving" parts, we mean changing a part's position on the 3D coordinate plane, without changing its orientation (or rotation angle). Moving basic bricks, plates, and most other elements in a "studs-up" (or vertical) configuration is easy in any of the LDraw-based applications, including MLCad. The purpose of moving parts within the workspace is to align them with other parts, creating a digital representation of your LEGO model. Remember, in the LDraw system, parts don't snap together like your real LEGO pieces do. You have to view the part from multiple angles to ensure it is in the proper place — and to make sure you don't have one part placed inside of another! MLCad's four view panes make this easy, whereas in the days of LEdit, you could only use one view at a time.

Chapter 3 gave you a head start and taught you this technique when you built the pyramid. At that point, you were simply moving basic bricks next to each other. To review this, place a part on the screen and move it on all three axes. You can use your mouse, MLCad's toolbar, or the keyboard in the three 2D windows. Remember, when using the keyboard, the Left and Right arrows move a part along the X-axis, Home and End correspond to the Y-axis, and the Up and Down arrows move parts along the Z-axis.

This is basic parts movement. At the coarse Grid setting, parts move at a half-stud increment. A stud is the knob or bump on top of a LEGO part (and it corresponds with an antistud inside of the sockets and tubes underneath a LEGO part). Why do they move at half-stud increments? They use it because standard LEGO models use the half-stud all the time — parts like the "Plate 1×2 with 1 Stud" make it possible (see Figure 5-2). We will examine other grid settings next.

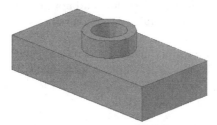

Figure 5-2: Part number 3794, Plate 1x2 with 1 Stud.

The 3D World: A Coordinate System

We've mentioned the 3D coordinate system a few times so far. Each part is assigned a set of coordinates inside of the file's coordinate system (which can be called the "world"). Every object in the world has a set of coordinates, so the software knows where it is placed. You may run into a few cases when you are using MLCad where you need to find out what a part's coordinates are, or perhaps you will need to manually set the part's coordinates. This is all accessible through the Enter Position and Orientation dialog as seen in Figure 5-3.

NOTE *Sometimes MLCad refers to this as the "Enter Position and Rotation" dialog. The two names mean the same thing in MLCad.*

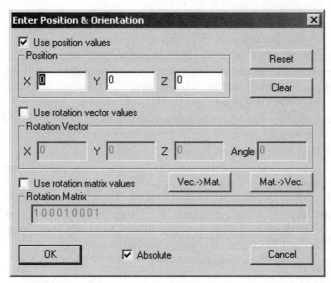

Figure 5-3: MLCad's Position and Rotation/Orientation dialog.

To access this dialog, click the icon on the toolbar, or right-click the desired part and select **Enter Pos + Rot** from the context-sensitive menu. In order to access this menu by either method, there must be a part currently selected. In the dialog you will see fields for entering a brick's position, rotation vector, and rotation matrix. For most building applications, you will only need to use the position values and not the other two values. Manually entering rotation values is useful for authoring parts, which we'll cover in Chapter 21, "Creating Your Own LDraw Parts."

If you want to tell a brick's location within the 3D coordinate system, look at the X, Y, and Z Position values. In the case of Figure 5-3, the brick is located at 0, 0, 0. You can also use this dialog to manually enter the position you would like the brick to move to. Clicking **OK** will move the brick to that position.

Grid Settings and Movement

You've heard us mention "Grid settings" a couple of times already, and you're probably curious as to what that is. The Grid setting is the most important element within MLCad (and other LDraw editors as well) when it comes to properly positioning parts. When we say *grid*, we mean a set of evenly spaced invisible parallel lines running parallel to all three axes. The size of the grid is determined by the grid's setting. When you move a part, the part *snaps* to the grid, meaning the program only allows the part to move in these specified increments.

This is a very helpful feature because as we stated earlier, LDraw parts don't click with each other; you have to learn to position parts by observing them at multiple view angles. The grid helps you place parts next to each other accurately, instead of struggling each time to align freely moving parts perfectly. The half-stud increment of the coarse grid enables you to line up your part easily with neighboring parts.

The smaller grid settings, Medium and Fine, help you when you are dealing with parts that don't line up with each other in half-stud increments. These types of parts include clips and bars, parts with studs that face sideways, hinges, Technic parts, and many other specialty parts.

Changing the Grid Setting

The Grid setting can be changed by using the Grid icons  on the toolbar.

Grid Coarse

The icon on the far left, selected in this image, is the Coarse Grid setting. This is the default selection and you should use this unless you need it to be finer. On the horizontal plane (X- and Z-axes) it moves in half-stud increments. In the vertical (Y-axis) it moves one third of a brick height, which is one plate. When you rotate the part on any axis, it rotates at a 90-degree angle.

Grid Medium

The middle button is Medium Grid. On the horizontal plane, it allows for movements of one-fourth of the width of a stud. When moving vertically, it is set at one-half of a plate height — requiring six steps to reach a brick height instead of three. The rotation increment is set at 45 degrees.

Grid Fine/Grid Off

The Grid icon furthest to the right represents Fine Grid. MLCad sometimes calls this "Grid Off" as well. This setting is used most often when you are creating assemblies on a model that uses hinges, or when you rotate a part to line up with a stud that is not facing up. With this setting, horizontal movements are made in one-twentieth of a stud width, vertical movements are one-eighth of a plate height (requiring 24 steps to reach a brick height), and by default, rotation in any direction is set in 5-degree increments.

Snap to Grid

If you have moved a piece using Fine Grid and want to return to using Coarse or Medium grid, MLCad's Snap to Grid feature can be of help. Sometimes it is easy to make the mistake of moving a piece while using too fine of a grid setting. Re-aligning a part you accidentally moved using fine grid with others aligned with coarse grid can be tedious. Snap to Grid takes the selected part (or parts) and aligns it with the closest grid increment. After snapping a part to the grid, you can proceed to align it with other parts easily. To use the Snap to Grid feature, click the ▦ icon, select **Edit > Snap to Grid**, or use the keyboard shortcut **CTRL + SHIFT + G**.

Get Acclimated

If you haven't already, go ahead and play with these grid settings to get a feel for what each one does. Take a few minutes to move parts around, rotate them, and attempt to align them — even at odd angles!

LDraw Units

The LDraw parts library has its own system of measurement, called *LDraw Units* or *LDU* for short. LDraw Units are very important to MLCad's Grid settings — they are the basis for deciding how far in a given direction to move or rotate each part. Each part file is made up of components that rely on this measurement system to represent each part accurately. James Jessiman put a good deal of thought into defining the LDraw Unit in relation to the dimensions of an actual LEGO brick. Figure 5-4 on the next page shows you how many LDraw Units compose bricks, plates, and studs.

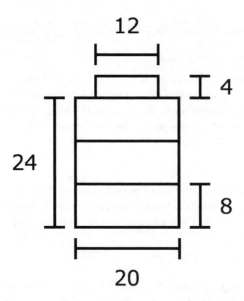

Figure 5-4: LDraw Units in relation to a LEGO brick.

The above diagram shows a 1×1 brick broken down into LDraw Units. While not essential by any means in a WYSIWYG environment such as MLCad, it is a good idea to commit these measurements to memory. Knowing these will help you understand how a brick relates to another brick you are placing next to it. LDraw Units will be discussed more in depth in Chapter 21.

Customizing the Default Grid Increments

You can change the increments each Grid setting uses in MLCad's settings. Go to **Settings > General > Change**. When the window pops up, click the **Step, Grid, Snap** tab. This is where the Grid settings are stored. Note that what MLCad calls "Grid Off" is the same as the setting we called "Fine Grid."

The numbers in the fields for X, Y, and Z of each setting are LDraw units. Note how they correspond to Figure 5-5. If you are new to the LDraw format, MLCad, and the Grid tool, do not change these values. It is a good idea to make a mental note of the location of these settings. After you have experienced using MLCad, you may want to change the defaults here to make it easier to build your model. The most useful feature here is probably the rotation angle on the Grid Off section — you can change your default angle to 1 degree or less, allowing for much more freedom on hinged parts.

NOTE *Undo? MLCad has a placeholder for an "undo" feature under Edit > Undo. However, this feature does not actually function. Be careful when moving and rotating parts, especially when using Fine Grid, since you cannot undo your most recent movement.*

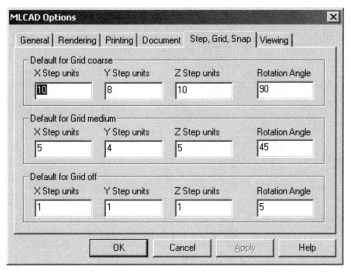

Figure 5-5: You can edit the Grid settings in MLCad's Options window.

Rotating Parts

This is perhaps the trickiest of all part positioning tasks. The model you are creating exists in a three-dimensional environment, just like your physical LEGO models. This time, however, you have to think carefully about exactly how your model's parts are placed. Rotating parts in the horizontal plane, or along the Y-axis, is a simple enough task. It gets more complicated when you start rotating along the other axes, and when you need to rotate one part along more than one axis.

The most convenient method of rotating parts is the icons on the toolbar, just right of the movement buttons (see Figure 5-6). A series of arrows following a circular path around a single bar illustrate clockwise and counter-clockwise rotations around an axis. A click of one of the buttons rotates the part one Grid increment. For Coarse Grid, that is 90 degrees.

Figure 5-6: Movement and Rotation commands on the Element Bar.

In order to see the effects of rotating a part easily, choose an asymmetrical part such as part number 2516, Space Chainsaw Body. First, start a new file, and then find the part in the Other Parts Space category and drag it onto the screen. Take some practice time now and rotate the part along each axis. Be sure to use only one axis at a time; this will help you avoid confusion. Notice how the part rotates at 90-degree increments.

For fun, try rotating the part on more than one axis at a time — click one axis, then another. See how many combinations you can come up with. Also, try to return the part back to its default orientation, shown in Figure 5-7 on the following page.

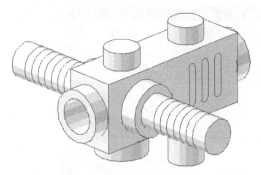

Figure 5-7: Part number 2516, Space Chainsaw Body, default orientation.

Center of Rotation

Each part has a default "center of rotation" that is defined by either the part's author or the LDraw.org parts administrator. When you rotate a part, it rotates around that specific point. Sometimes you may want to change the center of rotation to more easily position a part such as a hinge, or two Technic bricks joined with a pin. The default center of rotation on the part you are using may or may not be in the place you want it to be in. If it isn't in the place you like, you can take a few steps to move the point.

Understanding Center of Rotation

When you're changing the center of rotation in MLCad, realize you aren't actually changing a specific part's center of rotation. Instead you are creating a custom preset point in the 3D world of your LDraw file, which MLCad stores for reference, and can be used by any part you want to rotate. By default, MLCad uses the currently selected part's origin as the center of rotation. You can override this and tell MLCad to use your custom point for all parts in the file. You can define as many of these custom rotation points as you like. It is also possible to toggle between custom points, or return to using the selected part's origin as MLCad's center of rotation. This is all accessible via the Rotation Point Definition dialog window, which we will show you how to use in the following section.

Self-Centered

Do you really need to adjust the center of rotation to align parts? No. It becomes useful when you are making a model that needs to have its various components rotate or move with ease. If you are building a 3D model that you will never need to pose, it's not likely you'd need to use this technique. If you're building a model you want to animate, or you want to generate pictures of in many poses or positions, this will be quite useful to you. It really all depends on what you need your model to do.

Setting a Custom Rotation Point

For our example we will use two 1×6 Technic bricks joined by a pin, as seen in Figure 5-8. The default center of rotation for this part is on the top of the brick, in the middle, between the third and fourth studs. If you want to join the two via a pin in an axle hole, and rotate one on that axis, it will be very difficult to line up the parts as desired with the default rotation point. We need to set a custom rotation point inside the pinhole that we want to rotate around.

Figure 5-8: Our desired result: a 1x6 Technic Brick rotated around a single pin.

1. First, open a new file. Locate the 1×6 Technic Brick (**Technic Brick 1×6 with Holes**) and drag it into the workspace. Remember that you can center the part on the world's origin by dragging directly into the 3D workspace pane.

2. Find the **Technic Pin** on the parts list and properly align it with the last hole on your brick (Figure 5-9).

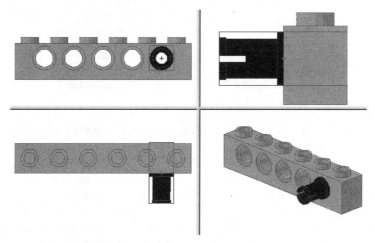

Figure 5-9: Pin alignment in a 1x6 Technic Brick.

3. Insert a second 1×6 Technic Brick and align it next to the first, as shown in Figure 5-10.

Figure 5-10: Two Technic Bricks with pin properly aligned.

4. Select the pin. We need to find the point we will use as the custom rotation point. To find this point, right-click the **Technic Pin** with your mouse, and choose **Enter Pos. + Rot.** . . . Note the pin's X, Y, and Z coordinates — in this case, if you centered your first beam by dragging it into the 3D window, it's 40, 10, and -10. This is our rotation point for the second brick. Write these numbers down and close the dialog by clicking **OK**. If your numbers differ from ours, write yours down and use them.

5. Click the toolbar icon 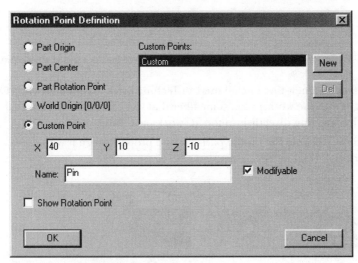 , or from the menu, choose **Settings > Rotation Point**. The Rotation Point Definition dialog appears (see Figure 5-11). Here is where you can select what rotation point MLCad uses for the model. Choose **Custom Point** among the buttons on the left. Enter the coordinates of the pin that you noted from the last step. Change the point's name to **Pin**, and click **OK**.

Figure 5-11: Rotation Point dialog.

6. Now you may rotate your brick around the pin. Choose the fine grid setting, and rotate the brick along the Z-axis.

By setting a custom rotation point as you just did, you told MLCad to use that specific point in 3D space for every single part you may select. In most cases, you don't want to do that; you only want to use a custom point for a specific connection. To go back to rotating parts as normal, return to the Rotation Point Definition dialog and select **Part Origin** as the rotation point. You will not lose the custom rotation point you just set; it will be saved and listed in the Custom Points list. To use that point again, select **Custom Point** and select the point **Pin** from the list.

Defining and Using Multiple Custom Points

You can define as many custom rotation points as you like. To do this, click the New button on the right hand side of the Custom Points list. The "Name" field will read "New Custom"; change this to your desired name. When you click OK and return to the Rotation Point Definition dialog, you should see your new custom point listed in the Custom Point list.

Other Rotation Points

There are also a few other rotation points you can use as MLCad's rotation point for parts. Besides Part Origin and Custom Point, you can define the rotation point as the center of a part, or at the origin of the world (0, 0, 0). There is also an option for "Part Rotation Point"; however, this does not differ from the Part Origin.

Showing the Rotation Point in the Workspace

You can select the checkbox **Show Rotation Point** if you want to see the rotation point in the workspace. When this box is selected, MLCad will display the rotation point in the form of a dark gray box in each of the 2-dimensional view panes.

Aligning Different Types of Parts

Now that you know how to place, move, and rotate parts, it's time to put your knowledge to good use. While it may be easy for you to line up bricks and plates in simple studs-up configurations, most LEGO models contain parts that require some degree of rotation — whether it is something as simple as placing wheels on a Minifig scale car or as complex as the landing gear assembly on a gigantic Technic fighter jet.

No Tissues Needed

Non studs-up construction is such an important part of LEGO-building that a group of die-hard fanatics coined the acronym "SNOT," which stands for Studs Not On Top. This affectionate but mucus-free phrase encompasses any brick or part with studs that are facing any direction but up. There are some builders who focus on building entire chunks of their models this way, to take advantage of the parts' geometries and come up with new and interesting shapes, or shapes that more accurately reflect the subject being modeled.

Truth is, there are many types of parts that require rotation. The LEGO Company has introduced countless specialty parts to the market over the last twenty years, which give builders a large sphere of creative possibilities. Learning how to align the LDraw versions of these parts can be rather tricky. Let's learn about some of the basic types of parts, and how to use the grid to align them.

Tires, Wheels and Axles

Wheels and axles are common parts, but lining them up with each other is pretty tricky. You need to use the Fine Grid setting to get the tire, wheel, and axle to line up with each other. Here are a few simple steps to accomplish this task.

1. Find the following parts and place them in the workspace.

   - Plate 2×2 with Wheels Holder Wide
   - Wheel Wide
   - Tyre Wide (Note: James Jessiman introduced tyres spelled with a "y." To a North American reader, this is the incorrect spelling. In Europe and in British-influenced English speaking countries, "tire" is spelled with a "y.")

2. Rotate the parts into their proper orientation, so the wheel and tire are ready to be lined up with the axle peg.

3. Move the wheel into place over the axle. Note that in some views, the two do not appear to line up when using the Coarse Grid setting. Set your grid to **Fine**, and use the keyboard to move the tire into place. (See Figure 5-12 on the following page.)

NOTE   *On the Fine Grid setting, it is easiest to use the keyboard or toolbars versus the mouse for all movements, as the increments are small enough that you are liable to slip a fraction of a stud away from your intended target. Make it a practice of using the keyboard while on Fine Grid.*

4. Switch your Grid settings back to **Coarse**.

5. Move the tire approximately around the wheel.

6. Switch your Grid settings to **Fine** again, and move the tire into place using the keyboard. (Remember, use the **HOME** key to move the tire in the vertical.)

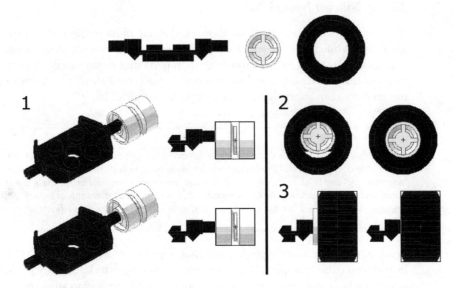

Figure 5-12: Aligning wheels and tires using Fine Grid and multiple views.

Congratulations! You have successfully assembled wheel parts.

SNOT Bricks

Another common case you will experience is the need to connect one brick to a horizontal-facing stud on another brick. As we said before, this type of technique falls under the term SNOT. There are many bricks that feature studs at 90-degree (or other) angles. Here is one example of that technique.

NOTE   *Make sure your Grid is set to Coarse before dragging new parts onto the workspace. It is much easier to work off of an even grid increment than off of small LDraw units.*

1. Find the following parts and place them in the workspace.

    - Brick 1×1
    - Brick 1×1 with Headlight

2. Determine which axis you need to rotate the brick on to meet up with the horizontal stud on the "headlight" brick. Rotate the brick so the antistud, or socket, aligns with the sideways-facing stud. Hint: Make sure you reset your 3D view by a **Right Click > View > 3D** in that pane. Next, compare the parts and their orientations to the axes on the toolbar icons.

3. Move the bricks as close together as you can on the Coarse Grid.

4. Switch to **Fine Grid** and finish aligning the pieces.

Clips and Bars

Another major part category has to do with clips and bars. These two categories encompass a wide range of parts. For the purpose of this section, we will group together parts with similar elements. Clips can be anything from minifig hands to wall mount brackets. Bars are standard LEGO-sized cylinders; they are present in minifig hand tools, Star Wars light sabers, fences and railings, and more.

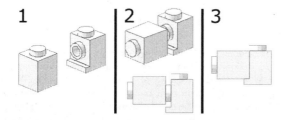

Figure 5-13: Aligning SNOT bricks using Fine Grid and multiple views.

The first image in Figure 5-14 shows a Plate 1×1 with Clip Vertical – Type 3 holding a Bar 4L Light Sabre Blade. Go ahead and bring these parts into the workspace and align them. You should notice that if your Grid is set to Coarse, they line up right away. This is because the parts' origins allow for easy alignment in this orientation.

Now grab a Plate 1×1 with Clip Horizontal. The only difference between this part and the previous is the clip is rotated 90°. With your Grid still set to Coarse, you will find these two don't line up when you rotate the bar. It needs to be moved into position using the Fine Grid (note images 2 and 3 of Figure 5-14).

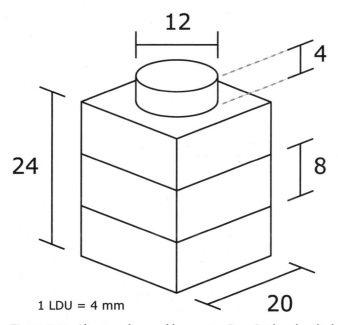

Figure 5-14: Aligning clips and bars using Fine Grid and multiple views.

Hinges

Hinges are relatively straightforward parts to align once you have mastered the use of Fine Grid. However, these parts are often used at arbitrary orientations, and this adds to the complexity of the alignment process. You should practice with a couple types of hinges to get the feel for aligning them after fine rotations.

Figure 5-15 demonstrates hinge alignment first using a Hinge Plate 1×2 with 2 Fingers, and its counterpart the Hinge Plate 1×2 with 3 Fingers. The second example uses the Hinge 1×2 Base and Hinge 1×2 Top. Practice these on your own.

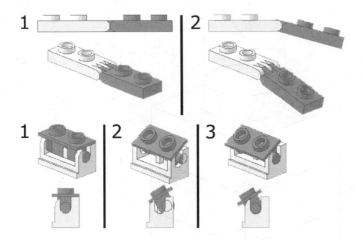

Figure 5-15: Aligning clips and bars using Fine Grid and multiple views.

Technic

The various Technic parts require heavy use of the Fine Grid. Gears go on axles, pegs and axles inside holes, and more. While you probably have the hang of Fine Grid positioning, why not go ahead and grab a few Technic parts to try out. In Figure 15-16, you see a Technic Brick 1×4 with Holes, a Technic Pin, Technic Axle 4, and Technic Gear 16 Tooth.

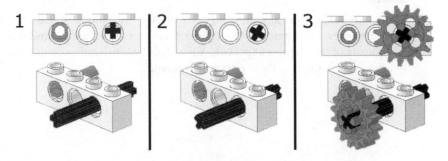

Figure 5-16: Aligning Technic parts using Fine Grid and multiple views.

Summary

In this chapter, we covered moving parts, rotating parts, and aligning parts within LDraw, and gave you examples of each task. You learned how to use the grid settings and how to set custom rotation points for your parts. The skills you learned in this chapter will equip you for building models later in this book and on your own in the future.

6

EXPLORING MLCAD

MLCad has many features beyond simple parts placement. Its powerful editing commands, program options, and integration of various automatic building generators make it the best choice for a Windows LDraw editor to date. MLCad even generates reports of the parts in a model on command — making it easy to know which parts you needed to build it.

MLCad's user interface is both simple and complex at the same time. You can quickly get around placing bricks from the parts list and orienting them, but learning advanced techniques can be tricky. To master the program, pay attention closely to the menus and icons. The English version of MLCad doesn't always make sense to native speakers. The program was written by an Austrian, Michael Lachmann, and is supported by a group of volunteers who test and make suggestions, using their own free time. Michael has recently offered his language documents to native English speakers to correct, so hopefully MLCad's English will be corrected in the near future. Despite the language drawbacks in the current version, you can become an expert at MLCad by practicing and by paying close attention to the various features we will discuss.

This chapter covers many important MLCad concepts that you will need to be familiar with before proceeding. A little bit of what we discuss here is review from the exercise in Chapter 3 and the lessons on placement in Chapter 5, but not much. Here we will cover the features that help you build models with efficiency, such as using the color palette, generating parts lists for your models, and using the context-sensitive menus. We'll save the higher-level concepts like sub-modeling, creating building instructions with Buffer Exchange and Rotation Steps, and using flexible elements for later chapters.

Using the Toolbars

Almost all of MLCad's common commands can be run through the buttons on the toolbars. Each command on the toolbars can be run through the menus above as well. We will list those commands when talking about each feature, but our focus will remain on running the commands by using the various toolbars.

You have already been introduced to a few toolbar features in previous chapters. In this section, we will skip the ones you've already learned and fill you in by teaching the rest of them. Once you have learned what these features do and how to use them, you will be well on your way to mastering MLCad.

Organizing Your Toolbars

MLCad allows you to organize your toolbars just as any other Windows application does. Remember from Chapter 3, you were able to drag the toolbars into place to set up MLCad like our screenshot? You can also show and hide different toolbars. Do this by right-clicking your mouse over a toolbar. A context-sensitive menu (as shown in Figure 6-1) appears that lets you check off which toolbars you want to show or hide.

Figure 6-1: Show or hide various toolbars by right-clicking over any toolbar in MLCad. This context-sensitive menu will appear, allowing you to choose which toolbars you want to see.

Toolbar Names

The following table shows MLCad's toolbars and their names, along with the commands they hold and the chapters that discuss them.

Image	Toolbar Name & Contents
	Toolbar File and Edit commands such as save, open, copy, and paste (Chapter 6).
	View Bar Modes and Grid settings (Chapters 3 and 6).
	Edit Bar Part, comment, step, rotation step, background image, clear, and Buffer Exchange commands (Chapters 6 and 12).
	Element Bar Movement, rotation, and position commands (Chapter 5).

Image	Toolbar Name & Contents
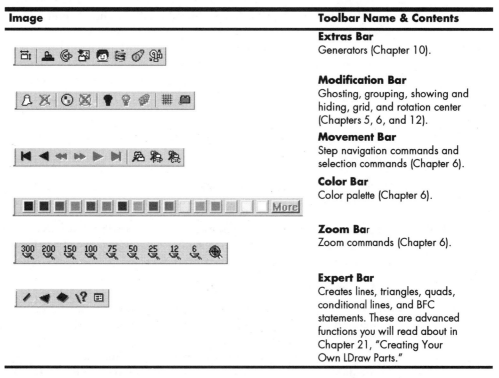	**Extras Bar** Generators (Chapter 10).
	Modification Bar Ghosting, grouping, showing and hiding, grid, and rotation center (Chapters 5, 6, and 12).
	Movement Bar Step navigation commands and selection commands (Chapter 6).
	Color Bar Color palette (Chapter 6).
	Zoom Bar Zoom commands (Chapter 6).
	Expert Bar Creates lines, triangles, quads, conditional lines, and BFC statements. These are advanced functions you will read about in Chapter 21, "Creating Your Own LDraw Parts."

Throughout this chapter and others, we will refer to various icons on these toolbars. We've already covered some on them in previous chapters, and will feature others later when explaining other concepts.

MLCad's Modes

MLCad has built in four different "modes" for the user to work in. The default mode is Edit Mode, where you can place parts on the screen to create your model. View Mode is for browsing through a model step by step. Zoom Mode lets you arbitrarily zoom with your mouse on the workspace. Move Mode changes your perspective of the model in each workspace pane.

Changing Modes

Each mode is selectable via buttons on the View bar.

Figure 6-2: MLCad's Mode buttons, left to right: View Mode, Edit Mode (active), Move Mode, and Zoom Mode.

View Mode

View Mode is used in conjunction with the navigation buttons in the Movement Bar. These buttons work in the same way as the play and skip buttons on a CD player (see Figure 6-3 on the next page). Press them to navigate your way through a model's construction step by step. If a model has no steps in it, pressing the buttons will not change how the model is displayed. The middle two buttons that look like fast-forward and rewind buttons are "Fast

Step" controls. Use them to jump ahead a predefined number of steps (the default is five steps). We'll show you how to change this value later in this chapter in the "MLCad's View Options" section.

View mode will not let you edit a model; you can only scroll between a model's steps. You can also access the controls in Figure 6-3 by using the Navigate menu. We will introduce you to the concept of inserting steps into your models later in this chapter.

Figure 6-3: The navigation buttons on the Movement Bar.

Edit Mode

Edit Mode is the mode you use to place and move parts. You're already familiar with working in this mode, so there's no need to discuss it further here.

NOTE *When you toggle between Edit and View modes, MLCad actually re-draws its window. Therefore, if you're working with MLCad maximized in Edit Mode and decide to switch to View Mode, MLCad may re-draw the window as a normal, non-maximized pane.*

Move Mode

Move Mode allows you to change which part of a model you are looking at. If you click and drag your mouse in one of the workspace panes, you will see the view scroll in the direction in which your mouse moves. Think of this like cropping a picture: If you are zoomed in close, the model is too large to fit in the pane. By moving the frame, or the crop box, you can see a different part of the model. This does not move the model in 3D space, but simply changes what you see.

Zoom Mode

Zoom Mode works in a manner similar to Move Mode, but when you drag your mouse within a workspace pane, it zooms in and out. Dragging the mouse up will zoom out, and dragging down will zoom in. This is useful if you prefer to choose a custom zoom level instead of using one of the Zoom bar presets.

Working with Parts and Lines

You've already learned the fundamental concepts behind the Parts List and Parts Preview Window, how to move parts in the workspace, and how to place, move, and rotate parts. Now we'll show you how to manipulate parts and lines when creating your model.

LDraw Files are Linear

Just as you build a physical LEGO model one piece at a time, you also assemble an LDraw model one piece at a time. An LDraw file is written line by line; each line in a model is either a part or a comment (we'll talk more about comments below). Therefore, the parts in each model file have a sequential order to them, by nature.

When building, you may find that you want to change the order of parts from time to time. For example, if you are designing instructions, you might mistakenly place a part that obscures the view of a part to be placed in your next step. Moving the part to a later step will allow the instructions to be read more clearly.

MLCad provides many options for editing LDraw files in a WYSIWYG environment. You can interact with the LDraw model file through the Model File Window, which is located directly above the workspace. This window displays all lines in your file — comments, parts, and meta-commands.

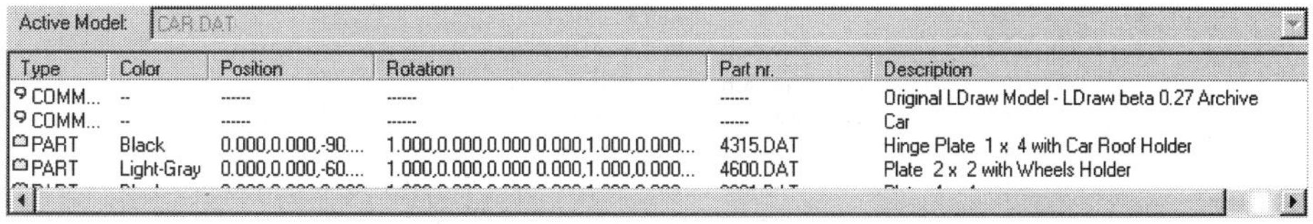

Type	Color	Position	Rotation	Part nr.	Description
COMM...	--	------	------	------	Original LDraw Model - LDraw beta 0.27 Archive
COMM...	--	------	------	------	Car
PART	Black	0.000,0.000,-90...	1.000,0.000,0.000 0.000,1.000,0.000...	4315.DAT	Hinge Plate 1 x 4 with Car Roof Holder
PART	Light-Gray	0.000,0.000,-60...	1.000,0.000,0.000 0.000,1.000,0.000...	4600.DAT	Plate 2 x 2 with Wheels Holder

Figure 6-4: MLCad's Model File Window, clearly showing the different lines of the model file.

Types of Lines

An LDraw file, at its core, is essentially all lines. There are different types of lines. Each type of line tells an LDraw editor or viewer (in this case, MLCad) what to display on the screen or what to do. Lines of a file can be parts, comments, steps (essentially special comments), and then lines (objects in LDraw), triangles, quadrilaterals, or optional lines. We'll go over the last four in Chapter 21, "Creating Your Own LDraw Parts," so don't worry about them for now! Right now we will focus on parts, comments, and the subset of comments, steps.

Parts

You've already experienced working with parts in this book; it was one of the first things we had you do in MLCad. A part is a type of line in the LDraw file, and you can see each part in the Model File Window. Note the description that MLCad displays at the end of each line in LDraw files, and you can tell which line represents which part.

In raw LDraw files, all part lines begin with a 1. All part lines reference another LDraw file, such as a part or another model, and keep track of its color, location, and rotation in 3D space. Each LDraw part is a separate file, residing in your own \LDraw\PARTS\ folder, that gets referenced in the raw file when you drag a part onto the workspace. Part lines look like this in the raw file:

```
1 15 -90 -7 0 0 1 0 1 0 0 0 0 -1 2909.DAT
```

The good news is, unless you're a parts author, you really don't need to know what all the numbers in the middle do! The last bit, 2909.DAT, is the actual reference to a part file.

Adding Parts with the Select Part Dialog

There's one more way of adding parts to your model, rather than just dragging them onto the screen. The Add Part button 🔲 brings you to the Select Part dialog (Figure 6-5 on the next page). This window allows you to scroll through the parts list and find the part you want to insert. It also allows you to insert a file as a part. For example, you can insert another LDraw model that you created, and have MLCad treat it like a part. This is called a *sub-model*. We will deal with sub-models in greater depth in Chapter 8, "Complex Models: Sub-modeling and SNOT."

While the Select Part dialog is a nice feature, you will probably want to stick to using the Parts List window and Parts Preview window until you are comfortable enough with the LDraw part names to select a part without a preview image.

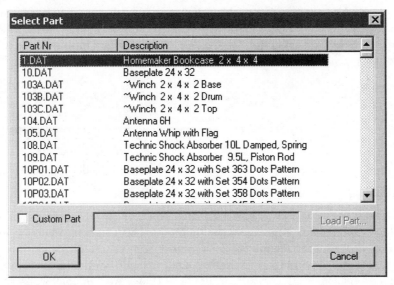

Figure 6-5: The Select Part dialog is useful when you know your way around the parts library.

Comments

Comments are just like the name suggests. You can insert a comment line into your model to note something for future reference or to clarify what you're doing, in the same way that programmers insert comments into their source code.

In raw LDraw files, all comment lines begin with a 0. Immediately following that leading 0 is the comment itself. So the following line is a comment, as seen in a raw LDraw file:

```
0 Hi, I am a comment!
```

Most comments have no effect on how the software functions. There are special comments called *meta commands* that MLCad or other programs use to perform special functions. Most of these are transparent to you, the user, because the program writes them itself. However, there will be a few extensions in this book we will need to teach you to create by hand.

Adding Comments to Your Model

The Add Comment button **abl** allows you to insert a comment line into your LDraw file. A dialog will pop up, allowing you to type in your comment. Upon clicking **OK** you will see your comment inserted in the Model File window at the top, and MLCad will highlight it as the active line.

NOTE   *When you insert a comment in MLCad, it appears as* 0 WRITE *(your text) in the raw LDraw file. MLCad inserts the word "WRITE" in each comment it creates. While this is acceptable if you're writing a message to yourself about your model file, it doesn't work with extensions. If you're ever inserting extensions by hand, remember you can edit out the "WRITE" portion using a text editor like Notepad or an LDraw editor like LDDesignPad (introduced in Chapter 10, "Creating and Using Flexible Elements," and covered in depth in Chapter 21, "Creating Your Own LDraw Parts").*

Steps

Steps are special types of comments that allow you to divide your model into individual steps, just like LEGO building instructions. The LDraw community doesn't quite consider them meta commands because they were a part of the LDraw package from the beginning. A *step* is essentially a marker that programs interpret as a dividing point in a model. A step tells a program that "every part before this point in the file should be drawn in this step, and every part after this point should not be drawn." Steps are very simple comments that look like this in the raw file:

---

0 STEP

Adding Steps to Your Model

The Add Step button ![Add Step icon] inserts a step marker into your file. This is how you tell MLCad where each step of your model's building instructions should be. For more in-depth discussion of building instructions, see Chapter 12, "Introduction to Building Instructions."

Editing Models in MLCad

The Model File Window can be used to change the order of parts, add parts in the middle of a file, and delete parts. These are important actions whether you are building or refining a model. It also allows multiple-line selections, grouping lines, and cut, copy, and paste functions.

Navigating a Model File

Because parts are placed in a linear fashion when building in an LDraw editor, there is always one line that is active; this line is the place you are editing in the model. We will call this the *Active Line*. In MLCad, you can select a line from the Model File Window by clicking on it. When you do this, MLCad highlights the line. You can click on other lines to select them as active, or even use your keyboard to navigate through active lines. To select the previous line as the active line using your keyboard, use the **Page Up** key. To select the line below, use **Page Down**.

Draw to Selection ![Draw to Selection icon]

When building a model, it is sometimes a good idea to work with the Draw to Selection option on; this way, MLCad will only render up to your active part. This feature gives you a good visual indication of where your active line is when you are building. If you have a complete model and turn this feature on, it only shows you the model up to the active line, like Figure 6-6 on the following page.

NOTE *If for any reason you scroll back in the model and insert parts in the "middle" of the sequence, you should notice your entire model is not rendered. Check your active line, scroll down in the model file window, and select the last part. Now all the parts you add will be inserted at the end of the file, instead of in the middle.*

Selecting Multiple Parts Simultaneously

Selecting more than one part at a time is very useful when editing a file. You may want to move an entire group of parts around, or you might even want to take that group and change where it appears in the sequence of parts in your file. You can select multiple parts in the workspace from the toolbar, in the workspace with your mouse, or by using a combination of the mouse and keyboard to select the parts in the Model File Window. The latter two methods are based on how Windows handles selecting multiple objects and files using both keyboard and mouse commands, so they should seem familiar.

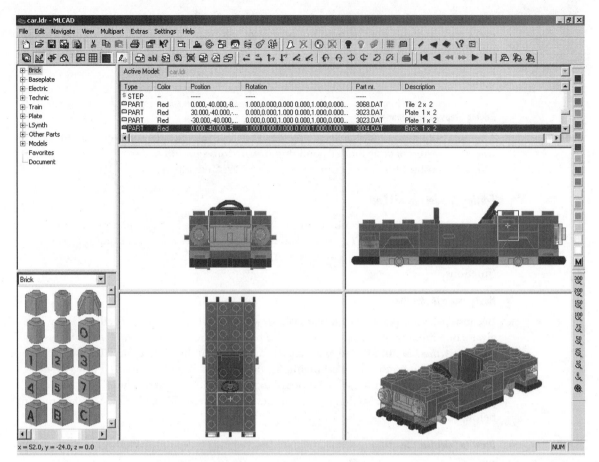

Figure 6-6: Toggling on Draw to Selection makes MLCad render to your selected part only: in this case, a 1x2 brick in the middle of the car. Note the active line in the Model File Window.

Selecting Multiple Parts on the Toolbar

There are three toolbar icons that allow you to select multiple parts simultaneously:

- **Select All**

  To select all parts, click this button.

- **Select Same**

  If you have a part currently selected, you can click this button to select all of the same type of part in your model.

- **Select Same Color**

  If you have a part currently selected, you can click this button to select all parts of the same color in your model.

Selecting Multiple Parts in the Workspace

There are two ways you can use your mouse to select multiple parts in the workspace.

For the first method, while holding down the **CTRL** key, click several parts in your model. (Note: Don't click parts in the 3D pane, just the three 2D panes.) You will see MLCad select the parts you pick in the workspace and highlight their corresponding lines in the Model File Window. Release **CTRL** and you can click one of the selected parts and drag them all around at once. Hold down **CTRL** again, click a part that's not selected, and MLCad adds it to the selection. Click a part that is not selected in this group, and it deselects the parts you just had selected.

NOTE   *If you want to select multiple parts but the parts you want to select aren't all visible in the same view pane, no worries — simply select some of the parts in one view with **CTRL** held down, then select the rest of the parts in another.*

The second method is a little bit easier to do, but it allows less flexibility. This method uses only your mouse, and does not require you to hold down a key to select multiple parts. Simply click your mouse and drag it across the screen in a 2D pane, as you see in Figure 6-7. When you release, all the visible parts wholly or partially covered by highlighting will be selected, as in Figure 6-8.

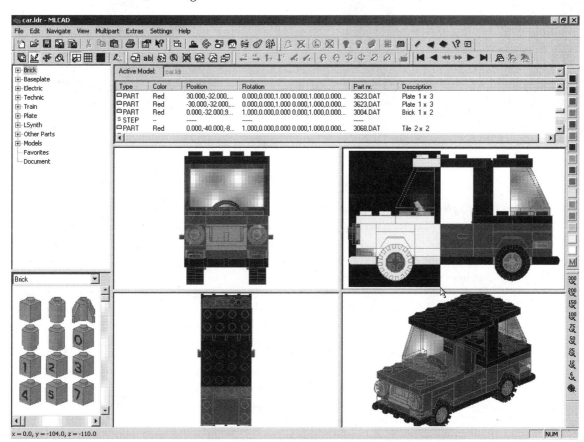

Figure 6-7: Selecting multiple parts by dragging the mouse.

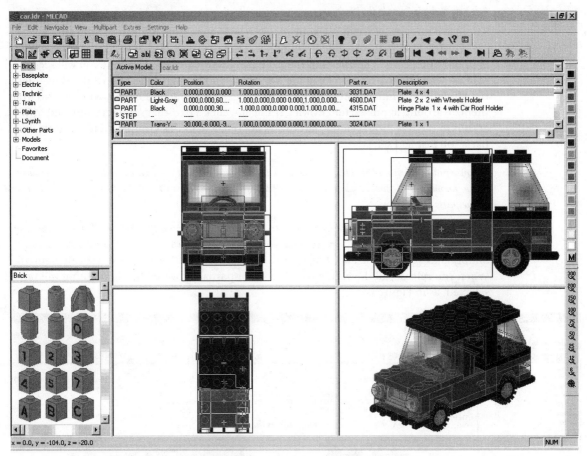

Figure 6-8: When you select multiple parts using the mouse, MLCad selects all of the parts either wholly or partially covered by the highlighting in Figure 6-7.

Selecting Multiple Parts in the Model File Window

There are two keyboard commands for selecting multiple parts in the Model File Window.

For the first method, hold down the **CTRL** key and at the same time click several lines. This will add each part you click to the current selection. Go ahead and give that a try; all of the lines you clicked will appear highlighted.

The second method uses the **SHIFT** key. Select a single line with your mouse, hold down **SHIFT**, then click another line several lines away from the first. By doing this, you are defining the endpoints of your selection. MLCad will select all of the lines in between.

Grouping and Ungrouping Parts

MLCad also allows you to bind multiple selected parts into a group. *Groups* take multiple parts and bind them together in a single unit. MLCad interprets a group as a single part, so when you rotate or move them, they act as one. They have a new, shared center of rotation, as though the parts were glued together as you see them.

Grouping is not the same as sub-modeling, which we will explore in depth in Chapter 8. In a sub-model, you build a group by starting a new model file, building the assembly you want to group, saving the file, and referencing your new file inside of your main model file. MLCad treats that assembly as one part because it uses one line to reference it.

Grouping versus Sub-modeling

Grouping is an MLCad-specific feature, defined by a meta-command. MLCad reads the meta-commands and interprets them as intended, but no other LDraw editor or viewer has been programmed to understand this. If a file containing a group were to be opened in another editor, such as LEdit for DOS or BrickDraw3D for the Macintosh, the "group" would not exist. Instead, the program would treat each part individually. The meta-commands MLCad used to create the groups would be seen as comments.

Creating a Group

To create a group, use the following steps:

1. Select the parts you want to include in the group.
2. Create a group from the parts by clicking the Create Group icon 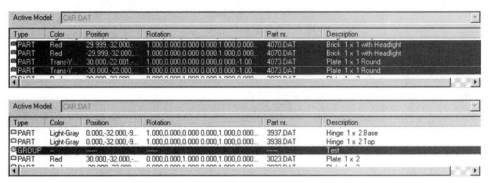, by selecting **Edit > Group > Create**, or by using the keyboard shortcut CTRL+G.

Get the feel for the behavior of the group you just created by moving and rotating it.

Figure 6-9: Selecting multiple parts and creating a group "Test."

Ungrouping

To ungroup your parts, select the group in the model file window and click the Ungroup icon. ⊠

Changing the Order of Parts within a File

Cutting, Copying, and Pasting are not the only ways to change the order of parts inside of an LDraw model. You can also use your mouse to drag and drop individual lines or groups of lines to a different location within the model file itself. To do this, select a part or group of parts. Next, drag the parts up or down in the file. You will notice your mouse cursor changes to an arrow with a small rectangle underneath it; this tells you that you have picked up the parts and are moving them. To complete the act, simply drop them in the place desired. For beginners, this can be tricky. Sometimes it is difficult dropping a part in the right location, because MLCad gives no indication (such as a small line) of where you are about to drop. Spend a little time getting used to moving parts with this method, and you will begin to find it quite useful.

Hiding Parts

MLCad allows you to hide parts without deleting them. This is especially useful if you need to see behind a part or a group of parts to build a portion of your model. You can use the Hide button 💡 to hide all parts you have currently selected. If you want to show an individual part, click the Show button. 💡 If you have multiple hidden parts and want to show them by clicking a single button, you can use Show All. 💡

MLCad does not save information on hidden parts with the LDraw file, so the next time you open your file, all parts will be shown.

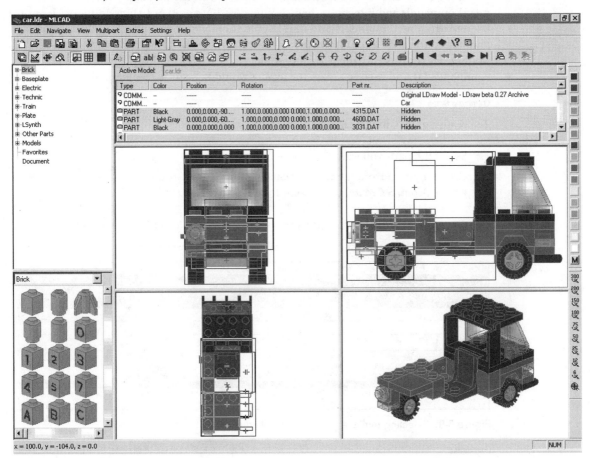

Figure 6-10: MLCad allows you to hide parts so you can see around them to work.

Managing Colors

Assigning colors to your parts is easy using MLCad's color palette. This is accessible through the Color Bar, which you can see in Figure 6-10 along the right hand side of the screen. While it's easy to change colors, there's a bit more behind how colors are assigned and interpreted in the LDraw format. There's also a lot of complexity behind how those colors are affected when you want to convert your model to POV-Ray to create nice renderings, which we will show you how to do later in this book.

LDraw Colors

The LDraw format offers three kinds of colors: solid (or opaque), transparent, and dithered. These are the "standard" colors. Because there are more LEGO colors than the original LDraw program defined, LDraw.org has defined additional "extended" colors. Most current programs recognize some or all of these extended colors, but they aren't completely standard — and the original LDraw program doesn't recognize them at all.

MLCad's Color Palette

To access MLCad's color palette, click the **M** button at the end (for "More" colors). If you happen to have your color palette along the top of your MLCad window, you will see the word "More" spelled out (see Figure 6-11).

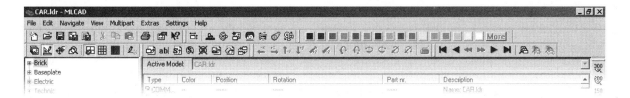

Figure 6-11: You can launch the MLCad Color Palette using the "More" button on the Color Bar when it is stored on top, or "M" when it is stored on the side.

If you are used to working with color palettes in image editors on your computer, you will quickly realize it is probably laid out differently than anything you have seen before. As you can see in Figure 6-12, the palette is arranged in rows of eight so that two rows give the full range of 16 standard solid colors. The scrollbar on the right-hand side lets you scroll through the available colors.

Below the color swatches (tiles of sample colors), MLCad displays the currently selected color along with its name. The grayed-out checkboxes indicate whether or not the selected color is transparent or dithered. Figure 6-12 shows Black selected, which is neither transparent nor dithered. You can enter the color's number in the Color Number dialog. Clicking on a color swatch will display the color's LDraw number. In this case, Black is color 0.

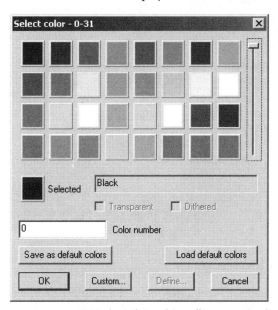

Figure 6-12: MLCad's color palette allows you to choose the original 16 solid colors along with extended ones.

Besides the 16 standard colors, this first "page" of colors shows two special color values, and MLCad's 11 extended solids.

The two special color values are 16 (first color, third row) and 24 (first color, fourth row). These colors are not often used in regular models, but they are very important in part files. They can also be useful in advanced modeling, when you might use sub-model files.

MLCad's 11 extended colors make up the remainder of rows 3 and 4. These colors are real LEGO colors, but are not included in LDraw.org's standard color set. The last three swatches on the page are "unused" colors. MLCad has no color definition for these values. You can use them in your models, but MLCad will render them as gray.

Transparent Colors

James Jessiman assigned transparent colors to numbers 32 through 47. A transparent color in the original LDraw/LEdit package was [Color Number] + 32. Transparent black, the equivalent of today's "smoke" gray windshield pieces such as in the Star Wars sets, would be color number 32 "Trans-Black." Clear is represented by color number 47 "Trans-White," which is white's color number (15) + 32.

To access the transparent colors in the color palette, take the scrollbar on the right and move it down one notch. Since the colors are laid out in rows of eight, and four colors per page, the first transparent color is the first swatch on the second page (see Figure 6-13). These colors are laid out in the same order the solid ones are on the first page. One catch is that not all the colors on the first page have transparent versions in MLCad. Only the first 16, plus Trans-Orange (a la Ice Planet 2002), are available. The first 16 are recognized by the original LDraw program, and are part of the standard color set. Trans-Orange corresponds to the extended color Orange, and its value is calculated the same way: Orange is color 25, and Trans-Orange is 25 + 32 = 57. Note that in Figure 6-13, the color is Trans-Orange, and the grayed-out checkbox "Transparent" is checked.

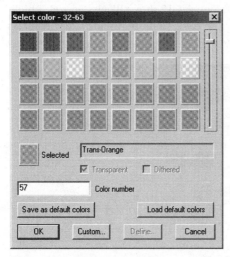

Figure 6-13: Transparent colors are accessible on the second page of colors in the MLCad color palette.

Manipulating the Workspace

MLCad has a few helpful options that allow you to manipulate the workspace while you are building. These are accessible via the context-sensitive menu in each workspace pane, and affect each pane differently. To access this menu, right-click in one of the workspace panes.

Add	▶
Change Color...	
Modify...	
Enter Pos. + Rot...	
Snap to grid	Ctrl+Shift+G
Visibility	▶
Wireframe	
Outline	
View Angle	▶
Zoom	▶
Scrollbars	

Figure 6-14: You can control how the workspace displays the model via each pane's context-sensitive menu.

Wireframe and Outline View

You can change any view pane to display a wireframe or an outline view of your model. Figure 6-15 shows an example of both — wireframe view on the left, and outline view on the right. Wireframe draws the vertices of each part; even on simple models like the car, it appears busy. Outline view displays what are called *bounding boxes*, or simple rectangular prisms that represent the extremities of each part. You can toggle either Wireframe or Outline view by right-clicking inside the desired workspace pane and selecting either **Wireframe** or **Outline**.

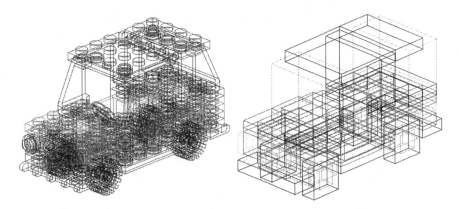

Figure 6-15: Wireframe view (left) draws the vertices of each part, where Outline view (right) draws bounding boxes.

Changing the View Angle

You can also change the view angle of each pane via the context-sensitive menu. This is useful if you want to look underneath, behind, or on the right-hand side of a model, because the default views are Front, Above, Left, and 3D. To change the view angle, right-click inside the desired workspace pane and select **View Angle > [desired angle]**.

NOTE *Sometimes when you are creating or editing large LDraw models, MLCad will run more slowly. This is usually because of the 3D pane and the computing power needed to keep rendering that view as you move parts. You can increase your performance on larger models by turning off the 3D view, either by selecting a different angle or by shrinking the 3D pane to the side so you cannot see it.*

Zooming In and Out

You can zoom in and out on a model either by using the Zoom Bar or by using the context-sensitive menu in the workspace (as a third option, the menu **Settings > Zoom** allows you to do the same). To get the most out of each view pane, select **Fit** from the context-sensitive menu, or click the ⊕ icon.

Using Scrollbars

If you are zoomed in, you can turn on scrollbars to help you navigate to different parts of your model. MLCad's view mode (discussed earlier in this chapter) does the same thing. Scrollbars allow you to scroll without changing the mode you are working in. To toggle scrollbars on or off, right-click inside the desired workspace pane and click **Scrollbars**.

MLCad's Reports

MLCad has several built-in report features that are designed to give you information about the model you are creating. It can tell you the dimensions of your model, generate a list of pieces used in each model (as well as save and/or print that list), and summarize the comments interspersed throughout your model as well.

These are very useful features if you ever want to use your 3D model practically. Because all of the parts are documented in the LDraw file, you can use MLCad's dimensions report to learn your model's physical measurements. Or, more importantly, should you ever decide to build someone else's model using the LDraw file they provide, you can easily find out which parts you need to collect to build it (you can also enter that model into BrikTrak and buy the parts needed off of BrickLink; more on that in Chapter 22, "LDraw and LEGO Bricks: Taking the Hobby Further"). An LDraw Unit (LDU) report feature would be useful, but MLCad has not (yet) implemented this.

Summarize Comments

To generate a summary of the comments in your LDraw model, select **Extras > Report > Comments. . . .** Choosing this menu command will bring up the dialog window seen in Figure 6-16. If you select the **Show old style comments**, MLCad will display the comments as it automatically inserts at the top of the file. You can also check **Show comments of parts** to display all of the comments in all of the parts your model uses. However, this feature provides quite a lengthy report and is not necessary for creating virtual LEGO models. The main drawback to this report is you cannot save a text file of the comments, nor can you print this out directly from MLCad.

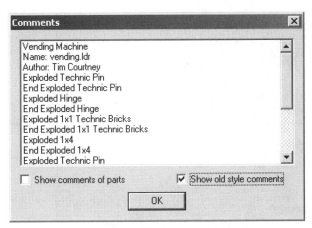

Figure 6-16: You can see all of your model's comments in one screen by using MLCad's Comments report feature.

Model Dimensions

To see a report of your model's dimensions, use the 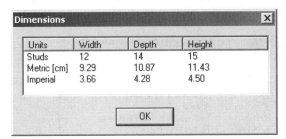 toolbar button, or select **Extras > Report > Dimensions . . .** As you can see in Figure 6-17, MLCad will give you a report of your model's dimensions in studs (LEGO units), centimeters (metric), or inches (imperial).

Figure 6-17: MLCad will give you the dimensions of your model in Studs, Metric, and Imperial units.

Parts List

Finally, possibly the most useful report feature in MLCad is the Parts List. You can tell MLCad to generate this report by selecting **Extras > Report > Pieces. . .** from the menu. Doing this brings up a dialog like the one in Figure 6-18. This displays (from left to right) the quantity of each part, the color of each part, the part number, and the part description.

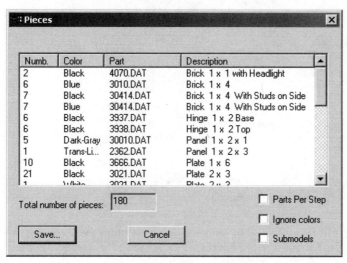

Figure 6-18: MLCad's most useful report displays how many pieces your model contains, including their color and their quantity.

You can also tell MLCad to show you the parts relative to each step, by checking the **Parts Per Step** checkbox (see Figure 6-19). This causes MLCad to add a column on the left, indicating the step number each part appears in. This also automatically sorts the parts list by step number, but you can sort by any column by clicking on the column header.

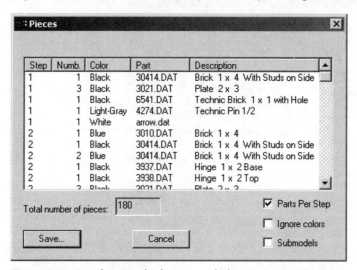

Figure 6-19: Sort the parts list by step with the "Parts Per Step" option.

Save a Text File

You can use the **Save . . .** button to save a text file of your parts list (another option is to use **File > Save Partlist**). After you've saved it, use Notepad or your favorite text editor to open the file and print out a hard copy. This is *very* useful when building a physical LEGO version of an LDraw model!

Saving Pictures and Printing

MLCad is able to print images of your model, print building instruction steps of your model, and save pictures of your model to your hard drive in BMP, JPG, or GIF format. These can be useful if you need images of your model for quick reference, or if you don't want to take the time to convert and render in POV-Ray. If you are using your LDraw files for projects and plan to render them in POV-Ray, saving images in MLCad can be useful for creating preliminary layouts of models before rendering.

Saving Pictures

MLCad's Save Picture(s) dialog (**File > Save Picture(s) . . .**) allows you to save BMP, JPEG, and GIF formatted images of MLCad output. As you can see in Figure 6-20, this feature can generate pictures either for every step, or a "snapshot" of the current view. Other checkboxes allow you to determine if it will generate pictures of your sub-models (to be covered in Chapter 8), or show the step number in the picture. The Central Perspective option is pretty nifty, and you can see what it does to the right of the screenshot in Figure 6-20. You can rotate the view in the preview window by dragging your mouse, just as you can on the 3D pane of the workspace.

The drop-down menus below the checkboxes allow you to set the image options. MLCad allows you to change your image resolution, or pixel dimensions of the image, in standard 3:4 aspect ratio increments up to 1600×1200. The Picture Format drop box allows you to choose between saving the file as a BMP, JPEG, or GIF.

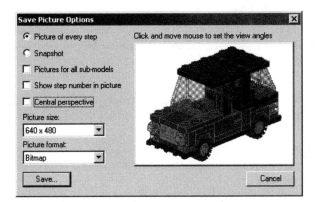

Figure 6-20: Save Picture Options window with Central Perspective off (left) and on (right).

After selecting your image options and clicking **Save . . .** MLCad will prompt you to choose a place to save the image as well as what to call it. If you are generating images of each step, it will save the images with the name you gave including a number (01, 02, 03, and so on) to distinguish the steps.

Adding a Background Image

You can also add a background image to be saved with your model's images, or to be printed out behind it. This adds a bit of life to your building instructions. You probably noticed that official LEGO building instructions have background images to them, and they improve the visual interest. It would be pretty boring to look at blank white pages of instructions, especially when working on a large model! You can see the results of MLCad's background image feature in Figure 6-21 on the next page.

Figure 6-21: You can add a background image to your model for MLCad to save with pictures, or to use when printing.

In this case, we used a simple gradient from an image editor. You can use any image you like. To add a background image, click the ▣ icon. Browse for your image and click **OK**. This will insert a BACKGROUND statement in your file immediately below the active line. Images must be in bitmap (BMP) format.

Be careful about where you position a background statement: MLCad will impose the background image you select on the step the statement appears in and all subsequent steps, but NOT in preceding steps.

NOTE *When you insert a background image into your project, you will not see it in the workspace. It is only visible when you open images you save, print, or when you look at the Print Preview.*

Image Size

Image resolution is an issue when using a background image in MLCad. Make sure that when you save images of your model, you are using a background image that shares the same dimensions with the image you are saving. So, if you are saving images at 1024×758, use a background image of the same size. If you use a smaller background image, MLCad will tile the background image you provide. Chances are you won't want that to happen!

If you look at a background image in the print preview, you will notice that it tiles your image. We recommend saving pictures and printing them through an image editor if you want to print background images, rather than using MLCad's print feature.

Gradient

You can use our gradient image, located under **C:\LDraw\VLEGO\CH06\gradient.bmp**. The file gradient800.bmp is a smaller 800×600 version of the same image.

Printing

MLCad's Print Preview option lets you see how your model will be laid out on the printed page before you print it. While you don't get results that are nearly as nice as you would if you were to create instructions through LPub (see Chapter 15, "LPub: Automate Building Instruction Renderings"), this feature is quick and effective. If your model includes steps, it will print out instructions; the numbers of each step are included in the upper left-hand corner, just like real LEGO building instructions. This method of printing building instructions does not allow for much flexibility. It has no page layout options, so this form of instruction printing should be restricted to simple models without sub-models. For more information on sub-models, see Chapter 8.

Print Preview

To see a preview of your model before printing, select **File > Print Preview** (see Figure 6-22).

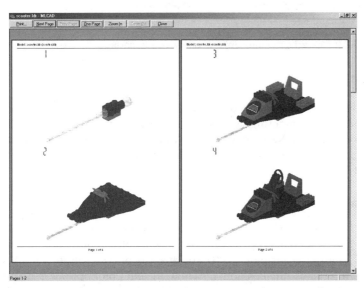

Figure 6-22: The Print Preview window gives you the option of viewing more than one page at a time.

Parts List

MLCad will print a parts list following the last step of your model. This is a useful addition because it can aid you in locating all the necessary pieces to build the model.

Printing Your Model

To print your model, select **File > Print**. This dialog is identical to the print dialog in other Windows applications.

Changing the Printing Options

You can change MLCad's printing options by selecting **Settings > General > Change . . .** and then the **Printing** tab in the MLCad Options window. This gives you access to the various elements of the printed page, as seen in the Print Preview window in Figure 6-22. Note that MPD files will be covered in Chapter 8.

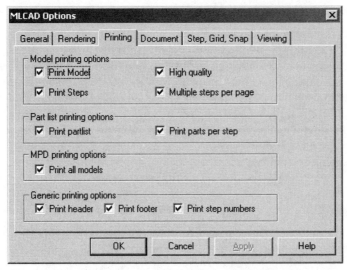

Figure 6-23: MLCad's printing options allow you to change the way MLCad prints your models.

MLCad's Options

We've already discussed some of MLCad's options, such as the Printing settings above; the General and Rendering settings in Chapter 3; and the Step, Grid, Snap settings in Chapter 5. There are two other tabs in MLCad's options to become familiar with, and they contain a few important controls. To access MLCad's Options window, select **Settings > General > Change . . .** from the menu.

Document Options

MLCad's Document options, accessible via the **Document** tab on the **Options** window, control how MLCad handles new elements being placed on the screen (see Figure 6-24). You can control the angle of the 3D workspace pane with the **Default 3D Rotation Angle** settings. You can use the Position, Orientation, and Color options to determine how MLCad treats new parts when inserting a part by either duplicating or using the **Select Part** dialog.

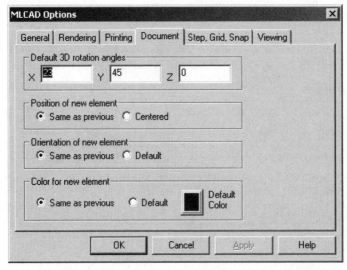

Figure 6-24: MLCad's document options allow you to set defaults for the 3D view and for new part placement and color.

Viewing Options

MLCad's Viewing options, as seen in Figure 6-25, allow you to control elements of View Mode.

When browsing through the steps of a model, MLCad can render the new parts for each step (or parts from previous steps) differently. You can change how new or old parts are treated on instruction steps by adjusting the **Added parts view type** drop-down menu. Leaving this option set to **Off** does not treat new parts in steps any differently than old parts. You can tell MLCad to highlight new parts when browsing instructions in view mode, or you can have the program gray out old parts or make old parts appear transparent. Graying out parts changes color parts to grayscale, whereas making parts appear transparent dulls them a bit in comparison to the current parts in a step.

MLCad allows you to define what your left mouse button does in the 3D pane when in view mode. You can either use it to rotate the model like you would in Place Mode, or draw the next step by clicking in the 3D pane.

Finally, you can change the number of steps MLCad will skip for "fast steps." We mentioned fast steps earlier in the chapter in the View Mode section. Here is where you can change how many steps MLCad skips when you use the fast step navigation buttons.

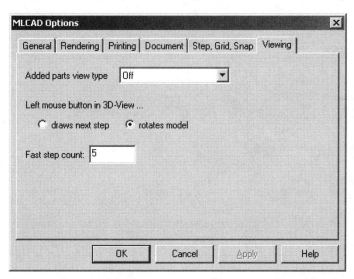

Figure 6-25: MLCad's viewing options allow you to change how models are rendered in View Mode.

MLCad's Help File

MLCad offers an HTML-based help system, with quick references to various features and key commands (we'll cover all of MLCad's key commands on the cover flap in the "Cheat Sheets" section). To access the help system, select **Help > Help Topics** from the menu. As you can see in Figure 6-26, the sidebar categories are collapsible, just as the Parts List window is. Here you can find informatio on any topic. This window is resizable as well. Contrary to the name on top, this file is not online; it is stored on your computer inside of your MLCad directory. MLCad's help file is also put together by volunteers, and as of this printing there are plans to clean up translations and translate this document into other languages.

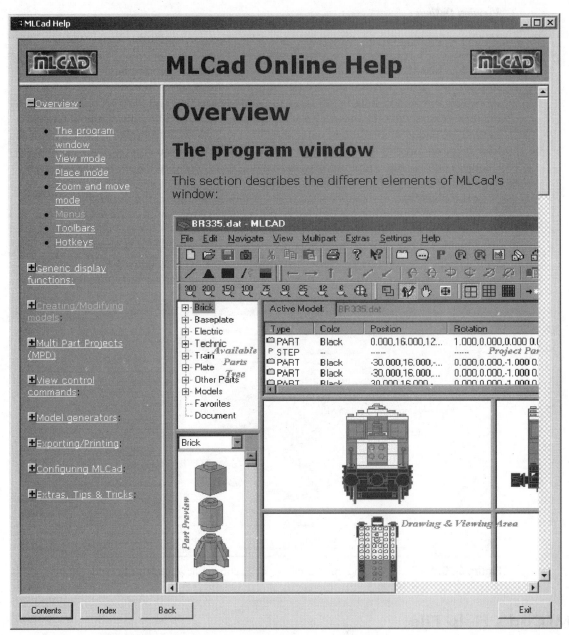

Figure 6-26: MLCad's help file features collapsible menus and pages describing the various features of the program.

Checking Your MLCad Version

To see which version of MLCad you are using, choose the About MLCad button 📇 on the toolbar, or select **Help > About . . .** As you can see in Figure 6-27, we include Version 3.00 in this book. You can check the MLCad website (www.lm-software.com/mlcad/) for newer versions. MLCad is a free program, and the author, Michael Lachmann, periodically posts updates to his website. It is a good idea to check there every once in a while for an upgrade.

Figure 6-27: Use the About MLCad button to find out which version you are running, and to find a link to the MLCad website.

Summary

In this chapter, we've taken you through the essentials of MLCad so you can be better equipped when using it to build your own virtual LEGO models. We've taught you about MLCad's modes and how the mode you're in determines how MLCad functions. We introduced you to the concepts of lines, parts, comments, and steps, and what each means to an LDraw model. You are now familiar with how MLCad and other LDraw-based programs handle color, and which elements of color in MLCad are standard LDraw features. Finally, you've learned about the usefulness of MLCad's reports and how to print your models using MLCad.

7

CAPTURING A SIMPLE MODEL: THE CAR

Now that you have learned how to use MLCad, it's time to put your skills to the test. In this chapter, you will be able to practice by building your own virtual copy of James Jessiman's Car (see Figure 7-1 on the following page). This model, like the Pyramid, is a rather simple design compared to many LEGO sets and custom creations that exist. Before you know it, you will be using the LDraw tools to design models that will blow these two models away! For now though, they're good practice. To see what the Car looks like in color, open the file **C:\LDraw\MODELS\car.ldr** in MLCad.

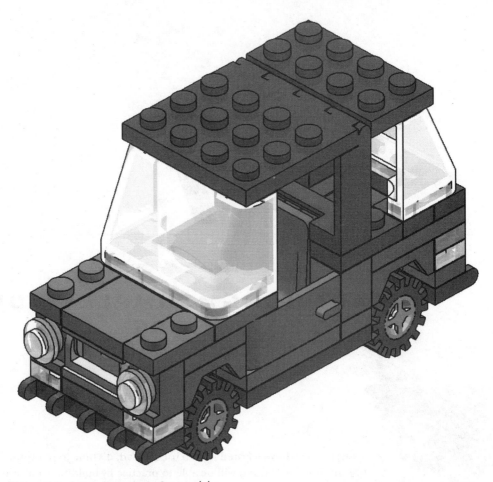

Figure 7-1: James Jessiman's Car model.

Building Instructions

Use these building instructions to create your own copy of the Car in MLCad. If you get
stuck finding part names, check the end of the chapter for a parts list. For a color version
of the instructions, refer to the companion CD-ROM file **\CH07\car.png** (BMP version also
available).

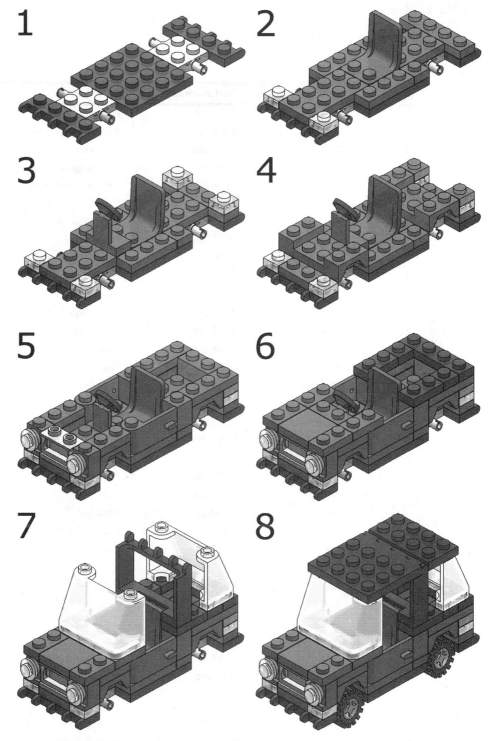

Figure 7-2: Building the Car.

Parts List

Here is the parts list for the Car model.

Table 7-1: Car Parts List

Quantity	Color	Part Number	Part Description
2	Red	3005.dat	Brick 1x1
2	Red	4070.dat	Brick 1x1 with Headlight
2	Red	3004.dat	Brick 1x2
2	Red	3788.dat	Car Mudguard 2x4
1	Red	3829.dat	Car Steering Wheel
1	Red	3822.dat	Door 1x3x1 Left
1	Red	3821.dat	Door 1x3x1 Right
1	Light Gray	3937.dat	Hinge 1x2 Base
1	Light Gray	3938.dat	Hinge 1x2 Top
1	Black	4213.dat	Hinge Car Roof 4x4
1	Black	4214.dat	Hinge Car Roof Holder 1x4x2
2	Black	4315.dat	Hinge Plate 1x4 with Car Roof Holder
1	Blue	4079.dat	Minifig Seat 2x2
2	Black	3024.dat	Plate 1x1
2	Red	3024.dat	Plate 1x1
2	Trans-Red	3024.dat	Plate 1x1
4	Trans-Yellow	3024.dat	Plate 1x1
2	Trans-Yellow	4073.dat	Plate 1x1 Round
2	Black	3023.dat	Plate 1x2
6	Red	3023.dat	Plate 1x2
4	Red	3623.dat	Plate 1x3
1	Black	3710.dat	Plate 1x4
1	Red	3710.dat	Plate 1x4
2	Light Gray	4600.dat	Plate 2x2 with Wheels Holder
1	Red	3021.dat	Plate 2x3
1	Black	3020.dat	Plate 2x4
1	Red	3020.dat	Plate 2x4
1	Black	3031.dat	Plate 4x4
1	Red	3068.dat	Tile 2x2
4	Black	3641.dat	Tyre
4	Light Gray	4624.dat	Wheel Centre Small
2	Clear	3823.dat	Windscreen 2x4x2

Summary

In this chapter, you gained valuable practice by testing your skills on a simple model. Hopefully, the practice will prepare you for the next few chapters on advanced virtual building techniques.

8

COMPLEX MODELS:
SUB-MODELING AND SNOT

Practicing simple building techniques is important when learning how to create LDraw models using MLCad (or any other LDraw editor), but it is only the beginning. In fact, most of the physical LEGO creations you build contain far more complicated connections than the car you just built. While it's pretty easy to snap two physical LEGO parts together at any angle, sometimes making the same connection work using LDraw isn't so simple.

LDraw users are hindered by the fact that parts don't actually "click" together in the system. Combine this with complicated models that have parts facing many different directions, and all of a sudden building in LDraw becomes very challenging! Think of all the rotating and fine movements you would have to do for each part at an odd angle to line them up. What if you ever wanted to move or rotate an entire chunk of a model? Fortunately, we have a solution for that.

This chapter teaches you a technique for managing these models — Sub-modeling — and explains how to use it with what we like to call SNOT (Studs Not On Top) construction and component-based construction techniques. With Sub-modeling, instead of inserting a part and rotating it, you actually insert another LDraw model into your creation like you would a part and orient it. With this technique, you can move entire chunks of your model at once with the same ease with which you move single parts.

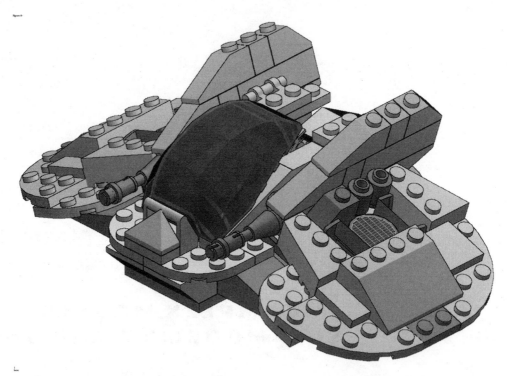

Figure 8-1: Jon Palmer's Scumcraft model.

We will use Jon Palmer's Scumcraft model as an example throughout this chapter because it makes a good design study for creating building instructions. Jon is a prolific builder who specializes in Space themed models. His most notable achievement is the Alphabet Project for From Bricks to Bothans (www.fbtb.net/jon/), where he created Star Wars style fighters derived from the letter designations not used in the movies. Visit Jon's website at www.zemi.net/lego/. This model was originally LDrawn for Jon by Jason S. Mantor.

Sub-models

Sub-models, as the name suggests, are models that serve as components of larger models. *Sub-modeling* is the technique of using an LDraw model that you have created as you would use a part, so the program moves and rotates all the pieces in the sub-model as one single unit. This feature makes creating complicated models very easy! Using Sub-modeling, you can think of a "complex" model as a collection of simple models assembled together, as shown in Figure 8-2.

How does this work? Because each LDraw file is made up of only plain text, and there is fundamentally no difference between the elements of a part file and the elements of a model file, they are interchangeable. Part files even use sub-parts, as well as smaller components called *primitives* (see Chapter 21, "Creating Your Own LDraw Parts," for more information). The LDraw file format relies on files referencing other files to conserve file size, increase portability, eliminate redundancy, and make models and parts easier to edit. In simpler terms, this translates into an easy way for you to build complicated models without a lot of hassle.

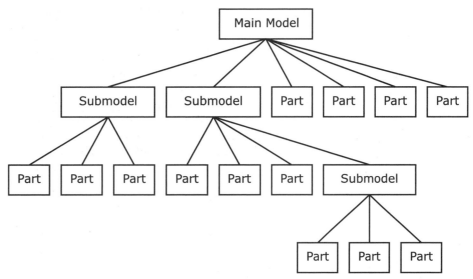

Figure 8-2: Sub-models are actually LDraw files referenced from a larger LDraw file, in the same way parts are referenced.

Two Storage Methods

There are two different storage methods for sub-models. Using the first method, you save individual LDraw files and reference one in another. The second method, called *MPD* (for Multi-Part DAT) allows you to have multiple LDraw models inside one file.

Individual Files

The first method of Sub-modeling is simply to save separate LDraw files, and use the file's name and extension in the place of a part number. LDraw editors and viewers scan for referenced LDraw files in the current file's directory, as well as the LDraw\PARTS\ and \LDraw\P\ directories. The P\ directory is for primitive files. If you are trying to reference an LDraw file outside of your model's current directory, or those three directories, you need to provide the full path. This, however, is not practical or wise, as you may reorganize your directories one day and end up breaking these links without realizing it.

If you are assembling a collection of many large and complex models in a single LDraw file with the intention of creating a scene or a very large creation, this is the method to use. It eliminates enormous Multi-Part files by splitting up the data. Also, if you are doing any manual editing of the files, it's a lot easier to find what you're looking for in a smaller file than in a larger one!

This method does have a distinct disadvantage, though. If you plan to send someone else your model, or if you are posting it online, you will need to send all of the sub-models together for the recipient to see the entire model. It is much easier to transport one file, and the MPD feature solves that problem.

MPD Files

MPD files actually place multiple LDraw *models* inside of one LDraw *file*. This eliminates the disadvantages you would have using separate files. MPD files are able to do that by using a line statement that declares each individual model:

```
0 FILE filename.ldr
```

LDraw editors read every line between the first 0 FILE statement and the next as one model. Also, it is standard to end a file with a 0 (or comment line), whether that is in an MPD file, or in a single-model LDR file. The text below is a sample MPD file. There are two sub-models: 2×4.ldr and 1×1.ldr. As you can see, 2×4.ldr references 1×1.ldr. When you view this file, you see the 2×4 brick with a 1×1 on top of it, as shown in Figure 8-3.

NOTE *Our use of the terms "model" and "file" may be confusing. When we talk about a model, we are referring to a group of parts assembled in a particular fashion to represent something — whether it's a complete object or a component of an object. When we talk about a file, we are referring to what LDraw-based programs read as a separate file. So, a model can be a file, and a file can contain multiple models.*

```
0 FILE 2x4.ldr
1 0 -20 0 0 1 0 0 0 1 0 0 0 1 3001.DAT
1 15 -10 -24 10 1 0 0 0 1 0 0 0 1 1x1.ldr
0
0 FILE 1x1.ldr
1 0 0 0 0 1 0 0 0 1 0 0 0 1 3005.DAT
0
```

Figure 8-3: The preceding LDraw code displays the bricks in this configuration.

As we said earlier, the name MPD stands for "Multi-Part DAT," which is a reference to the early days of the LDraw community when all LDraw files had the ".dat" file extension. LDraw.org has not (yet?) changed the multi-part extension to reflect the move to LDR.

This method's advantage is its portability. One file can contain all the data of a complex model, so it is easy to move around. Also, MPD files keep a lower profile when you are browsing your folder; instead of seeing many files for one model, you only see one.

Using Sub-models

Now that you have learned about what sub-modeling is, here are some techniques for putting it into practice — from inserting a model as a part to storing sub-models in an MPD file.

Creating a New MPD File

To create an MPD file (by adding a sub-model to your current file), select **Multipart > New Model** from the menu bar. MLCad will prompt you to enter a sub-model name, and give you the option of entering a description. Upon clicking **OK**, you will notice the parts you have already placed in your model disappear. They are not gone — they are just not shown because you are now editing the sub-model you just created.

NOTE *If you're creating a sub-model of an assembly that rotates on a hinge, try placing the hinge at the origin by dropping it in the 3D view pane. This will help reduce the need to set a custom rotation point, since MLCad uses a sub-model's origin as its center of rotation.*

Inserting a Sub-model

To insert a sub-model into your model file, open the Select Part dialog in MLCad by clicking the ⛏ icon and using the **I** (for insert part) key on your keyboard, or by right-clicking **Add > Part** in the workspace. In the Select Part dialog, check the box for **Custom Part** near the bottom, as shown in Figure 8-4. Then type a name or browse for your desired sub-model.

Remember that you must first create the sub-model before you can reference it. You can reference either saved LDR files or the name of a sub-model in the current file you are working in. For example, if you just created a sub-model named "sub-model.ldr" by selecting Multipart > New Model, you could insert that through the Custom Part field in the Select Part dialog by typing in **sub-model.ldr**.

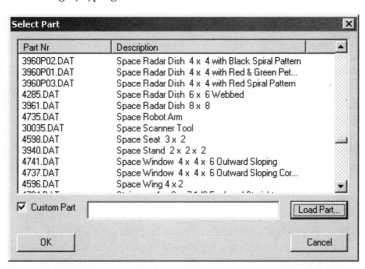

Figure 8-4: The "Custom Part" checkbox in the Select Part dialog allows you to enter a filename of your choice.

You can enter the name of a file in your current directory, the name of a part or primitive in \LDraw\PARTS\ or \LDraw\P\, or you can browse for a file in a different directory by clicking **Load Part . . .** Also, most importantly, MLCad will recognize a filename of a sub-model if you are currently editing an MPD file. We do not recommend using parts from different directories, unless they are parts or primitives located in PARTS\, P\, or S\.

After you have inserted a sub-model using the Select Part dialog, you may now manipulate it as you would any other LDraw part. Go ahead and try this out; select a sub-model and try moving and rotating it (try this using **C:\LDraw\VLEGO\CH08\fig8-3.mpd** from our earlier example).

Managing MPD Files in MLCad

MLCad also has options for you to edit and manage your MPD files, accessible from the Multipart menu.

Navigating Between Sub-models

You can only edit one sub-model at a time. To change the sub-model you are currently editing, select your desired sub-model from the **Active Model** drop-down list immediately above the File View Window, as shown in Figure 8-5 on the next page.

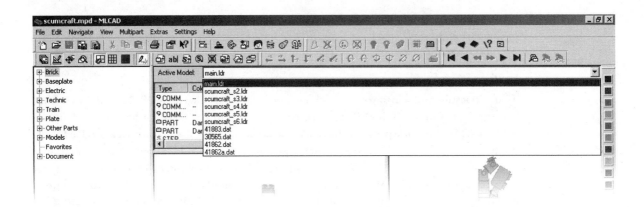

Figure 8-5: The "Active Model" drop-down menu allows you to navigate between sub-models in a Multi-Part file.

Importing a Model

MLCad allows you to import an individual LDraw file into your Multi-Part file. Simply put, you browse for an LDraw file on your hard drive, and MLCad incorporates it as a part of the file you are working on. This is useful if you want to assemble models into a scene, or if you want to use an unofficial part in your model, and at the same time make sure other LDraw users will be able to see it. Also, if you have previously built LDraw models using individual files, you can insert them into your MPD this way.

To import models, click **Multipart > Import Model** on the menu bar. MLCad will prompt you to browse for a model on your hard drive. Clicking **OK** will insert the model's contents into your current file as a separate sub-model.

Exporting Sub-models

You can also export all of your sub-models to individual .ldr files at once. To do this, click Multipart > Export Model. You are prompted to select a directory to export your models to. The dialog MLCad provides does not allow you to create new directories, so if you want to save to a directory that does not exist, create the directory in Windows Explorer first, before trying to export.

Removing a Single Model

You can remove a model from your MPD file by selecting **Multipart > Remove Model** from the menu bar.

CAUTION   *If you want to remove something from your MPD file but retain it, DO NOT use this option. It will delete the sub-model permanently, and it cannot be recovered.*

Sequencing Sub-models

MLCad allows you to change the order your sub-models appear in the MPD file through a simple dialog box. Click **Multipart > Sequence** to open this window, as shown in Figure 8-6. To change where a sub-model appears in the file, select the sub-model and use the Up and Down buttons to move it.

Figure 8-6: The "Model Sequence" window allows you to change the order of your sub-models in the MPD file.

Saving an MPD File

When you save your creation, it will automatically save as an .mpd file instead of an .ldr file.

NOTE   *You can place a sub-model inside of a sub-model, and LDraw tools will recognize it. Just be sure to include each sub-model (no matter where it is referenced) inside the MPD file, or inside the model's directory. This technique is called* nesting, *a term derived from computer programming, hierarchical lists, outlines, and so forth.*

Construction Techniques

Now that we have taught you how to use sub-models and MPD files, here is an explanation of construction techniques that rely heavily on using them.

Component-Based Construction

The idea of component-based construction is more of a frame of mind than a formal construction technique. Your model may have large sections with studs mostly oriented in the same direction, but when building, it is convenient to think in terms of component parts. For example, Jon Palmer's Scumcraft model (C:\LDraw\VLEGO\CH08\ scumcraft.mpd) creates a sub-model for the canopy/engine mounting point, which attaches to the main model in a studs-up orientation. The model is shown in Figure 8-7 on the following page.

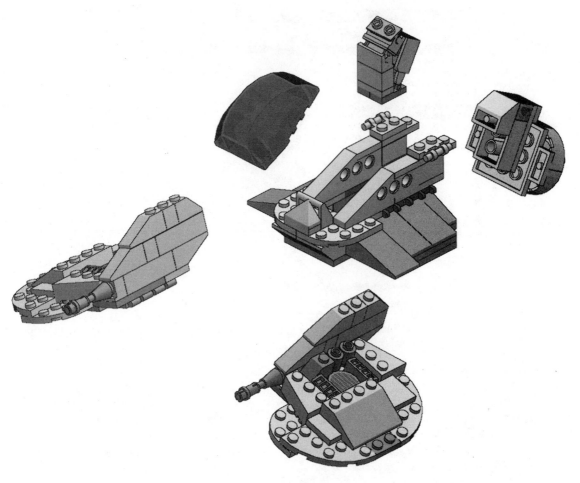

Figure 8-7: Jon Palmer uses component-based construction in his Scumcraft model.

Recall the mentions about functional versus static building? This is where it all comes together. In general, think of your model in components. Whether these components are moveable or not does not matter. In the case of the Scumcraft, Jon sub-modeled the mounting point because it becomes mostly hidden in the construction and contains a few parts that would be difficult to attach if it was built in place in the model. The building instructions for the Scumcraft even treat this as a sub-assembly (more on "sub-model" versus "sub-assembly" in Chapter 12, "Introduction to Building Instructions").

SNOT Construction

We briefly touched on the idea of "SNOT Construction" in Chapter 5. *SNOT* is an acronym first coined in the online LEGO community, which stands for "Studs Not On Top." This is a quick way of saying a model's studs don't face straight up and down, which is the typical orientation of LEGO parts. People build models where groups of studs don't face up and down — they may face to one side or another, upside down, or any orientation in between. Any use of this technique is commonly referred to as "SNOT," which is illustrated in Figure 8-8. Mastering this technique is nothing to sneeze at, and its artful use is highly regarded in the online community.

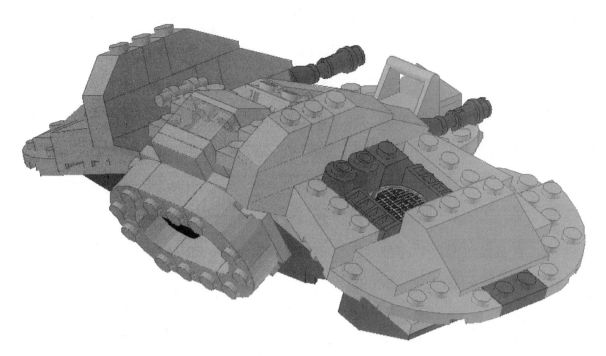

Figure 8-8: The Scumcraft engine uses SNOT Construction. Its studs do not always face "up."

SNOT and Sub-modeling

If you plan to SNOT (the online community uses the term as a verb as well as a noun) more than a couple parts, or if you want your SNOT assembly to be able to rotate easily, you need to create a separate sub-model for it. This way your LDraw editor groups the parts as a single unit, and you don't have to rotate each one individually, or spend time painstakingly aligning each one using fine grid movements.

Build Studs-Up

When you are building a SNOT component as a sub-model, build it studs-up. As you've already experienced, it is far easier to align bricks that all face studs-up than it is to align bricks that face any other direction. You can rotate your entire SNOT component at one time, in the main model.

Rotating SNOT Components

After you insert your SNOT component into the main model by using MLCad's **Select Part** dialog, you are free to move it as a single unit. You may notice that SNOT assemblies rotate rather awkwardly if you don't set the rotation point yourself. This is because MLCad sets the default rotation point to the part origin, as shown in Figure 8-9 on the next page. The part origin is the point within a part file that defines its location on the 3D coordinate plane. If you look closely at how sub-models rotate, you will notice that this origin is sometimes outside of its bounding box! (A border indicates the bounding box, which identifies the part that you have selected; the bounding box outlines the extremities of the part.) This behavior is rather strange because the point that MLCad chooses does not appear to be the origin of any part on the sub-model itself.

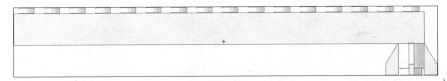

Figure 8-9: MLCad sets the rotation point at the part origin by default. This can be awkward when using sub-models. Note the rotation point, indicated by a black box above the selected sub-model.

Set a Custom Rotation Point

To solve this issue, do two things. First, check the box for **Show Rotation Point** in the Rotation Point dialog (![toolbar icon] on the toolbar). Once you click **OK**, you will see the rotation point defined by a small square with crosshairs running through its center. Next, set a custom rotation point for your model using the procedures outlined in Chapter 5.

Be careful to select the point on the part from which you want it to rotate, as shown in Figure 8-10. Use the base of the studs where you will be attaching this part to the main model, or use the center of rotation of a hinge part. Now, you will be able to rotate and align the sub-model with ease.

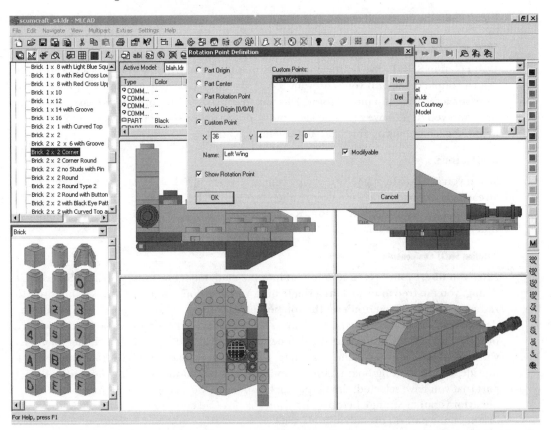

Figure 8-10: Setting a custom rotation point allows you to rotate a sub-model with ease.

Summary

This chapter taught you the concept behind Sub-models — LDraw models that serve as components of larger models. They are referenced inside a model file in place of a part, and you can manipulate them as a single unit. We covered the two storage methods: individual LDR files and MPD (Multi-Part DAT) files. MLCad offers robust MPD support, allowing you to easily manage LDraw files that contain sub-models. The component-based and SNOT construction techniques provide a reference point for you as you go on to create your own complex LDraw files. As you experiment on your own, remember, a complex LDraw model is nothing more than several simple models that have been put together.

9

MINIFIGS, SPRINGS, RUBBER BELTS, AND MORE

In this chapter, we'll introduce you to MLCad's generators. These tools vary from construction aids (such as the Minifig, Spring, and Rubber Belt generators) to semi-useful add-ons to the building experience (Fractal Landscape, Rotation, and Mosaic). Though they each have different purposes, we've grouped them together in this chapter because of what they share: Each generator takes individual components and automatically constructs something new with them.

These features are timesavers. Before the Minifig Generator, you would have had to assemble minifigs by hand out of the various component parts. This was very time consuming and required complex rotations along with heavy use of the fine grid setting. Now, with the Minifig Generator at your disposal, creating and posing minifigs is a breeze (and fun, too! Take a look at our minifig friends in Figure 9-1 on the following page).

Figure 9-1: Minifigs are just one thing you can create with MLCad's generators.

NOTE *Organizing these generators (along with the flexible elements in Chapter 10, "Creating and Using Flexible Elements") has been a difficult task. From a construction perspective, springs and rubber belts belong grouped with the hoses, wires, and treads we'll discuss in Chapter 10. However, the way you go about creating them using the available tools is totally different. We've chosen to group the MLCad tasks together, and separate out the non-MLCad tasks of generating the flexible elements from the next chapter. While there is some overlap in subject matter between this chapter and the next, the methods for creating these elements are completely different.*

MLCad's Spring and Rubber Belt generators take the complicated task of assembling new parts on the fly and simplify it into an easy-to-understand interface. If it weren't for the MLCad Spring Generator (which originated as the stand-alone application Spring2DAT by Marc Klein but is not covered in this book), creating compressed springs would be nearly impossible for most users.

Creating Minifigs with the Minifig Creator

For many fans, minifigs are the crowning touch to a LEGO creation. The personality they add to LEGO models is priceless. Now, thanks to MLCad's Minifig Generator, creating minifigs in LDraw is painless.

Fun Fact:

Minifigs are indeed central to LDraw. The original program got a jump-start early on in the LEGO community by way of the Minifig World Tour. In early 1997, a small group of online friends from all over the world organized and sent two minifigs, whom they named Jill and Gary, to various LEGO fans around the world. A part of the project was a community idea book, and when the participants wanted to share their building instructions, James introduced them to LDraw. The DOS software quickly took off, and the original L-CAD (for "LEGO CAD") email list was started. Ironically (or maybe not so), the minifigs were sent off on their journey by Jacob Sparre Andersen in Denmark, who later became LDraw.org's server admin, and made their first stop at author Tim Courtney's house in the Chicago suburbs. They also made their way to author Steve Bliss' residence, where they took a two-year hiatus (by way of being forgotten) before resurfacing and making their way back to the World Tour organizer, Tamy Teed.

Using the Minifig Creator

To use the Minifig Creator, click the 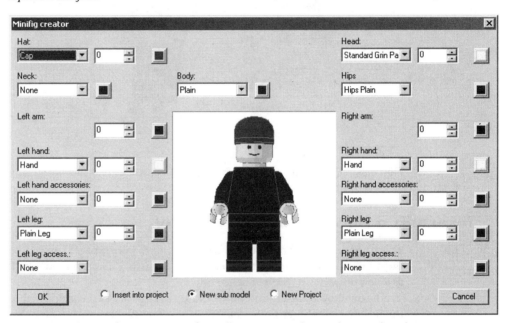 icon. The layout (as you can see in Figure 9-2) may seem complex at first, but take a closer look: Each component is logically arranged around the preview image. Along the left-hand side of the image, you control the parts on the left side of his body; on the right, you control the right-hand components. Above the diagram are the components that lie in the center: hat, head, neck gear (air tanks, quivers, and so on), body (torso), and hips.

Use the drop-down list boxes to choose the item a minifig is holding or wearing. Use the box to the right of that to rotate each component by using the scroll buttons or by entering an angle. The color swatch to the right of each entry gives you access to the Color Palette to change the color of each item.

NOTE *The MLCad Minifig Creator layout is based on the minifig generator that Leonardo Zide has had available in LeoCAD (www.leocad.org) for quite some time. LeoCAD is a CAD program that maintains its own proprietary parts library but is based on the LDraw parts library. LeoCAD can import and export LDraw files.*

Figure 9-2: The Minifig Creator interface allows you to select and manipulate the various components of your minifig easily.

Rotate Your Minifig

You can use your mouse to drag the minifig around and spin him or her in different orientations, just like the 3D pane in the workspace. This is useful if you are placing objects on the neck, or if you're rotating body parts and accessories that may conflict with each other. Remember that there is nothing stopping two objects from occupying the same space in the LDraw system!

Our Custom Minifig

To give you a chance to see how the Minifig Creator works, we created our own custom minifig. Figure 9-3 shows how we selected each object and rotated it using the rotation dialogs. Also, the color was simple to change via the color palette, accessible by clicking on each respective swatch. You can find our minifig under **C:\LDraw\VLEGO\CH09\minifig.ldr**.

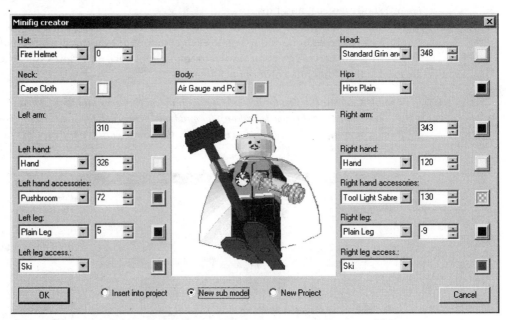

Figure 9-3: Our minifig uses a wide range of accessories and shows how easy it is to rotate parts.

Inserting a Minifig Into Your Model

There are three ways to insert your minifig into the workspace: You can insert the parts of the minifig directly into your project, insert them as a new sub-model, or insert the minifig as a new project.

If you choose "insert into project," MLCad will insert the individual pieces straight into the file. Be careful that you don't "break" your minifig by moving his or her parts, especially if you're working in Fine Grid! (Because MLCad's Undo feature does not function, realigning them could be time consuming.)

If you choose to insert the minifig as a new sub-model (the recommended method), MLCad will prompt you for the sub-model's name and description. Once you have entered that information, MLCad will make the minifig the active model. You can switch back to the main model via the **Active Model** list box above the file view window.

CAUTION   *If you choose to insert the minifig as a "new project," be sure to save your model first. If MLCad detects that your model has not been saved, it will prompt you to save before inserting the minifig. When it inserts a minifig as a new project, it does not close your current file and open an "untitled.ldr" file for your minifig; instead, it erases your current data and places the minifig in the file you were editing. If you select the "save" button after this, it will overwrite your previous work! For this reason, we do NOT recommend using the "new project" method of creating minifigs! You should either insert them directly into your project or create a new sub-model.*

Experiment!

Have fun experimenting with the Minifig Creator to see how many different combinations you can come up with. Remember that minifigs can have hooks for hands, wear skis and various hats and helmets, and so on. If a minifig part that you want to use isn't available, just remember that volunteers are responsible for the LDraw part libraries. There are so many printed elements in existence; it will take quite a while for the community to catch up with what's available. At present, there's a good selection of minifig parts, so you can create a variety of figures from all different themes.

NOTE *A fun semi-related website is Suzanne Rich's Minifig Generator, available at http://www. baseplate.com/toys/minifig/. This online toy enables you to scroll through different minifig heads, torsos, and legs, and to mix and match them.*

Creating Rubber Belts with the Rubber Belt Generator

Rubber belts are used to tie Technic pulleys together. MLCad has a simple, two-pulley rubber belt generator that is very easy to follow and very useful. Creating rubber belts in MLCad is a matter of taking a few simple measurements and inserting them into the generator interface.

NOTE *This generator also gives you the option to make your belt editable when you insert it into the project, by way of the **MLCad Part** feature. The Minifig Creator does not offer this.*

In the next chapter, we will teach you another method of creating rubber belts. Kevin Clague's LSynth is capable of creating the parts by inserting pulley parts as "constraints," adding a few simple comments, and generating a new file. LSynth's distinct advantage over MLCad's rubber belt generator is that it can create belts that wrap around more than two pulleys, as illustrated in Figure 9-4. Beyond that, the two tools have quite a few differences between them. The LDraw system does not allow for actual flexible elements, so these must be simulated using several segments that create the image of a rounded part. Also, LSynth uses many more segments than the MLCad rubber belt generator does, though it offers more flexibility. It's up to you to decide which one is the most appropriate tool for your particular project.

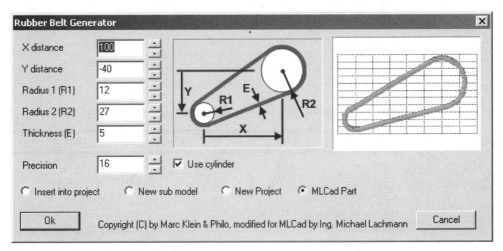

Figure 9-4: MLCad's Rubber Belt Generator.

The Rubber Belt Generator requires you to enter the distance between the center of the two pulleys (X and Y), the radius of each pulley, and the thickness of the rubber band. It allows you to determine the number of line segments used when the belt wraps around the pulleys, via the Precision field. This generator also gives you several options for how to insert the rubber belt into your LDraw file.

Creating Your Rubber Belt

To create your rubber belt, you must find the values for your specific arrangement of pulleys. We have created a sample pulley setup (shown in Figure 9-5), located under **C:\LDraw\VLEGO\CH09\pulley.ldr**. You may open the file and follow along.

Figure 9-5: We created our rubber belt using this sample model.

Finding the Radii

Let's start by finding the easiest values, R1 and R2. As you saw previously in Figure 9-4, these values refer to the radius of each pulley. In Table 9-1, you can find the radius measurements of the most common parts used for pulleys, courtesy of Philippe Hubrain (http://philohome.free.fr/rubberbelt/rubberbelts.htm).

Table 9-1: Use the information in this table to find the diameter of common pulleys.

Image	Description	Part #	Radius
	Technic Pin	3673	8
	Technic Bush ½	4265C	8.5

Table 9-1: Use the information in this table to find the diameter of common pulleys.

Image	Description	Part #	Radius
	Electric Technic Micromotor Pulley	2983	13
	Technic Wedge Belt Wheel	4185	30.5
	Technic Pulley Large	3736	46.5

Notice that we used the Technic Wedge Belt Wheel and the Technic Pulley Large for our sample model. With our sample **pulley.ldr** file open, launch the Rubber Belt Generator (the ✏ icon, or **Extras > Generators > Rubber Belt**) and enter the proper radii in the fields. For R1, use **30.5**; for R2, use **46.5**. These are the radii of the two pulleys we've selected.

Finding the X and Y Distances

Now you need to find the X distance and the Y distance. As the diagram in the Rubber Belt Generator shows, these numbers represent how far the centers of the pulleys are from each other. You'll need to measure the distances between the two to find these values.

Jaco's Tip for Measuring Distances

As we were writing this book, Jaco van der Molen showed us how he measures in LDraw. He inserts a part that does not exist, and MLCad displays crosshairs in place of a part for the cursor. Then he can take note of where that point is located using the Enter Pos & Rot dialog (⊞), or by counting how far he moves the part.

To use Jaco's tip, press the **I** key or click the insert part icon (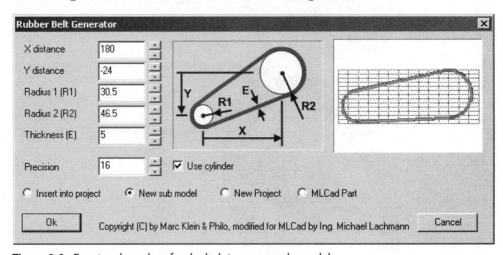), select **Custom Part**, and key in **0.dat**. Now you have created crosshairs to use for measurement.

It's relatively easy to measure the distances in our sample.

To find the X distance, notice that counting along the X-axis (horizontally) there are 10 LDU per keystroke (or one-half of a stud). Center the crosshairs in the hole directly below the large pulley, then move the cursor with the arrow keys until it is centered on the smaller pulley, counting the number of keystrokes along the way. Multiply this by 10, and you have your X distance value.

To find the Y distance, remember from Chapter 5 that a full brick height is 24 LDU. Again, center your cursor on the hole below the large pulley; this time, use the **HOME** key to move it up. You should have to press **HOME** three times to reach the center of the large pulley. Each keystroke moves the cursor one plate height, so three strokes is one brick height. The Y distance value is 24 LDU. In the LDraw world, positive Y actually goes down, so you'll need to enter a **-24** in the Y distance field for it to be accurate. You can see the values being entered in the Rubber Belt Generator in Figure 9-6.

Figure 9-6: Entering the values for the belt in our sample model.

Thickness

The Thickness value refers to the actual thickness of the rubber belt. 5 LDU, the default, approximates the new-style colored Technic rubber belts that come in many Technic and MINDSTORMS kits. You can adjust this value up or down (some of the older rubber belts were thinner than this).

Precision

The Precision value refers to the number of line segments the Rubber Belt Generator will use to construct the round edge around each pulley. The higher the level of precision, the more line segments are used, and the more realistic the belt appears (see Figure 9-7). Next to the Precision field is a checkbox for **Use Cylinder**. With this checked, the generator uses a cylinder to construct the belt; with it unchecked, the belt is square-shaped on the edges.

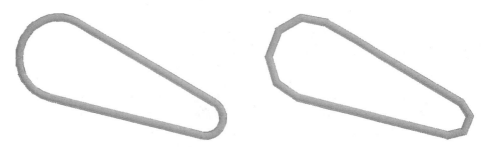

Figure 9-7: Precision effects the number of line segments used for the round edges. On the left is a belt with a precision of 16, on the right a belt with a precision of 4.

Inserting the Belt Into Your Model

Once you've entered all of the values for your belt, you are ready to generate it. There are four ways MLCad can insert the rubber belt into your model. You can select which you would like to use from the radio buttons near the bottom of the generator window.

MLCad Part

The best way to insert a rubber belt into your project is as an MLCad part. Even though this option has its pros and cons, its superiority lies squarely in that it is the only method that allows you to edit your rubber belt after you are finished generating it. This also groups all the subcomponents of the rubber belt into one unit in MLCad's file view window. Were it not for this feature, you could not edit rubber belts after they had been generated.

The disadvantage of using MLCad Parts is that they are MLCad-specific. Other editors *will not* display the belt as a group; instead, each individual cylinder that makes up the belt and the curve of the belt will appear in a separate line.

When you insert a rubber belt as an MLCad part, there is a trade-off — you gain the ability to edit your belt, while you lose the LDraw standard part grouping that sub-modeling offers. Once you have inserted an MLCad part, you cannot ungroup the components.

TIP *To avoid having these parts appear ungrouped outside of MLCad, you can create your rubber belt inside a blank sub-model. Simply create a new sub-model via **Multipart > New Model**, then use the Rubber Belt Generator to insert an MLCad Part inside your sub-model. You can then switch back to the main model, insert the sub-model, and manipulate your belt into place. However, this adds another layer to your creation, and we only suggest doing this if you need to use the belt as a group outside of MLCad. If you only plan to edit this file within MLCad, there is no need to use this technique.*

New Sub-model

The second best way to insert your rubber belt into your model is to create a new sub-model. This method allows you to move your belt as one unit, though it does not allow you to edit your belt after you generate it. If you select this method and choose **OK**, MLCad will prompt you to save your file, and then ask you for the name of your sub-model. Name the sub-model something like **belt.ldr** and click **OK**.

Just as when you created the minifig, the rubber belt will become the active model. If you are using our sample file, select **pulley.ldr** from the **Active Model** list (if you are using your own file, select your own main model from this list). To insert the belt into the main model, use the Insert Part icon or the **I** key, select **Custom Part**, and key in **belt.ldr**. Now move the belt into place.

Insert Into Project

As mentioned earlier in the Minifig Creator section, this option inserts all of the individual lines the generator creates directly into your project. However, unless you specifically need the belt inserted this way, *avoid this method* because it will clutter up your main model. Most of the time, you want your main model to remain as free of clutter as it can be. With this method, you cannot edit your belt via the Rubber Belt Generator after you have generated it once. Also, all of the individual part components (not parts themselves, seeing that this is generated on the fly) will be in your model. What a mess!

New Project

As with the Minifig Creator, this feature does not work the way it should in MLCad 3.00. In theory, it should prompt you to save and then insert the belt in an entirely new file. Instead, it prompts you to save and then replaces the contents of your file with the new belt. It does not (as the name would suggest) close the file you were working in and put the belt in a new file. Instead, it hijacks your current file, deleting all of your work. Be careful here, so you don't lose valuable data by saving your new belt! (See the warning in the "Inserting a Minifig Into Your Model" section on page 120.)

In addition to this flaw, be sure to note that using this method will not group your files, and will not allow you to edit your belt after you have generated it.

Insert the Belt

Once you've chosen how you will insert the belt into your project, click **OK**. Depending on the method you have chosen, MLCad may prompt you to save your project. If it does, we recommend you save your work before generating the belt. Figure 9-8 shows the result.

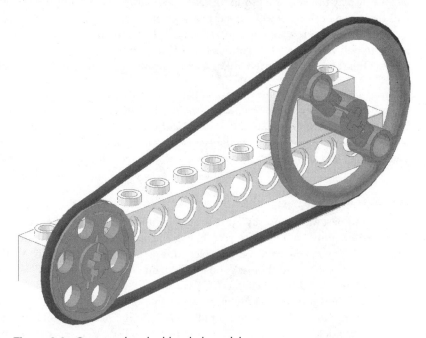

Figure 9-8: Our completed rubber belt model.

Creating Springs with MLCad's Spring Generator

MLCad also features an easy to use Spring Generator, originally created by Marc Klein (http://marc.klein.free.fr/spring2d.html) as a separate application. This feature is useful for creating Technic shock absorbers that are compressed (not fully extended).

There are shortcut parts for the two types of shocks that are fully extended, but if you want to create shocks that appear compressed, you need to assemble them manually, including creating your own spring. This means you need to take careful measurements for the spring length and plug it into the Spring Generator. Take a look at MLCad's Spring Generator as shown in Figure 9-9.

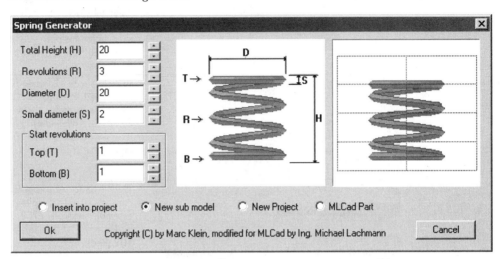

Figure 9-9: MLCad's Spring Generator allows you to create custom springs.

The Spring Generator requires a few pieces of information about the spring you are creating. It needs the total height of the spring ("Total Height"), how many revolutions the spring coils between the top and the bottom ("Revolutions"), the total diameter of the spring ("Diameter"), and the diameter of the wire itself ("Small Diameter"). It also allows you to set the number of revolutions on each end ("Start Revolutions").

Types of Shock Absorbers

There are currently three different types of shock absorbers available in LDraw, as described in Table 9-2 on the following page: The first is the small (6.5L) Technic shock; the second is the dampened shock; and the third is the large Technic shock. They all feature different size springs. In real life, the dampened shock part has a cylinder of air with a gasket that helps soften the compression, and the spring itself stays on the outside of this airtight area.

Table 9-2: Shock Absorbers Available in LDraw

Image	Description	Size
	Technic Shock Absorber 6.5L. Parts 731 (Cylinder) and 732 (Rod). Unfortunately, the Rod file has the spring included so you cannot customize your spring on this file without editing the actual part. There is also no complete assembly shortcut for this part; you have to assemble the Cylinder and the Rod manually.	Spring Diameter: 14 LDU
	Technic Shock Absorber 10L Dampened. Parts 32181C02 (complete assembly shortcut), 32181C01 (Cylinder), 32183 (Rod with Gasket), and 108 (Spring).	Spring Diameter: 20 LDU
	Technic Shock Absorber 9.5L. Parts 2909C01 (complete assembly shortcut), 2909 (Cylinder), 110 (Spring), and 109 (Rod).	Spring Diameter: 24 LDU

Creating Your Spring

To walk you through creating a custom spring, we will show you how to create a spring for our compressed shock absorber as shown in Figure 9-10. You can open the file and follow along. It is located under **C:\LDraw\VLEGO\CH09\shock.ldr**.

Finding the Total Height (H)

The Total Height tells the generator how long the spring is. Use Jaco's 0.dat trick to measure the distance between the flanges on the shock; this is the length of the spring. Enter **0.dat** in the **Custom Part** field of the Insert Part dialog. Now, align the crosshairs with the flange on the rod as shown in Figure 9-11. Use the fine grid to place it there if necessary.

Now that you've aligned the crosshairs, make sure the grid is set to coarse, and move it to the left using the arrow keys (not the mouse) to measure the distance between the two flanges. You will notice it does not align with the second flange evenly. Count the number of keystrokes it takes you to get as close as possible to the flange (without going past it) and then multiply that number by 10 (we got 60). Now set the grid to fine and count how many keystrokes it takes to reach the flange on the cylinder (we got 6). So, the total length (note: not "Total Height") of the spring is 66 LDU. Write this down.

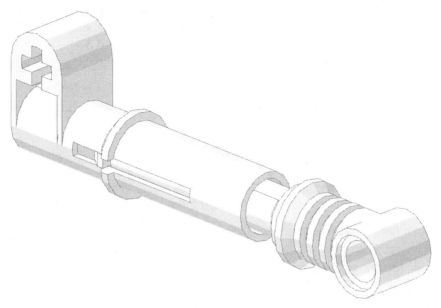

Figure 9-10: Our sample shock absorber.

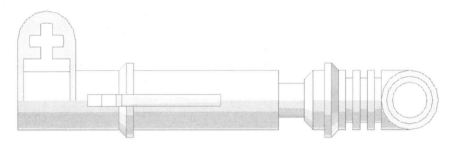

Figure 9-11: You may need to align the crosshairs you create using the 0.dat technique using the fine grid settings. Note the crosshairs are perfectly aligned with the flange on the rod.

Diagram Error

Now this gets tricky. In MLCad 3.00, the diagram you see on the Spring Generator is in error. It falsely represents the Total Height as the total length of the whole spring, including the Start Revolutions. In fact, Total Height represents the length in LDU of the Revolutions portion of the spring. The Start Revolutions, those tighter coils at the beginning and end of the spring, add to the spring's overall length.

To fix this, you need to take your total length value and subtract from it the number of total start revolutions on your spring, multiplied by the Small Diameter. In our case, the total length is 66. To make a spring that looks like the actual part, there are three Start Revolutions on each end. The Small Diameter, the thickness of the wire itself, is 2 LDU; so, you would take 6 (your total number of Start Revolutions, 3×2 for the top and bottom), multiply that by 2 (your Small Diameter), and know that you need to subtract 12 from 66 to get 54. You're now very close to the number you need for the Total Height field.

There's only one problem with the above formula. The Spring Generator seems to shrink the Total Height 2 LDU smaller than one would think. So, add 2 LDU back to the length of the spring, and its total length from end to end is now 56 LDU. Go ahead and try this out on the spring yourself, generate it, and you can see the difference.

Finding the Number of Revolutions (R)

The next measurement you need to take is that of the Revolutions. This value indicates how many times the wire wraps around the shock along the length of the spring. In this case, all you need to do is look at the physical LEGO part or the LDraw spring shortcut (part number 2909c01) for the fully extended version of the shock. Count the number of times the wire crosses a straight line down the length of the spring. Start at one end and count your way towards the other. The real Technic shock absorber we're talking about has about 13 revolutions, so enter **13** in the Revolutions field.

Finding the Diameter (D)

We have given you the diameters of the three shocks in the table earlier in this section. This particular spring needs a diameter of 24 to fit on the shock pictured.

Finding the Small Diameter (S)

Spring Generator calls the diameter of the wire itself the "Small Diameter." The standard for this type of shock is 2 LDU. You can leave this value alone; you can change it if you want to insert springs that use a different diameter wire.

NOTE *In real life, the small Technic shock (6.5L, parts 731 and 732) has a smaller diameter wire than the 9.5L shock we're assembling. In LDraw, however, it is still represented by a 2 LDU Small Diameter.*

Start Revolutions (T and B)

We mentioned Start Revolutions earlier when we talked about the trick in finding the Total Height. Again, Start Revolutions are the tightly wound coils at the beginning and end (or Top and Bottom as the Spring Generator says) of the spring. Because the physical springs have three start revolutions at each end, enter **3** for the T and B values.

Inserting the Spring Into Your Model

As with inserting belts, you have four choices for inserting your spring into the model. Since you probably want the flexibility to edit the spring, choose MLCad part and generate the spring. Once you've inserted it, take a couple of seconds to use the Fine Grid and align it properly with the shock absorber as shown in Figure 9-12.

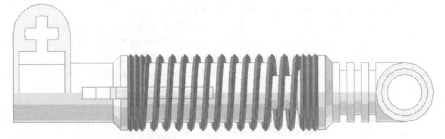

Figure 9-12: Our completed spring and shock absorber.

Creating Rotation Models with the Rotation Model Generator

MLCad's Rotation Model Generator is useful for creating what the math world calls a "surface of revolution" out of virtual LEGO pieces. In layperson's terms, this is the equivalent of a potter's throwing wheel. You can define a cross-section of a LEGO model, and the Rotation Model Generator will take that cross-section and rotate it along an axis, generating a sculpted round version of it.

Figure 9-13 shows a sample rotation model. Imagine an axis running vertically through the center of this. To define this model, we took a cross-section in MLCad's Rotation Model Generator, and it created this construction out of it.

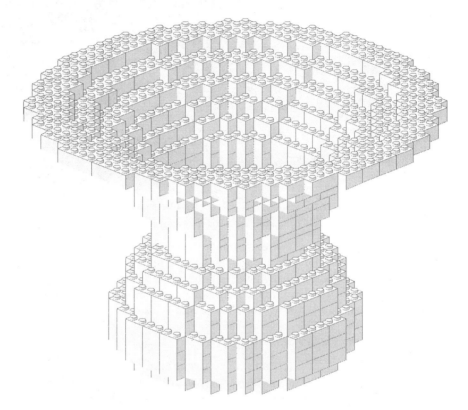

Figure 9-13: You can create a surface (or a solid) of revolution with MLCad's Rotation Model Generator.

In the LEGO community, there is an entire genre of building called sculpture. Like the name suggests, these LEGO fans build large sculptures out of LEGO bricks. Even though basic LEGO parts are square, they still find ways to make these models appear rounded by staggering them in a certain arrangement. The LEGO Company's "master builders" are specialists in this genre. The Rotation Model Generator can be seen as an aid to LEGO sculptors, who want to create rounded objects with square bricks. MLCad gives the best arrangement of these parts, and then it is up to the builder to construct the model.

Creating a Rotation Model: An Apple

We will teach you how to use the Rotation Model Generator by creating a model of an apple. To begin, launch the Rotation Model Generator using its toolbar icon: 🔄 This will open the Create rotation model dialog as shown in Figure 9-14 on the next page.

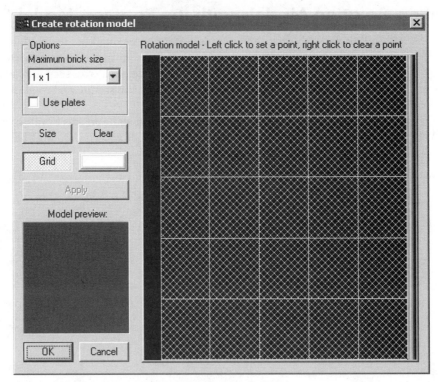

Figure 9-14: MLCad's Rotation Model Generator dialog.

The grid you see on the right hand side of the dialog is where you plot your cross-section. On the left, you have the option of changing the maximum size of bricks used, using plates instead of bricks, changing the size of the grid, and changing the color of the pieces used to generate the model. You can also toggle to show or hide the grid if you desire (by pressing the depressed **Grid** button), but we find it easier to create a cross-section with the grid on.

Changing the Grid Size

For our purposes (creating an apple), the current grid is too small. We need to change the size of the grid, and can do so by clicking the **Size** button on the right. Go ahead and click it. The **Rotation object size** dialog opens, as shown in Figure 9-15.

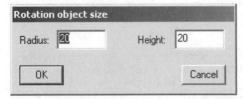

Figure 9-15: You can change the size of the grid you use for defining your rotation model with this dialog.

We like to have the most area available to work with when creating a rotation model, so increase both the radius and the height to **20**. MLCad only allows up to 20 for each field. Click **OK** and you will see the changes reflected in the **Create Rotation Model** dialog.

Changing the Maximum Brick Size

We also want to change the Maximum Brick Size. MLCad can use different sizes of bricks together when creating rotation models. You need to tell MLCad the largest size brick you will allow it to include in the model it generates. For example, if you are creating plans for a sculpture that you are going to build, it is a good idea to crank up the maximum brick size.

You can tell the generator to use up to a 2×4 brick; this will allow MLCad to overlap bricks on different layers. Were you to leave the maximum brick size set to 1×1 (as it is by default), you would not see how bricks should overlap.

Changing the Color

With the enlarged grid and maximum brick size set, we're almost ready to begin creating the apple. Before we start plotting out the points for it, it's a good idea to change the color. The default color in the Rotation Model Generator is white — and we know apples aren't white! Click the white color swatch to the right of the **Grid** button to bring up the color palette. Choose red (or light green if you prefer) for the color.

NOTE *You can also select a color, plot a few points, and change the color again to plot points in a different color. We will create the apple's stem this way.*

Plotting the Points

Now that we have set the color, let's create the apple. Figure 9-16 shows you our apple's cross-section. You can plot the points by clicking each square in the grid you want to fill. When you are plotting the points on the grid, be sure to note that the axis is on the right-hand side. For this reason, our apple's core is on the right, and the outside of the skin faces left.

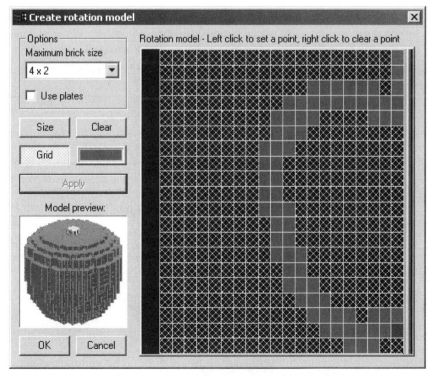

Figure 9-16: Here is the cross-section of our apple, with the preview image generated in the lower left.

Using Plates

You can use plates instead of bricks in your rotation model. To do this, check the **Use Plates** checkbox near the top left. When you check this box, the grid will change to reflect the dimensions of plates more accurately. When you use plates, you still can only have 20 for the height value, so it decreases the size of what you can create in one rotation model. In our situation, using plates would make a rather squatty apple, so we have chosen to use bricks.

Overlap

If you plan to use this feature to create a diagram for building a physical sculpture, it is a good idea to double or triple up on the thickness of your cross-section when plotting points. This way, MLCad can generate parts overlapping, just as you would need to do in real life to ensure the structural stability of your creation.

NOTE *While MLCad will use different sized parts to generate the rotation model, and these parts will overlap, overlapping is not perfect. We have noticed some structural weaknesses by observing the model MLCad generates. When you go to build the physical LEGO model, you may have to make some changes to the parts used.*

Using Multiple Colors

In our model, we chose to use green and brown as well as red. To do this, simply plot all the points you wish in one color, use the color swatch to access the color palette, change the color, and return to plotting points in a different color. We used green for the stem (the three squares in the upper right corner) and brown for the base (the one square in the lower right corner).

Generating a Preview

The dialog can generate a preview image of your rotation model before you click OK, giving you the chance to make tweaks before actually generating it. When you make changes to the grid, the **Apply** button will become clickable. Click it to generate the preview image. If you have not made any changes since the last time you clicked **Apply**, the button will appear grayed out.

Because you cannot go back and edit your rotation model once you have generated it, it is wise to use this feature before generating your model.

Generating Your Rotation Model

When you are done plotting points and are satisfied with the preview image, click the **OK** button to generate the rotation model. MLCad will place the model in your current file as individual LDraw parts. Figure 9-17 shows our completed apple. You can view the apple file on the CD-ROM under **C:\LDraw\VLEGO\CH09\apple.ldr**.

When you generate your rotation model, MLCad automatically inserts a step marking each layer of bricks. This is convenient for printing out building instructions of your rotation model, to be turned into a LEGO sculpture. Browse the instructions in View Mode, or print them out and spread out on the floor with your bricks!

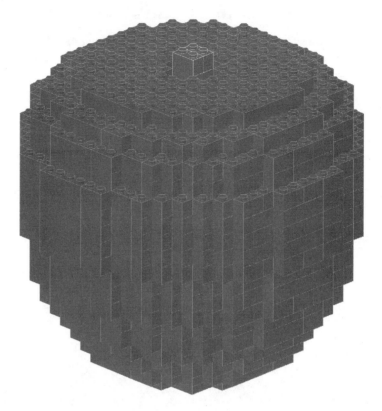

Figure 9-17: Our completed apple, as seen in MLCad.

No Substitute

The rotation model generator can be used as an aid for creating sculptures, but it is no substitute for sculpting a model with physical bricks. If you haven't already noticed, the rotation model generator does not always create perfect objects for transforming into LEGO sculptures. For example, our apple looks a bit chunky compared to what you may see coming out of a LEGO model shop. Still, it's a good start. You can always take these generated designs and refine them further to create more realistic looking rounded objects.

Sculpting in LEGO is an art, and a skill you acquire over time. It's difficult to replace that human skill with a computerized generator, but this tool can help you get a start with some basic techniques.

NOTE *The Rotation Model Generator is limited to 20 bricks/plates in maximum height, but if you are creating taller cylindrical models, you can create them in sections. It may be a good idea to plot out the entire cross-section on graph paper first before using the generator. Also, because MLCad inserts ungrouped individual pieces into your model when generating these, create a new sub-model for each segment of rotation model so that you can easily move sections as a single unit.*

For those of you who are curious, we found mathematical information about surfaces of revolution at this web address: http://mathworld.wolfram.com/SurfaceofRevolution.html.

Creating Mosaics with MLCad's Mosaic Creator

MLCad's Mosaic Creator is a helpful little tool that lets you create virtual LEGO mosaics based on image files stored on your computer. The mosaic is an art form that is popular among the adult LEGO fan community.

People such as Eric Harshbarger (www.ericharshbarger.org/lego/) have created large and stunning mosaics. You can also design and purchase your very own LEGO mosaic online at http://shop.lego.com. The availability of 1×1 plates in the various shades of gray (as well as black and white) has added to the popularity of mosaics in the LEGO community.

When browsing Eric Harshbarger's website, be sure to check out the New York Mosaic and the San Francisco Mosaic, which are examples of the enormous traditional mosaics (LEGO plates, studs-out) that MLCad's Mosaic Creator can generate. For non-traditional mosaics, see Calista (made from Modulex lettered tiles) and the Mad Hatter (SNOT construction).

Creating a Mosaic: Red Futuron Minifig

To teach you how to use the Mosaic Creator, we will create a mosaic of a red Futuron minifig, such as the one shown in Figure 9-18. To get started, launch the Mosaic Creator from within MLCad by clicking the 🖼 icon, located just to the left of the Minifig Creator icon. The Mosaic Creator dialog appears, as shown in Figure 9-19.

Figure 9-18: A Futuron minifig mosaic generated with MLCad's Mosaic Creator.

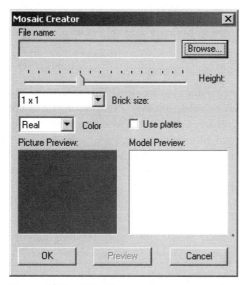

Figure 9-19: MLCad's Mosaic Creator dialog.

Loading an Image

The first step is to load an image. We have stored an image of the red Futuron minifig on the CD-ROM under **C:\LDraw\VLEGO\CH09\futuron.bmp**; load it now by locating it with the **Browse** button. When you find it, you should see a preview of the picture and of the model in the two lower panes of the dialog (see Figure 9-20).

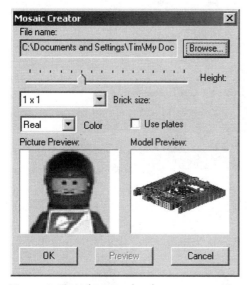

Figure 9-20: When you load an image in Mosaic Creator, it displays a preview of the image you are using, as well as a preview of the model it will generate.

Image Tips

When working with your own pictures, there are a couple of important points to keep in mind:

- Mosaic Creator can only load images in BMP format.
- Each pixel in your image is equivalent to one stud. This means the mosaic created by a 48×48 image like the Futuron minifig will cover a 48×48-stud LEGO baseplate. Because images of any significant size (such as the standard monitor resolutions of 640×480, 800×600, 1024×768, 1280×1024, and up) contain hundreds of thousands and even millions of pixels, we suggest using only small images (under 100 pixels square) for the best performance.

NOTE *The image preview display is slightly distorted. This is probably a result of stretching the image for the preview, since it is only 48×48 pixels. MLCad does not appear to reflect these errors in the final result.*

Setting the Height

Mosaic Creator allows you to set the height (in bricks or plates) of your mosaic by manipulating the slider bar all the way to the left for a height of one, or to the right for a height of 16. Setting a height of one generates a flat, solid image as you saw previously in Figure 9-18. However, if you set the height to greater than 1, mosaic creator will create a rather unique sunken relief effect based on the pixel color, as you can see in Figure 9-21. Black and darker colors do not reach very high in the mosaic, but the lighter colors can tower up to 16 bricks high.

Figure 9-21: When you set the height of a mosaic greater than 1, Mosaic Creator generates a rather unique sunken-relief effect. This image was generated using File > Save Pictures, with Central Perspective selected.

For simple mosaics, use a height of **1** to get a nice solid image. Though setting the height higher may be fun, it does not reflect traditional LEGO mosaics. (Who knows, though? Perhaps with the help of this tool you could create a new raised/sunken-relief style of LEGO mosaics.) For now, we will stick to using a height of 1 when creating our Futuron minifig mosaic.

NOTE *We have included the file used to generate this image under C:\LDraw\VLEGO\CH09\mosaic-high.ldr. After taking a look at the file, you can probably imagine being able to create some rather neat relief models by manipulating images and their colors.*

Setting the Color Option

Mosaic Creator lets you choose between three color settings: Real, Fake, and None. These names indicate the results you will get in your mosaic. "Real" approximates the actual colors of the image, as nearly as possible using the range of available colors (as discussed in Chapter 6). "Fake" color creates bands of solid colors when the height is greater than 1, a sort of topographical map effect, which is similar to how it handles the bricks it places in the Fractal Landscape Generator in the next section. "None" creates a mosaic out of entirely gray bricks. (This is pointless when setting the height to 1, but it can create some neat effects when you set the height higher.) For our mosaic, we will use the **Real** color setting.

NOTE *You may find once you've generated a mosaic that it helps to clean up the stray colors by hand, selecting each part and changing its color on the color palette.*

Changing the Brick Size and Using Plates

You can change the maximum size of bricks allowed in the mosaic via the **Brick Size** drop-down list. If you select 4×2 (the same as a 2×4 brick), Mosaic Creator will use any brick between a 1×1 and a 2×4 in size. It will use larger bricks when allowed by the color pattern.

You can tell the Mosaic Creator to create your mosaic using plates by checking the **Use Plates** checkbox. When "Use Plates" is checked, a one-layer mosaic will be created entirely of plates, whereas a multi-layer mosaic will be composed of bricks and plates.

Generating a Preview Image

To see a preview of your mosaic after adjusting the settings, click the **Preview** button at the bottom center of the window. A preview image will appear in the lower right pane. This is useful while tweaking, and before actually creating the mosaic.

Generating the Mosaic

Once you are satisfied with all of the settings for Mosaic Creator, click **OK** to generate your mosaic. Depending on your image size, this may take a few seconds.

Enjoy your mosaic!

NOTE *You may want to manually edit some of the color selections to achieve a more consistent pattern.*

Creating Fractal Landscapes with MLCad's Fractal Landscape Generator

MLCad offers a fast and easy way to create landscapes with its Fractal Landscape Generator. This very interesting tool can be useful for creating a rough brick landscape, or when creating scenes with your models.

Before generating a landscape out of LEGO parts, the generator first computes a non-LEGO based landscape using semi-random triangles meshed together. Next, it takes the generated landscape and computes it into LEGO pieces. The landscape can be made to fit a single 32×32 baseplate, or a group of 32×32 baseplates.

As with the two generators we discussed previously, you can control the height and the size of bricks used. To see the results of one of our Fractal Landscape Generator attempts, check out Figure 9-22.

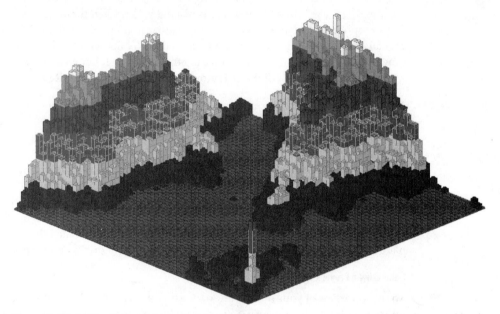

Figure 9-22: MLCad's Fractal Landscape Generator can create unique brick landscapes like the one pictured here.

Using the Fractal Landscape Generator

To begin using MLCad's Fractal Landscape Generator, click the 🏛 icon. You should see the "Fractal landscape generator" dialog appear, as shown in Figure 9-23.

MLCad will automatically generate a random fractal landscape upon launching the dialog. You can then tweak the settings to change the composition and limited characteristics of the landscape, but you cannot design the landscape yourself.

To see a preview of the brick version of the fractal landscape seen in the top right pane, click the **Preview Model** button; it will display a preview image like the one shown in the bottom right pane.

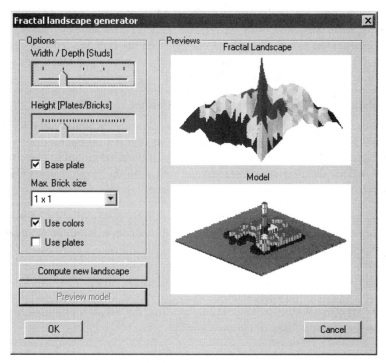

Figure 9-23: MLCad's Fractal Landscape Generator dialog.

While the current landscape in the top pane of Figure 9-23 looks decent, the brick version is seriously lacking. Changing the settings on the left can give you a much better looking landscape, such as the one shown previously in Figure 9-22.

Changing the Width/Depth and Height

You can change the characteristics of possible fractal landscapes by adjusting the width/depth and height settings on the top left of the dialog.

By default, MLCad generates the landscape within a 16×16 area, or half of a standard baseplate. When the Width/Depth (or more accurately Area) slider bar is all the way to the left, it will restrict the brick landscape to an 8×8 area. The third setting allows the landscape to fill one baseplate, the second four square baseplates (2×2) and the far right setting sixteen square baseplates (4×4). However, the greater the area you choose for your landscape, the more time MLCad will have to take to compute each brick. Just as with the mosaics, if you use a large area, you could be using tens of thousands, if not hundreds of thousands of bricks.

You can adjust the height slider to tell MLCad how many bricks tall you want your landscape to be. When the slider bar is all the way left, it will use a height of one brick; all the way to the right will generate landscapes up to 30 bricks tall. Each time you change one of these two settings, the generator will re-compute the landscape.

Other Options

There are also options to toggle baseplate usage, toggle plate usage, change the maximum brick size, and to toggle the use of colors. These settings are similar to the ones we discussed when covering the previous two generators.

To select the maximum size brick Fractal Landscape Generator is allowed to use, use the **Max. Brick Size** drop-down list. Unchecking the **baseplate** checkbox (checked by default) will eliminate the blue baseplate(s) from the final model. Unchecking **use colors** will generate an all-gray brick landscape. Finally, checking **use plates** will add plates into the mix along with bricks, giving a more detailed landscape.

Computing a New Landscape and Generating Your Final Landscape

In the event that the current landscape the generator has computed is not to your liking, you can tell it to create an entirely new random fractal landscape by pressing the **Compute New Landscape** button. You can see a preview of the final model by pressing **Preview Model.** When you are satisfied with the landscape shown on the preview, press **OK** for MLCad to generate the final brick model in the workspace.

NOTE *Because we're dealing with potentially tens of thousands of pieces here, MLCad may take a while to respond when working in your new model. This is natural due to the size of the model, so be patient, and have fun with fractal landscapes!*

Hose Generator

Finally, MLCad has a flexible hose generator that can be useful for creating the classic Space flexible hose, as pictured in Figure 9-24.

Figure 9-24: A recent version of the classic LEGO Space hose, with tabs on the ends. The original hose of this design did not include the tabs.

However, we've found a somewhat roundabout way of generating the same hose, which requires a little extra work but which also requires you to enter fewer points. We've chosen to discuss this method in Chapter 10, since this chapter focuses solely on generators available inside MLCad. MLCad's hose generator can be laborious to the new user. Instead of teaching it, we have decided to teach what we feel is ultimately an easier method. In the next chapter, we provide a brief mention and description of MLCad's Hose Generator as well.

Summary

In this chapter, we've introduced you to MLCad's array of generators and how to use them. We've taken you step-by-step through creating your own minifigs, rubber belts, springs, rotation models, mosaics, and fractal landscapes. You've learned about all of the options of each generator and how to manipulate them to create your desired effect.

By now you should be well equipped to use these tools for your own creations, and we encourage you to play with them and be creative.

10

CREATING AND USING FLEXIBLE ELEMENTS

Flexible parts, such as hoses, pneumatic tubing, and string, are somewhat challenging to create using LDraw tools, but they are essential to many LEGO projects. Figure 10-1 on the following page shows an example. Over the last few years, we have seen a handful of solutions to this problem, but they all add a layer of complexity to the LDraw building experience. We have chosen two programs to assist us in creating flexible elements: Kevin Clague's LSynth and Orion Pobursky's Bezier Curve Plugin for LDDesignPad.

The chief problem with creating flexible elements (such as the hoses shown on the crane in Figure 10-1) in the LDraw system is this: The LDraw file format has no built-in way to handle flexible elements. Flexible elements are simulated by using many small static parts that, when aligned together, give the appearance of a flexible part.

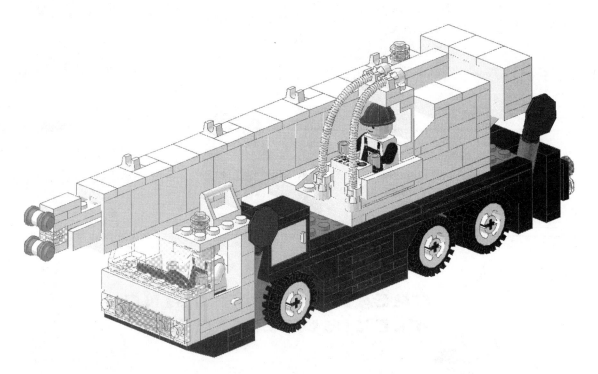

Figure 10-1: Flexible elements like the hoses in this crane add detail to an LDraw model, but often it can be difficult to understand how to create them.

Why doesn't the LDraw system handle this? We're using tools based on a grassroots community of volunteer contributors. Within that community, methods for adopting standards to solve these types of problems have not developed to the point where file format extensions are commonplace. Fortunately, within this group of free tools there are solutions like LSynth to work around these limits.

Tubing, Wires, Fiber Optics, Belts, and More with LSynth

LSynth, by Kevin Clague (http://users.qwest.com/~kclague/), is a program written to aid in creating bendable parts. Elements such as flexible Technic tubing, electric wires, tank treads, and ribbed tubes (you can see these elements in Tables 10-1 and 10-2) naturally have endless possible configurations because of their flexibility. LSynth paves the way for flexible elements to be created easily. It is the first robust solution that doesn't require number crunching on the part of the user.

NOTE *The name "LSynth" is derived from the action the program performs on an LDraw file. LSynth is a "bendable LEGO part synthesizer." It takes individual segment components of each type of flexible part and "synthesizes" them to form a curved path, creating what looks like one single flexible element.*

When it comes to creating these flexible parts, LSynth does most of the work for you. In this section, we will introduce you to the various flexible elements LSynth synthesizes. We will also teach you how to place LSynth's guide ("constraint") parts, which LSynth uses to guide the creation of the flexible element. Finally, we will show you how to run LSynth to synthesize the flexible elements.

Learning these techniques requires you to pay special attention to the text. You need to clearly understand how to place comments in specific places in your file and learn the characteristics of the constraint part files. Despite the slight learning curve, the capabilities this tool adds to the array of LDraw tools is phenomenal.

Parts LSynth Creates

LSynth divides the parts it creates into two major categories: Hose and Band. These categories reflect the characteristics of the types of parts they include. The LSynth documentation describes hoses as "parts that start in one place and snake their way to another place." *Hoses* are non-looping flexible parts. *Bands* are closed loops, and their shape is determined by their constraints, which are, as the documentation puts it ,"pulleys, pins, gears, and wheels."

Tables 10-1 and 10-2 list the parts in each category, each part's LSynth keyword, and information on the constraint parts each is used with. Note the part keywords, and use these tables to refer back to each part's keyword when you are creating your own flexible elements in LSynth.

NOTE *In the future, you can refer to these tables if you are ever unclear about which parts can be used to constrain certain types of flexible elements. We will cover the concepts behind using constraint parts in a later section.*

Table 10-1: "Hose" Style Flexible Elements

Image	Description	LSynth Keyword	Constraint Part
	Electric Cable This is the standard electric cable used to connect two "Electric Bricks" (part number 5306). These parts can be seen on the end of MINDSTORMS sensors, and are used in LEGO train sets to connect the track power wires to the Speed Regulator. These parts are always black in real life.	ELECTRIC_CABLE	LSynth Constraint Part for all "hose" style elements
	Technic Fiber Optic Cable Used with the Fiber Optic System. These parts are always clear (transparent) in real life. The end piece for the fiber optic cable is LS30.dat, and LS30c.dat is a completed shortcut of the cable.	FIBER_OPTIC_CABLE	

Table 10-1: "Hose" Style Flexible Elements

Image	Description	LSynth Keyword	Constraint Part
	Technic Axle Flexible Available in some Technic sets. LSynth does not automatically include the endpoints: Custom Part Number LS40.dat.	FLEXIBLE_AXLE	
	Pneumatic Hose Works with the Pneumatic System.	PNEUMATIC_HOSE	
	Ribbed Hose Available in the MINDSTORMS Robotics Invention System 1.x and other Technic sets.	RIBBED_HOSE	LSynth Constraint Part for all "hose" style elements
	Flex System Hose Used in conjunction with the Technic Flex System: The Flex System cable travels through the center of this hose. This is also the rigid tubing available in many Technic sets.	FLEX_SYSTEM_HOSE	

All of the hose-style parts use the same constraint part to guide their direction. The LSynth Constraint Part is reminiscent of an arrow; the pointed end signals the flow of the hose from the starting point to the ending point. The notch on the side indicates the alignment of its central axis: For example, electric wires, which are not round, are oriented along the notch.

Table 10-2: "Band" Style Flexible Elements

Image	Part	LSynth Keyword	Constraint Part(s)
	Technic Link Chain Available in many Technic sets.	CHAIN	Tooth Gears
	Technic Link Tread This is not a very common part, but very useful for Technic scale models.	PLASTIC_TREAD	Tooth Gears
	Rubber Tread These belts are widely available in MINDSTORMS sets.	RUBBER_TREAD	Technic Tread Sprocket Wheel
	Rubber Band This accomplishes the same task as the MLCad Rubber Belt Generator, but also allows the belt to loop around more than two points.	RUBBER_BAND	Pulley, Wedge Belt Wheel, Bushings, and so forth

Adding LSynth Constraint Parts to MLCad

Before you start using LSynth, you will need to add a category to MLCad's Parts List for quick access to constraint parts. By doing this, you will be able to easily drag constraint parts into the workspace via the Parts List pane. We will take you through this process below.

Editing the LSynth Parts

First, you need to edit three LSynth parts to include the "LSYNTH" keyword so that MLCad will recognize them when you configure the Parts List. To do this, follow these steps:

1. Open Windows Explorer and then open the **C:\LDraw\PARTS** folder.
2. Arrange the folder contents by name by right-clicking a blank space in the pane your files are stored in and selecting **Arrange Icons > By Name**.
3. Locate the three part files: **LS00.dat**, **LS30.dat**, and **LS40.dat**.
4. On the first line of each part, add the term **LSYNTH <part description>** after the first 0. Specifically, you should edit the first line of each file to look like this:

LS00.dat:

```
0 LSYNTH Constraint Part
```

LS30.dat:

```
0 LSYNTH Technic Fiber Optic Cable - large end
```

LS30.dat:

```
0 LSYNTH Technic Axle Flexible - end piece
```

5. Save each file and close it.

Creating a New Parts List Category

Now create a new parts list category to house the three LSynth parts, as follows:

1. On the MLCad toolbar, select **Settings > Groups > Group Configuration . . .**
2. In the two fields at the bottom of the "Parts Lists Configuration" window that appears, add the following (see Figure 10-2 for a screenshot of this window):
   - Name: **LSYNTH**
   - Search: **<LSYNTH**
3. Click **OK** to save the configuration and close the window.

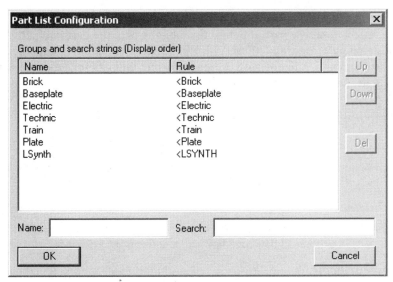

Figure 10-2: The MLCAD Parts Lists Configuration Window. Once you have completed the preceding steps, you should see an LSynth category in your window similar to the one pictured here.

Updating the Parts List

Before the parts will display in the new category you just created, you need to update the master parts.lst file. To do this from MLCad, select **File > Scan Parts** (alternatively, you could type **C:\LDraw\mklist.exe -d** in a DOS command prompt). The mouse pointer will turn into an hourglass for a minute or two, while MLCad searches the approximately 2,000 parts in the library. When MLCad finishes, a small dialog will appear (see Figure 10-3) telling you it has found new parts and asking whether you want to write a new parts.lst file. Click **Yes**.

Figure 10-3: MLCad will prompt you to write a new parts.lst file once it has found new parts. Click Yes.

Now check to see that your category works properly. In the Parts List window, click the plus sign before "LSynth," and you should see the three parts listed.

Using LSynth

There are two techniques for creating flexible parts in LSynth. Not surprisingly, they correspond with the two types of existing parts: hose and band. We will teach you how to create each type of part separately, and lead you through creating several specific parts in LSynth.

Creating Hose-Style Parts

To create a hose-style part, you must first plot the path the part will follow. The LSynth constraint part (Figure 10-4) used to plot the path has a pointed end, indicating the direction of flow. The notch on the side indicates the part's orientation, so you can tell, for example, how the electrical wires will lie (as you'll soon see).

Figure 10-4: LSynth's constraint part for hose-style flexible elements is pointed, indicating the direction of flow, and notched, to indicate its orientation.

Example: an Electric Cable

We'll use an electric cable as our example as we describe the necessary steps to create hose-style parts. First, open the file **C:\LDraw\VLEGO\CH10\e-cable.ldr** in MLCad (see Figure 10-5). This file contains two sets of electric cables. One is ready to be synthesized through LSynth, while the other is not. You must follow along in the file mentioned above in order to learn the material in this section.

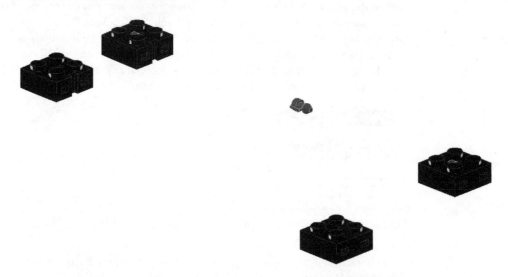

Figure 10-5: e-cable.ldr provides an example of a ready-to-synthesize electric cable (top) for reference, and a second electric cable setup with no constraints or LSynth commands (bottom).

Placing Constraints

With the file open in MLCad, you will notice three green constraint parts in the first electric cable. (We made these parts green to make them easier to spot.) There is one constraint part at each end, and a single constraint in-between the cable ends. This is a good configuration for a simple electric cable. (Pay close attention to our discussion of constraints in the following sections.)

Constraints' Direction of Flow and Linear LDraw Files

In Chapter 6, we discussed the linear nature of LDraw files. This concept is critical when placing LSynth constraints. Remember, in an LDraw file, parts are placed in a defined order, corresponding to the order in which they appear in a file.

As you saw previously in Figure 10-4, the "pointed end" of the constraint points to the right. When LSynth synthesizes the file, the cable (or other hose-style part) will "flow" in that direction. LSynth interprets each hose-style part as having a beginning and an end. The part begins at the first constraint in the file and ends at the last constraint in the file.

In the File View pane above the workspace, you can see the three constraints in e-cable.ldr, identified by the description "LSYNTH constraint part." Note that the first constraint in the file (buried within the leftmost end piece) is the starting point of the electric cable. The constraint point between the two ends is the second, and the third constraint is buried in the rightmost end piece.

When LSynth reads the LDraw file, it notes that the cable *starts flowing at the first constraint mentioned after the SYNTH BEGIN command* (seen in the e-cable.ldr file), and *stops flowing at the last constraint mentioned before the SYNTH END command* (we discuss these commands in the next section, "Entering the Necessary LSynth Commands"). Whenever you are placing constraints in a model of your own, remember this critical fact. If you do not place your constraint parts in the proper order, your part will "flow" erratically as LSynth creates electric cables by guiding the cable through constraints in the order they appear in the LDraw file. In essence, LSynth does what you tell it to do, so make sure you're using the proper order of constraints!

Refer to Figure 10-6 to understand the phenomenon that we described previously. The electric cable on the left used properly ordered constraint parts. The first constraint was located in the leftmost electric brick, the second was in the middle, and the third was in the rightmost electric brick. In the cable on the right, the first constraint was the middle one, followed by the constraint in the leftmost brick, and finally the one in the rightmost brick. You can see how the cable is improperly arranged, as the components that compose it start in the middle, travel back to the leftmost brick, and double back to the rightmost.

Figure 10-6: Proper order of constraints (left) and the results of improperly ordering constraints (right).

Orientation of Constraints

For electric cables and flexible Technic axles, be observant of how your constraint part is oriented. As you can see in Figure 10-7, the tab on the side of the constraint can be rotated to different orientations. When you have parts such as the electric cable that is not perfectly round, or the Technic flexible axle that has the profile of a cross, this tab indicates how the part should be oriented.

If you orient your electric cable using the leftmost constraint that is shown in Figure 10-7, it will appear to stand on end. If you use the orientation of the rightmost constraint, your cable will lay flat. Finally, if you angle your constraint like the one in the middle, the cable will appear at an angle. This applies to Technic flexible axles also — one of the points on the cross profile will follow the notch on the constraint.

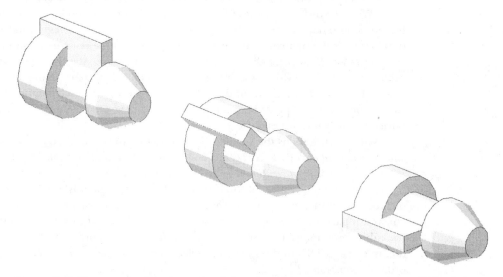

Figure 10-7: You can use the orientation of the tab on LSynth constraint parts to tell LSynth how to orient your electric cable or Technic flexible axle.

NOTE *If you were to synthesize a file like the one in Figure 10-7 for an electric cable, you might expect LSynth to gradually twist the cable as it flows along. We tried this, and LSynth produced some rather choppy results. There were breaks in the cable, and no gradual twisting motion. Make sure all of your constraints are oriented the same direction to avoid this problem.*

Entering the Necessary LSynth Commands

Before you can generate a flexible element in LSynth, you must enter commands in the form of comments surrounding your constraint parts. Once you have your constraint parts in order, follow these steps to enter the LSynth commands:

1. Select the part immediately preceding the first constraint part, and make it the active line in the File View Window.
2. Open the Comment dialog by clicking the **ab|** icon.
3. Enter **SYNTH BEGIN <KEYWORD> <COLOR>**, where <KEYWORD> is the LSynth Keyword of the type of element you want from Table 10-1 or Table 10-2, and <COLOR> is the color number. Click **OK** to insert the comment. If you are creating the second electric cable for Figure 10-6, for example, you will want to insert **ELECTRIC_CABLE** for <KEYWORD>.

NOTE *Finding the Color Number*
To change the color of your flexible element, you need to manually enter the color number in the SYNTH BEGIN command as shown above. The color of your constraint parts has no bearing on the color of your flexible element. To find your color's number, refer to the color chart on the back flap of this book.

4. Select the very last constraint part in your sequence, and make it the active line.
5. Repeat Step 2 and open the **Comment** dialog.
6. Enter **SYNTH END** and click **OK.**

Figure 10-8 shows you the commands properly placed preceding and following the constraint parts. The steps outlined above apply to any flexible element you want to create. Keep this in mind when you are using LSynth in the future: The program will not synthesize flexible elements through your constraint parts if you do not use these steps!

⌑PART	Black	-190.000,0.000,...	1.000,0.000,0.000 0.000,1.000,0.000...	5306a.DAT	~Electric Brick 2 x 2 x 2/3
⌑PART	Black	160.000,0.000,1...	-1.000,0.000,0.000 0.000,1.000,0.00...	5306a.DAT	~Electric Brick 2 x 2 x 2/3
⌾COMM...	--	------	------	------	SYNTH BEGIN ELECTRIC_CABLE 0
⌑PART	Green	-190.000,4.000,...	0.000,-1.000,0.000 1.000,0.000,0.00...	LS00.dat	LSYNTH constraint part
⌑PART	Green	-10.000,4.000,1...	0.000,-1.000,0.000 1.000,0.000,0.00...	LS00.dat	LSYNTH constraint part
⌑PART	Green	160.000,4.000,1...	0.000,-1.000,0.000 1.000,0.000,0.00...	LS00.dat	LSYNTH constraint part
⌾COMM...	--	------	------	------	SYNTH END
⌑PART	Black	-190.000,0.000,-...	1.000,0.000,0.000 0.000,1.000,0.000...	5306a.DAT	~Electric Brick 2 x 2 x 2/3
⌑PART	Black	160.000,16.000...	-1.000,0.000,0.000 0.000,1.000,0.00...	5306a.DAT	~Electric Brick 2 x 2 x 2/3

Figure 10-8: LSynth requires commands in the form of comments before and after the block of constraint parts you use to create your flexible elements.

Creating Multiple Flexible Elements in One LDraw File

Because LSynth only reads constraints between the SYNTH BEGIN and SYNTH END commands, you can easily create multiple flexible elements in one LDraw file. Remember to group all of the constraints for one particular flexible element segment in one place. Then place the appropriate SYNTH BEGIN and SYNTH END commands before and after *each group of constraint parts*. You can even create two different types of flexible elements in one LDraw file, by using the appropriate LSynth Keyword for each respective set of constraints.

NOTE   *When you're creating more than one flexible element in one file, it may be a good idea to change the color of your constraint parts for different flexible elements. The constraint part's color has no bearing on the color of the final element, since you enter the color number in the SYNTH BEGIN command. Using different colors for different elements helps you distinguish visually which constraints belong to which part.*

Synthesizing a Flexible Element

Before you can synthesize your flexible element(s), you must save the file in MLCad. Once you have saved the file, open LSynth. As you can see in Figure 10-9, the interface is very simple. To complete the process and synthesize your flexible elements, simply browse for your file in the **Input File** field. LSynth will synthesize the flexible elements in your file and write to a separate file. In the **Output File** field, you can specify the filename you want the output file to have, and you can also browse for a directory in which to save the file. If you don't specify a directory, LSynth will write the output file to the same directory the input file is located in.

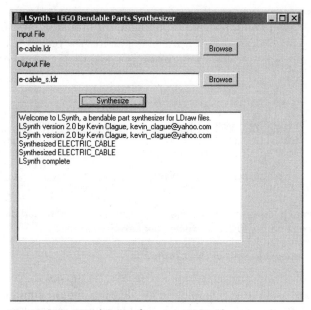

Figure 10-9: LSynth's interface is very simple.

When you are done selecting your input and output files, click the **Synthesize** button. LSynth will display the actions it has completed in the status window below. In Figure 10-9, we clicked Synthesize, and it read two electric cables in our e-cable.ldr file (after we had placed the constraints for the second set of cable ends). When you create other flexible elements, it will display those specific keywords in the status window.

NOTE   *The LSynth status window does not work in Windows 98.*

Creating Your Own Electric Cable Step by Step

Here are a few clear steps to take to create your own electric cable in our e-cable.ldr file. Follow along and create a flexible electric cable using LSynth:

1. Because you want to create a separate electric cable, begin by selecting the last line of the file from the **Model File Window** before placing any constraints. Doing this will allow you to insert your constraints after the two endpoints you are connecting, as well as separate them from the constraints for the first cable.

2. From the **Parts List Window**, select the **LSynth Constraint Part**, and drag it onto the workspace. Move the constraint so it is inside the cable end part. To help you place the constraint inside, try making the endpoint temporarily clear to place the part, such as in Figure 10-10. Be sure the pointed end of the constraint faces towards the other endpoint!

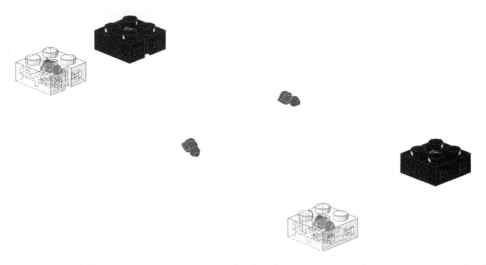

Figure 10-10: Making opaque parts temporarily clear helps you see to place constraints inside of them when needed.

3. Duplicate your constraint, and move it in between the two endpoints. Angle it slightly clockwise and down using the Fine Grid to guide the cable towards the other end. Be sure to check all four of the view panes for alignment while doing this.

4. Duplicate this constraint, re-align it to face straight and level, and move it to the inside of the second endpoint. Be sure each constraint has the pointed end facing in the same direction.

5. Place the proper LSynth commands before and after the block of constraint parts. Use **SYNTH BEGIN ELECTRIC_CABLE 0** before, and **SYNTH END** after the group.

6. Save the file. Synthesize the file in LSynth, and open the output file in MLCad to view your results. Figure 10-11 shows our results.

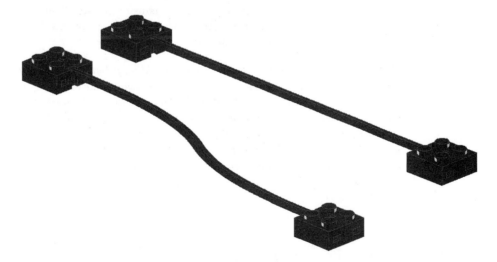

Figure 10-11: e-cable.ldr after being synthesized in LSynth.

Now you've successfully created your first flexible element. If you are unhappy with the results, you can open the previous file easily through the **File** menu (the bottom of the menu lists recently opened files), reconfigure the constraints, save, and re-synthesize your elements.

Additional Examples of Hose-Style Parts

Here are some more examples of hose-style parts. We have included the LDraw files on the CD-ROM for you to use.

Pneumatic Hose

We've created a ready-to-synthesize example of a pneumatic hose in use. You can access the file from **C:\LDraw\VLEGO\CH10\pneumatichose.ldr**. While we aren't connecting the hose to any actual Technic pneumatic parts, it is useful for other items as a decoration, as illustrated by Figure 10-12.

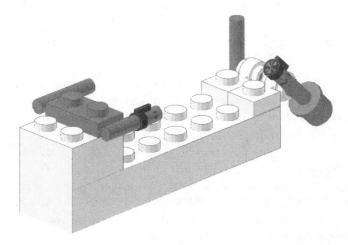

Figure 10-12: An example model, set up to create a pneumatic hose using LSynth.

Notice how, in Figure 10-12, the center of the constraint parts aligns perfectly with the center of the cylinders. Keep this in mind when creating your own flexible elements. Check out Figure 10-13 to see what this file looks like synthesized.

NOTE *When rotating two or more parts together such as the Minifig Torch and LSynth constraint in Figure 10-12, or the Technic Angle Connector #1 and Technic Axle Pin in Figure 10-14, you can group-select them. MLCad uses the rotation point from the first part you select. If you were to move multiple parts such as the two that are shown in Figure 10-12, first select the part you want all of the parts to rotate around. Next CTRL + click the additional parts. Then rotate the group. Note that the way the Radar Gun and LSynth constraint are currently arranged, you cannot CTRL + click the constraint. Move the constraint outside the bounding box of the Minifig Torch, group-select, rotate them as desired, and move the constraint back into place.*

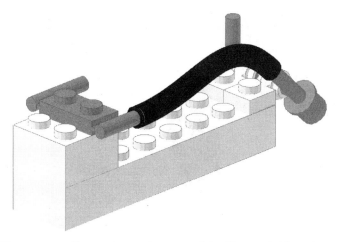

Figure 10-13: The synthesized version of the file, available in the same directory as pneumatichose_s.ldr.

Ribbed Tube

We've also created a ready-to-synthesize ribbed tube. The file is available under **C:\LDraw\ VLEGO\CH10\ribbedtube.ldr**. These tubes are common in some Technic and MINDSTORMS sets. See Figures 10-14 and 10-15 for the setup, and Figure 10-16 to see what the synthesized file looks like.

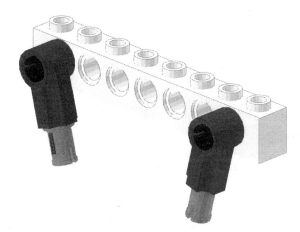

Figure 10-14: ribbedtube.ldr, set up to create a ribbed tube in LSynth.

Remember, when aligning constraint parts inside of other parts, you should make the pieces temporarily transparent so you can see to align your constraint. We've made the Technic pins transparent in Figure 10-15 as an example.

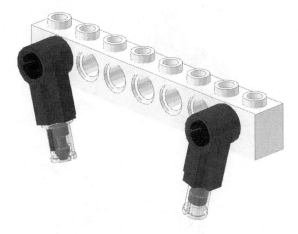

Figure 10-15: Make your parts temporarily transparent so you can see to align the constraints inside of them.

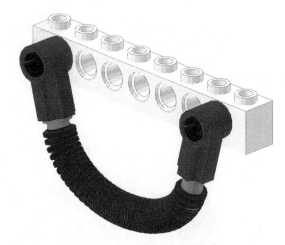

Figure 10-16: The synthesized ribbed tube, available in the same directory under ribbedtube_s.ldr.

File Size Warning

One known problem with the current version of LSynth is the large files it creates. When you synthesize large hose-style parts, and even rubber bands in the belt-style part category, LSynth uses hundreds of tiny segments to create the illusion of one flexible element. The large files it creates can make it difficult for MLCad to edit the file efficiently, and that may slow down the editing process if you are working with a post-synthesized file.

LSynth adds SYNTH SYNTHESIZED BEGIN and SYNTH SYNTHESIZED END comments before and after the line segments. It might be a good idea to select the first comment, scroll to the last comment, and group-select using the **SHIFT** key and a mouse click; then, create an MLCad group for the synthesized parts. Even though MLCad may still render the model slowly as you edit, you can still manage the parts reasonably. You should note that adding an MLCad group also increases your file size because it adds one comment line for each line in the group in the file. These comments are invisible to you in MLCad, but are visible in a text editor or an LDraw file editor such as LDDesignPad.

Creating Band-Style Parts

An entirely different technique exists for creating band-style parts. For starters, you don't need to use the LSynth constraint part the way you had to for hose-style parts. Here you use standard LDraw parts that LSynth recognizes as constraints. It will wrap a band-style part around these parts, such as a rubber band, chain, rubber tread, or plastic tread, just as you would do in real life (see Figure 10-17).

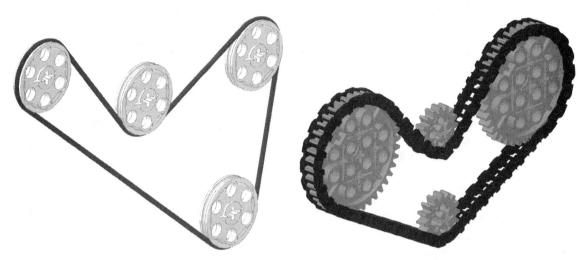

Figure 10-17: Band-style parts include rubber bands (MLCad refers to these as *rubber belts*) and chain. LSynth can create complex rubber band configurations where MLCad's Rubber Belt Generator cannot.

LSynth has three distinct advantages over MLCad's Rubber Belt Generator, which we discussed last chapter: you don't have to take any measurements to create your belt; it can create belts around more than two pulleys; and LSynth allows you to cross the rubber belt on the return loop. This creates a "figure eight" arrangement, which in real life allows the two linked pulleys to rotate in opposite directions. However, we decided to discuss MLCad's generator in the previous chapter for the sake of completeness.

Placing Constraints

Placing constraints for band-style parts is much easier than placing them for hose-style parts. Simply pick the gear or pulley you want your part to wrap around and place it in the model. You'll recall that Table 10-2 lists the parts you can use as constraints for chain, plastic treads, rubber bands, and rubber treads. Refer to this table when you are placing constraints in your projects.

Direction of Flow and Proper Order of Constraints

You should still pay attention to how you order these constraints in the file. As we discussed in the previous section, recall that LDraw files are linear. We learned from our experimentation that LSynth band-style parts have definite start and end points, even though they synthesize to create a continuous part or group of parts. Also, the band-style part starts flowing counter-clockwise from the first constraint in the group. This tidbit will be important later on when we discuss using more than two pulleys on a rubber band.

If you are using four or more constraints for any belt-style part, be careful how you arrange them. Because the flexible element travels in a counter-clockwise direction from the first constraint you place, be sure to arrange all of your constraints in a circular fashion, traveling counter-clockwise. If you don't do this, you will find some rather interesting results, as shown in Figure 10-18 on the next page. Also note that if you wrap your constraints clockwise from the first constraint, for example making the fourth constraint on the left in Figure 10-18 the second, and the second the fourth, you will also achieve undesirable results. Be sure to wrap your constraints in a counter-clockwise direction!

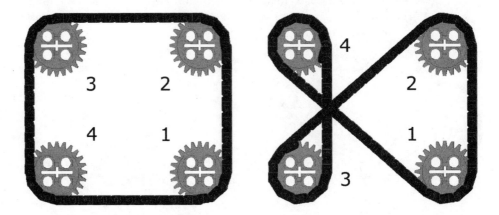

Figure 10-18: The proper way to order constraints in a file (left) and an improper way (right).

Aligning Constraints Properly

The physical limitations of the chain, rubber tread, and plastic tread parts in real life apply some limitations on the placement of constraints. These parts must be placed in line with each other so that the flexible element does not have to contort itself to fit in the grooves. Essentially, this means the grooves of all of the constraint parts must be aligned with each other, in the same orientation, and in the same plane. You can see an example of this phenomenon in Figure 10-19. The two top sets of rubber bands have incorrectly aligned constraints and therefore have broken rubber bands; the bottom is correct.

Inserting Begin and End Commands

Band-type parts require SYNTH BEGIN and SYNTH END commands just like hose-type parts. Refer to Table 10-2 for the different keywords for band-type parts. An example of a command would be:

```
SYNTH BEGIN RUBBER_BAND 0
```

This produces a black rubber band. Insert these as comments just like we taught you for hose-style parts, with SYNTH END at the end of a group of constraints.

Using Special Commands for Rubber Bands and Chains

When using more than two constraint parts, you may want to specify which direction LSynth wraps a flexible element around a particular constraint. LSynth wraps a belt-style element counter-clockwise by default, as you saw in Figure 10-18. In this image, the belt starts somewhere near the bottom right corner of the gear marked 1, and flows counter-clockwise from that point. That assumed, the gear and chain arrangement on the right hand side illustrates clearly that these parts flow counter-clockwise by default.

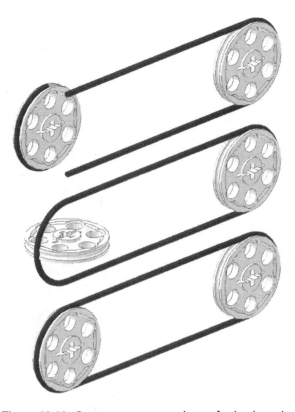

Figure 10-19: Constraint parts must be perfectly aligned with each other in 3D space and be of the same orientation to function properly. Not doing so will produce broken or improper rubber bands.

What if you want to wrap a chain or rubber band around a gear or pulley in a clockwise direction? What if you want to create a figure eight with your rubber band to reverse the direction of a pulley? LSynth offers the INSIDE, OUTSIDE, and CROSS commands for that purpose. These commands are inserted strategically among constraint parts to tell LSynth how to route the rubber band or chain around a constraint, or to criss-cross rubber bands to form a figure eight. You can see these commands in Table 10-3.

Table 10-3: Rubber Band and Chain Commands

Command	Effect	Used with
INSIDE	Constraint is located inside of the loop	RUBBER_BAND, CHAIN
OUTSIDE	Constraint is located outside of the loop	RUBBER_BAND, CHAIN
CROSS	Crossover (figure eight)	RUBBER_BAND

Using the Commands

You issue these commands like the SYNTH BEGIN and SYNTH END commands. An INSIDE command would be written "SYNTH INSIDE," an OUTSIDE command "SYNTH OUTSIDE," and so on.

Inside and Outside with Chain

Figure 10-20 shows a chain wrapping inside and outside a set of four gears. To achieve this behavior, we used a few commands, which are visible in Figure 10-21. To see the file for yourself, open **\CH10\chain.ldr**.

Figure 10-20: This chain wraps around four gears. All of the gears except for the 16-tooth on the top are INSIDE of the loop. The 16-tooth gear is OUTSIDE.

Type	Color	Position	Rotation	Part nr.	Description
COMM...	--	------	------	------	UNOFFICIAL MODEL
COMM...	--	------	------	------	SYNTH BEGIN CHAIN 0
PART	Light-Gray	-10.000,24.000,...	0.000,0.000,-1.000 0.000,1.000,0.00...	3649.DAT	Technic Gear 40 Tooth
COMM...	--	------	------	------	SYNTH OUTSIDE
PART	Light-Gray	-10.000,34.000,...	0.000,0.000,1.000 1.000,0.000,0.000...	4019.DAT	Technic Gear 16 Tooth
COMM...	--	------	------	------	SYNTH INSIDE
PART	Light-Gray	-10.000,24.000,-...	0.000,0.000,1.000 1.000,0.000,0.000...	3649.DAT	Technic Gear 40 Tooth
COMM...	--	------	------	------	SYNTH INSIDE
PART	Light-Gray	-10.000,114.000...	0.000,0.000,1.000 1.000,0.000,0.000...	4019.DAT	Technic Gear 16 Tooth
COMM...	--	------	------	------	SYNTH END

Figure 10-21: INSIDE and OUTSIDE commands guide the belt-style chain around the gear constraints.

Our loop of chain (Figure 10-20) starts off with the large 40-tooth gear in the upper left corner. That is reflected in Figure 10-21's view of the model file. The second gear, a 16-tooth, is on the outside of the loop, so it requires a SYNTH OUTSIDE command. The third gear, a large 40-tooth again, is on the INSIDE, and finally, the last 16-tooth gear is also on the INSIDE.

NOTE   *When working with chain, you may find it does not mesh perfectly with the gear teeth because LSynth does not give you precise control over the positioning of the links.*

Inside and Outside with Rubber Bands

The same principle is true for rubber bands. Figure 10-22 shows an example.

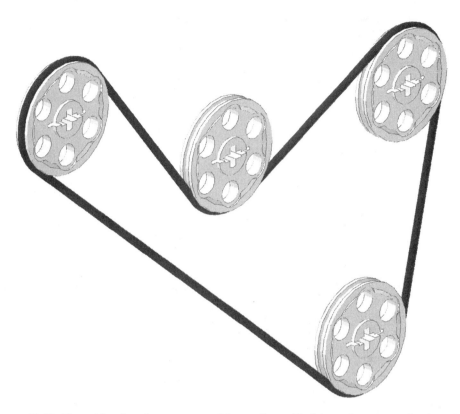

Figure 10-22: This rubber band wraps around four pulleys. All of the pulleys except for the center one are INSIDE of the loop. The center pulley is on the OUTSIDE.

In the preceding example (Figure 10-22), the first pulley is the bottom one. To snake the chain inside and outside of the pulleys, we use the commands that are shown in Figure 10-23. To see the file for yourself, open **C:\LDraw\VLEGO\CH10\rubberband.ldr**.

Type	Color	Position	Rotation	Part nr.	Description
⚲ COMM...	--	------	------	------	UNOFFICIAL MODEL
⚲ COMM...	--	------	------	------	SYNTH BEGIN RUBBER_BAND 4
▱ PART	White	0.000,0.000,0.000	0.000,0.000,-1.000 0.000,1.000,0.00...	4185.DAT	Technic Wedge Belt Wheel
⚲ COMM...	--	------	------	------	SYNTH INSIDE
▱ PART	White	0.000,-130.001,...	0.000,0.000,-1.000 0.000,1.000,0.00...	4185.DAT	Technic Wedge Belt Wheel
⚲ COMM...	--	------	------	------	SYNTH OUTSIDE
▱ PART	White	0.000,-130.001,-...	0.000,0.000,-1.000 0.000,1.000,0.00...	4185.DAT	Technic Wedge Belt Wheel
⚲ COMM...	--	------	------	------	SYNTH INSIDE
▱ PART	White	0.000,-186.001,-...	0.000,0.000,-1.000 0.000,1.000,0.00...	4185.DAT	Technic Wedge Belt Wheel
⚲ COMM...	--	------	------	------	SYNTH END

Figure 10-23: INSIDE and OUTSIDE commands guide the belt-style rubber band around the pulley constraints.

The only difference between this loop (Figure 10-22) and the loop of chain (Figure 10-20) is the start constraint. In this example, the start constraint is the bottom pulley; the chain starts off on the far right with the big 40-tooth gear. You can also see in Figure 10-23, after taking the change into account, that the commands we used are identical for the same reason.

Cross with Rubber Bands

Finally, LSynth allows you to cross rubber bands on a return loop to create a "figure eight." Figure 10-24 illustrates what we mean.

Figure 10-24: This rubber band crosses itself to form a figure eight.

Using the CROSS command is simple. Insert two pulleys as constraints, and encapsulate the second constraint with SYNTH CROSS commands, as shown in Figure 10-25.

Type	Color	Position	Rotation	Part nr.	Description
COMM...	--	------	------	------	Untitled
COMM...	--	------	------	------	Name: figure-8.ldr
COMM...	--	------	------	------	Author: Tim Courtney
COMM...	--	------	------	------	Unofficial Model
COMM...	--	------	------	------	SYNTH BEGIN RUBBER_BAND 4
PART	Light-Gray	0.000,-16.000,9...	0.000,0.000,-1.000 0.000,1.000,0.00...	4185.DAT	Technic Wedge Belt Wheel
COMM...	--	------	------	------	SYNTH CROSS
PART	Light-Gray	0.000,-16.000,-5...	0.000,0.000,-1.000 0.000,1.000,0.00...	4185.DAT	Technic Wedge Belt Wheel
COMM...	--	------	------	------	SYNTH CROSS
COMM...	--	------	------	------	SYNTH END

Figure 10-25: The CROSS command is used on either side of the second constraint to achieve the figure eight.

Synthesizing a Flexible Element

To synthesize one of these elements, recall the method we taught you in the same section under hose-style parts. Use the LSynth interface to select a file and synthesize it.

Flexible "Space" Hoses with LDDesignPad

One flexible element LSynth does not create is the "Space" hose, as shown in Figure 10-26. To solve this, we can use a program called LDDesignPad, and its Bezier Curve plugin. This program, written by Carsten Schmitz (http://www.m8laune.de), is a text editor that contains LDraw-specific functions. This editor is designed specifically to create and edit raw LDraw files. It also has a wonderful built-in plugin system, and several others have written their own plugins for the program. Orion Pobursky (http://ldraw.pobursky.com/) created a plugin to generate the classic flexible hose part that was introduced in the classic Space line in the 1980s.

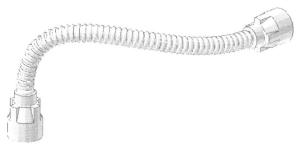

Figure 10-26: Flexible "Space" hose.

NOTE *According to the Free On-line Dictionary of Computing, a "plugin" is a file used to extend or enhance the operation of an application. You use plugins in your web browser: for example, Flash and Shockwave.*

As we said at the beginning of this book, LDraw files are plain text. This makes them easy to edit (assuming you know what you're doing, of course!). The process for creating hoses in LDDesignPad is simply this: Create your hose endpoints in MLCad, save and open your file in LDDesignPad, highlight the parts of the two endpoints, and run the Bezier Curve Generator. Once you save and reload your file in MLCad, you will have a complete hose.

NOTE *If you want to create a straight hose, you can use the Complete Assembly Shortcut part number 73509B, "Hose Flexible 8.5L with Tabs."*

Creating Your Hose

Here are the steps to create your own flexible hose.

Inserting Endpoints in MLCad

We created our hose using this example model, **C:\LDraw\VLEGO\CH10\hose.ldr** (see Figure 10-27). Go ahead and open this file in MLCad so you can create your own hose by following our instructions. We will teach you how to place a hose between the "headlight" brick (1×1 with the stud on its side) and the 1×1 plate with the two horizontal studs.

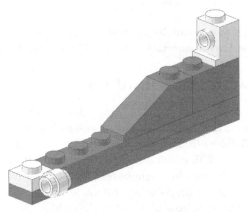

Figure 10-27: Our sample model for creating a hose.

After you've opened the file, you need to put the hose ends in place. We used part number 750, "Hose Flexible End 1×1×2/3 with Tabs," but if you prefer, you can use part number 752, which is the version without tabs. The tabbed version is the new style hose, whereas the version without tabs is the original.

Take one type of flexible hose end and align it with the studs as shown in Figure 10-28. Note that the hose you are about to create will be the same color as the endpoints you use here. We used black to make it stand out easily in our illustrations.

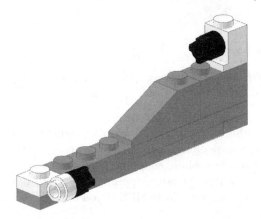

Figure 10-28: Add black flexible hose ends to hose.ldr.

Save the file to your hard drive with the hose ends you just inserted.

NOTE *This hose part has a fixed length in real life, and the LDraw part mimics that. If you ever place the hose ends too far apart from each other, the Bezier Curve plugin in LDDesignPad will not generate the hose for you.*

Generating the Hose with Bezier Curve Plugin

Now it is time to use LDDesignPad's Bezier Curve plugin to generate your hose. Launch LDDesignPad and open the file you just saved. It will open a text file that looks like this:

```
0 Untitled
0 Name: hose.ldr
0 Author: Tim Courtney
0 Unofficial Model
0 ROTATION CENTER 0 0 0 1 "Custom"
0 ROTATION CONFIG 0 0
1 7 0 0 0 0 0 1 0 1 0 -1 0 0 3460.DAT
1 7 0 0 90 1 0 0 0 1 0 0 0 1 3024.DAT
1 15 0 -8 -70 0 0 -1 0 1 0 1 0 0 4081B.DAT
1 7 0 -8 20 0 0 1 0 1 0 -1 0 0 3460.DAT
1 7 0 -32 80 0 0 1 0 1 0 -1 0 0 3004.DAT
1 7 0 -32 50 1 0 0 0 1 0 0 0 1 4286.DAT
1 15 0 -56 90 1 0 0 0 1 0 0 0 1 4070.DAT
1 0 0 -46 68 1 0 0 0 0 -1 0 1 0 750.DAT
1 0 20 -6 -50 -1 0 0 0 0 -1 0 -1 0 750.DAT
0
```

This is the raw LDraw file. To be able to use the plugin, first you must select the bottom two lines with "750.DAT" at the end, just as you would select some lines of text in any text editor. Remember that this is the number for the hose end part. Selecting the text tells the plugin which parts it is using to create the Bezier curve.

Once you have selected the two lines, choose **Tools > Plugins > Generate Bezier Curves** from the menu. The window that is shown in Figure 10-29 will appear.

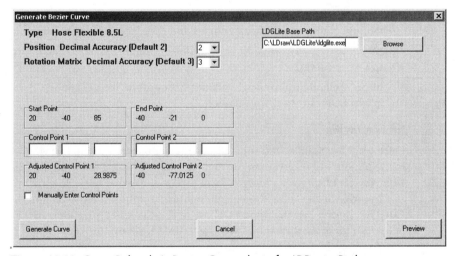

Figure 10-29: Orion Pobursky's Bezier Curve plugin for LDDesignPad.

Click the **Generate Curve** button. The window will close, and you will notice many more lines were added to your file. These lines represent each rib in the hose, all aligned to curve perfectly. Now save the file and reload it in MLCad. (Note that MLCad does not have a reload feature; simply open another model and then open **hose.ldr** again, or you can choose **Extras > Filecache > Reset** from the menu and open **hose.ldr** again.) Now you can see the hose you just created (see Figure 10-30).

One thing you might want to do is check out how the physical hose part behaves when you add it to the same arrangement of LEGO bricks in real life. This will help you judge if the curve generated is true to life or not. To see our completed file, check out **C:\LDraw\ VLEGO\CH10\hose-complete.ldr**.

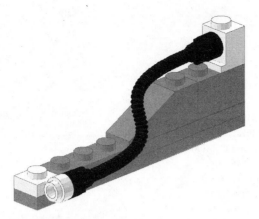

Figure 10-30: Your new hose, courtesy of the Bezier Curve plugin.

Adjusting the Curve Manually

As you probably noticed in the plugin, you can adjust your curve manually by entering control points. This is definitely an advanced feature and requires some knowledge of Bezier curves to be used effectively. If you are dissatisfied with the curve the plugin generated, you can adjust the control points to make the hose flow more true to life.

Bezier Curves

Check out these websites for information on Bezier curves:

http://www.math.ucla.edu/~baker/java/hoefer/Bezier.htm: This is an interactive demonstration of how Bezier curves and their control points function.

http://astronomy.swin.edu.au/~pbourke/curves/bezier/: A mathematical approach to Bezier curves.

http://folk.uio.no/fredrigl/technic/ldraw-mode/bezier-curve/: Bezier curves in Fredrik Glöckner's LDraw-mode.

Previewing Your Hose

Once you have familiarized yourself with Bezier curves and start playing with control points in the plugin, you may want to preview your hose before generating it. This is possible uwith Don Heyse's LDGLite, a cross-platform program capable of viewing LDraw files.

In order to take advantage of this feature, you must tell the plugin where to look for LDGLite. Because we included LDGLite in the CD-ROM's installer, all you have to do is point it in the right direction. LDGLite is already set up for you.

To set up LDGLite and preview your hose, follow these steps:

1. With your endpoints selected, launch the Bezier Curve plugin by selecting **Tools > Plugins > Generate Bezier Curves**.
2. Underneath **LDGLite Base Path** (see Figure 10-31), select **Browse**.
3. ldglite.exe should be installed at **C:\LDraw\Programs\LDGLite\ldglite.exe** (unlike the figure below). Find it, and select **Open**. This updates the Base Path field. Now, LDGLite will generate previews on demand.
4. To preview your curve, click the **Preview** button in the lower right hand corner. The plugin will take a second to process this, but when it is done you will see an image of the hose you are about to create.
5. To create the hose, click **Generate Curve**.

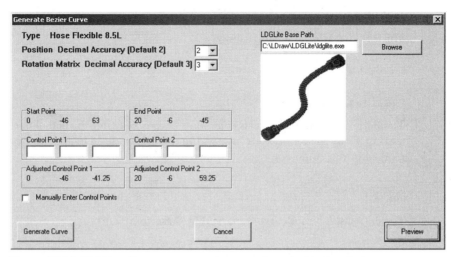

Figure 10-31: LDGLite allows you to preview your hose inside the plugin.

Cleaning Up Your File

When you generate a Bezier Curve to create a hose, the plugin adds many lines to your file. As we mentioned earlier, each line represents one rib in the hose. It is sometimes inconvenient to edit an LDraw file in MLCad with all of those lines, because you still have to deal with them in the File View Window. To get around this, you can create a separate sub-model for your hose, move it into the model, and save it as an MPD file. This way, your main model only sees one line to represent your hose, and you can move the entire hose around easily without the risk of breaking it.

Editing by Hand

Here is your first lesson in editing an LDraw file by hand. It may seem intimidating, but once you learn what the different elements are, it's not so difficult. All you need to do is identify a few elements of the file, and insert some text. We will teach you more about editing files by hand in Chapter 21, "Creating Your Own LDraw Parts." Follow along with our description of a few basic elements, and you should do fine. We also provide our completely edited hose MPD file so you can see one done right.

Elements of an LDraw File

In Chapter 6, you were introduced to the basic elements of an LDraw model file, parts, and comments, and learned what they looked like in MLCad. (There are a few more elements, but those have to do with creating parts; we will touch on those in Chapter 21 when we discuss parts authoring.)

LDDesignPad conveniently colors different types of lines (and different elements of each line) so that you can easily pick them out. This is another one of its wonderful features. By default, LDDP colors comments blue, where with parts it colors the various elements differently so they're easily identifiable. Most importantly though, the first character in a comment line is a 0, where the first element of a part is a 1.

Comment

The first character in a comment is always a 0.

```
0 Author: Tim Courtney
```

Part

The first character in a part is always a 1.

---

1 7 0 0 0 0 0 1 0 1 0 -1 0 0 3460.DAT

Elements of a Bezier Curve

The Bezier curve plugin inserts comments before and after the actual parts it inserts, making it easy to identify your hose within the LDraw file. The beginning of the curve is marked by this comment:

---

0 Begin Bezier Curve

The end of the Bezier curve is marked by these comments:

```
0 Start Point (0 -46 63) Control Point 1 (0 -46 -41.25)
0 Control Point 2 (20 -6 59.25) End Point (20 -6 -45)
0 Number Of Segments: 50 Curve Length: 130
0 Curve created using the Generate Bezier Curve plugin for LDDesignPad
0    by Orion Pobursky based on code by Fredrik Glockner
0 End Bezier Curve
```

The first two lines are useful for information keeping purposes, to have a record of the elements that make up your hose. The credit lines can be safely deleted if you like. Leave the End Bezier Curve comment in as a marker.

Creating a Sub-model for the Hose

To create a sub-model for your hose, you need to add some elements to the file, and then place the sub-model in MLCad. Follow these steps:

1. At the top of the file, add the following line:

---

0 FILE main.ldr

> This is the syntax for creating an MPD file. By inserting an **0 FILE** statement, you are telling LDraw compatible software that the following lines should be considered a part of this file. By declaring more than one file within one LDraw file, you have created an MPD, or Multi-Part DAT file.

2. Find the comment "0 Begin Bezier Curve." Place your cursor before the 0 and hit **ENTER**, creating a blank line. On the new line, key in:

```
0
0 FILE hose.ldr
```

> The first "0" is to provide a space between the two sub-models.

3. Save your file as **hose-edited.mpd**. Open the file you just saved in MLCad.

When you open your file in MLCad, you will notice the hose does not appear. You need to add it as a sub-model, as we taught in Chapter 9. Click the [icon] icon or press the **I** key to insert a new part. Check the **Custom Part** box, and key in **hose.ldr** in the field. Click **OK**. MLCad will place the hose inside your file. You can also find your hose in the Document category (beneath Favourites) where you can drag it into your model just as you would add a new part. Now all you need to do is re-align it with the parts it connects to, and you're done! You have now successfully generated your hose and stored the hose parts in a separate sub-model.

If you would like to have a look at our completed MPD file, check out **C:\LDraw\ VLEGO\CH10\hose-edited.mpd**.

NOTE *If you are creating your own MPD files by hand when creating your own models, consider this. With our example, the hose elements were the last parts in the file. Sometimes you may want to insert a hose in the middle of your LDraw file. In order to successfully create your MPD, you need to cut and paste all of your hose elements to the end of the file before inserting the MPD comments. If you do not do this, parts of your model will be included in the hose.mpd sub-model because they appear after the 0 FILE comment.*

MLCad's Flexible Hose Generator

MLCad also features a hose generator (accessible through the [icon] icon). However, we found this generator more difficult to use than the LDDP plugin (see the generator in Figure 10-32). It requires you to manually find and insert the beginning, ending, and control points. We found it doesn't follow the same principles as the interactive Bezier curve page at http://www.math.ucla.edu/~baker/java/hoefer/Bezier.htm, so that also adds a level of confusion. If you're up for the challenge, play with this generator and figure it out.

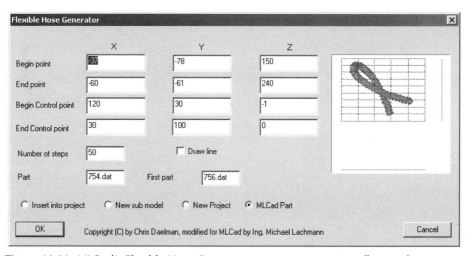

Figure 10-32: MLCad's Flexible Hose Generator requires you to manually enter beginning, ending, and control points.

LDraw-Mode

As an alternative, you can also try Fredrik Glöckner's LDraw-mode for Emacs at http://www.math.uio.no/~fredrigl/technic/ldraw-mode/index.html.

Summary

In this chapter, you learned how to create many different types of flexible elements using LSynth and LDDesignPad. Because the LDraw file format does not support flexible parts, these elements have to be simulated by aligning many smaller component parts together to form the appearance of a bend or curve. This would be nearly impossible to do without the help of the tools outlined in this chapter.

11

MANAGING YOUR MODEL FILES

As you start to use the LDraw tools more and more to document your models, your Models folder will probably start to get cluttered. In order to help yourself stay organized when you build, you should consider developing a folder structure. When you place all your work in one directory, it's easy to become overwhelmed with the number of models you have — especially when you start using sub-models! The best way around this is to stop the clutter before it starts. This chapter contains our suggestions for keeping your models organized.

Develop a Folder Structure

You can keep your Models folder clean by developing a folder structure. A good method is to keep each model in a separate folder. This may not be a big deal when you are making simple models, using only one file and not sub-modeling at all. When you start using sub-models more often and drawing complex models with a lot of sub-models, it will clutter your Models folder fast. The best way to avoid confusion about which files belong to which model is to keep each in a separate directory.

Themed Folders

If you build several different themes, create top-level folders with the theme names, like "Castle," "Space," "Town," and "Trains." If you're building a group of models in a custom theme, you could name a folder accordingly.

Other People's Models

The wonderful online LEGO community gives us the opportunity to download some amazing creations by other people. To avoid confusing things that other people have built with your own creations, why not create a folder specifically for that purpose? Years ago, some people in the LDraw community asked that very question. Many LDraw users have a folder under Models called "Authors." Within that folder, you can create individual directories for each person: "Tim Courtney" (who rarely publishes an LDraw model he's made), "Steve Bliss" (who has published maybe one model in his years of writing parts), "Ahui Herrera" (who publishes a bunch of models), and so on.

Develop a File Naming Convention

Most of the time, you'll probably be using MPDs to create models with sub-models. However, in the case where you make a file (or a scene, for that matter) that has multiple .ldr files, consider adopting a naming convention. When you keep each model in its own directory, you can repeat the same naming convention across all of your models, without worrying about overlapping the names. For example, if you built a crane, you might name your files main.ldr, superstructure.ldr, boom.ldr, and boom_extension.ldr.

Example: LDraw.org Official Model Repository

In 1999, when LDraw.org was founded, the members of the community discussed a standard for creating LDraw representations of Official LEGO models. The intention was to collect the files, created by individuals in the community, and host them on LDraw.org as a repository of models, called (not surprisingly) the Official Model Repository (OMR). The system has not yet been put into place on LDraw.org, but the file naming convention is published on the website. Here is that system, to give you a bit of a peek at a developed naming convention.

The following is reprinted from http://www.ldraw.org/reference/specs/omr.shtml. If you are planning to create official models to contribute to the site, you should check the page for any updates that may have happened since this book was published:

Official Model Repository Filenames and Headers

The Official Model Repository has developed standards for file headers and file names. This is for consistency between models and easy indexing by a script and a database. Here is outlined the standard for the filenames and headers.

Headers

The file header must be concise and at the same time display all the information needed for the model/sub-model. Here is the official file header for the Repository:

Main Model Header

```
0 7140 X-Wing Fighter
0 Name: main.ldr
0 Author: Tim Courtney
0 LDraw.org Official Model Repository
0 http://www.ldraw.org/repository/official/
```

Sub-model Header

```
0 7140 X-Wing Fighter, Main Fuselage
0 Name: m-2a.ldr
0 Author: Tim Courtney
0 LDraw.org Official Model Repository
0 http://www.ldraw.org/repository/official/
```

Minifig Header

```
0 7140 X-Wing Fighter, Luke Skywalker Mini-figure
0 Name: mf-1.ldr (see below [1])
0 Author: Tim Courtney
0 LDraw.org Official Model Repository
0 http://www.ldraw.org/repository/official/
```

Terms

Here are the terms used in the Repository along with their contextual meaning:

- Set: Display of every model and sub-model (in essence the whole box).
- Model: Individual model in set. For example, Service Vehicle and X-Wing Fighter on 7140. Includes such models as cars, airplanes, helicopters, etc.
- Sub-model: Sub-assembly (in instructions) of a model. Also used as the sub-assemblies that make up rotating parts of the model.
- Minifig: Specific to a mini-figure sub-model within a set.

Filenames and Storage

Filenames for the Official Model Repository files are restricted to the MS-DOS 8.3 compatible name length. Each set in the OMR is packaged into an MPD file. This MPD file can be extracted to create another subdirectory for the particular set (see diagram below). Within that are several files that make up the set.

The set subdirectory uses the set number for a name, and contains an optional modifier and the two-digit year. This will be accomplished by placing these directory instructions within the MPD file that contains the entire set. All of this will fall under a sets/ subdirectory under the models/ directory.

```
          Four digit TLG [LEGO Group] set number
          |
          | Optional modifier to the year [2]
          | |
          | | Two digit year - year set was released.
          | | |
models/sets/XXXXZ-YY/
```

Within the set's subdirectory, the following filenames will represent the following aspects of the set/models.

- main.ldr: Main instructions, including all models that can be built at once. Basically, this is just a display of m1.ldr through m(x).ldr
- m-1.ldr: First model in main instructions
- m-1a.ldr: First sub-model of first model

- m-1aa.ldr: First component of first sub-model of first model
- m-1ab.ldr: Second component of first sub-model of first model
- m-2.ldr: Second model of main instructions
- alt.ldr: Alternate instructions, display file
- a-1.ldr: First model of alt. instructions
- a-1a.ldr: First sub-model of first alternate model

To use further levels of sub-models, one could continue adding on letters to the end until a total of eight characters in the main filename has been reached.

Minifig Sub-models

Minifig sub-models will be named in order of appearance in the instructions.

- mf-1.ldr (7140) Luke Skywalker minifig
- mf-2.ldr (7140) Biggs Darklighter minifig
- mf-3.ldr (7140) Mechanic minifig

Optional Modifier to the Year

The optional modifier to the year comes into play when there are more than one set in the specific year that bear that set number. This usually takes place with the multi-packs of the late 80s and early 90s, but also occurs in other sets. For example, set 1974, a multi-pack:

```
                        Optional modifier to the year
                         |
C:\LDRAW\MODELS\SETS\1974A-90\MAIN.ldr Flyercracker U.S.A.
C:\LDRAW\MODELS\SETS\1974B-90\MAIN.ldr Smuggler's Hayride
C:\LDRAW\MODELS\SETS\1974C-90\MAIN.ldr Star Quest
```

Each separate set/model that was released with that specific set number in that year gets a modifier in its directory. The files within the directory still follow the standard naming system. Typically the first set in a pack, or the first set to be released with that number, would get A as the modifier, the second B, etc.

This system of headers and naming files will be implemented on the LDraw.org Official Model Repository for the redistribution of official sets to LDraw users. Each individual author of official models used in the OMR will receive credit for the model in the header and on the LDraw.org website. The OMR does not intend to detract from the individual author or the author's website, but seeks to distribute LEGO models to LDraw users free of charge.

(End of reprinted material.)

Summary

In this chapter, we discussed the benefits of organizing your Models folder and using a file naming convention. As you build your own models, take our thoughts and use them as a guide to develop your own structure. Realize that different structures work for different people, and be sure to create one that works the best for you. We also cited LDraw.org's Official Model Repository as an example of a well-developed naming convention.

12

INTRODUCTION TO BUILDING INSTRUCTIONS

Custom building instructions can be a very useful tool when you are trying to share your models with others. Using the techniques we outline in this chapter, you can prepare your LDraw file for creating instructions for your own models that look just like the ones produced by the LEGO Company! In this chapter, we will teach you several useful techniques to prepare your LDraw files for creating building instructions. Also, we will teach the MLCad features Rotation Step, Buffer Exchange, and Ghost, which are critical in creating effective instructions.

We will revisit Jon Palmer's Scumcraft from Chapter 8 in greater depth, and use it as an example of how to create custom building instructions. The Scumcraft MPD file is located on your hard drive under **C:\LDraw\VLEGO\CH08\scumcraft.mpd**.

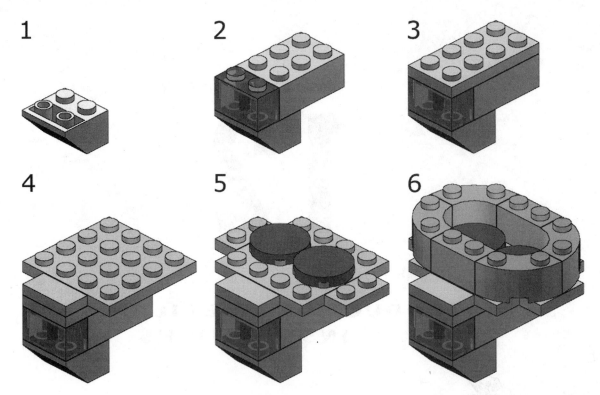

Figure 12-1: With some patience, it is not difficult to create building instructions comparable to those the LEGO Company produces. The steps in this image show you how to build the Scumcraft's engine. They were created using MLCad, L3P/L3PAO, and MegaPOV, and laid out in an image editor.

NOTE *You can use this chapter along with Chapter 15, "LPub: Automate Building Instruction Renderings," and Chapter 18, "Post-Processing Your Building Instructions." These are the three key chapters for learning how to create, render, and finish building instructions using LDraw tools.*

Static vs. Functional Building

In Chapter 8, we discussed how to apply functional building techniques in LDraw. When creating building instructions, especially for models with complicated sub-assemblies, you will want to take a functional (versus static) approach to documenting your models. The various components of your model should be separate files (stored in a single .mpd file or in multiple .ldr files) so you can easily create sub-assembly instructions for them. The functional building technique is useful when you need to rotate entire sub-assemblies in an instruction step.

> **Sub-Models and Sub-Assemblies**
>
> In Chapter 8 we talked about Sub-models — models that serve as components of larger models. What is the difference between a sub-model and a sub-assembly? Often times, there is no physical difference. When talking about building instructions, we want to refer to a component group of parts as a "sub-assembly" instead of a "sub-model," because we are in the process of building. The word "assembly" allows us to think about a model in terms of the most logical way to put it together. Sub-models, on the other hand, are helpful tools for creating complex LDraw files, whether or not those files are destined to become building instructions.

Building Instruction Theory

Building instructions are designed to clearly communicate the steps needed to construct a model; they should use minimal words, if any. Anyone should be able to understand building instructions, no matter what language they may speak. It's important to keep this in mind when creating your own building instructions because you want your audience to understand each individual step perfectly. There are exceptions, of course, in which you will want to use some amount of text, but if you design instructions correctly, these exceptions will be few and far between.

To that end, there are many important things to consider. When creating your own instructions, you should strive to create a step-by-step building process so clear that the builder feels totally comfortable with the model he or she is putting together. As with driving a car or turning on a light switch, if it is working properly, you don't notice the system that is behind what makes something work.

The Building Instruction Process

Before you can create your own building instructions, you should have built your model out of physical LEGO pieces. Some people, such as the custom kit makers of the Guild of Bricksmiths™ (www.bricksmiths.com), spend a lot of time creating building instructions for their models. Guildsman Steve Barile has written a page worth of notes on the process he used to create his Freight Train Instruction Book. You can read these notes on his site at http://www.bricworx.com/1001-how.php.

NOTE *We will revisit Steve's production process in Chapter 18, when we discuss the post-production of building instructions. For now, we will discuss modeling techniques to optimize your LDraw file for creating instructions.*

In order to share an outside perspective with you, we asked Jake McKee, creator of the Building Instruction Portal (www.bricksonthebrain.com/instructions/), to share his instruction creation process with us. Jake takes a very integrated approach to creating building instructions. He usually jumps back and forth between building and revising a model using physical LEGO pieces and documenting the model in LDraw, until a design is finished.

Jake starts a project in plastic. He designs his model until he gets a solid "first draft." Next, he takes the design to LDraw and mocks up the creation, making changes to the design that make sense to him along the way. For instance, in a particular model he may find that using a 1×4 plate, rather than two 1×2 plates, helps reduce the parts count and strengthens the design at the same time.

Once the model is built in LDraw, he will use the built-in MLCad sort function (Edit > Sort, by position, Ascending) to sort parts from bottom to top. He then goes through the LDraw file, adding steps at logical points. Based on these rough instructions, Jake builds another version of the model using physical LEGO bricks. This tells him if the changes he made in the LDraw file are "buildable," and gives him an idea of how his steps flow. He repeats this process until he feels his design is finished. When he is happy with the design, he uses the same sorting technique to create the final steps. Jake then polishes his instructions using sub-models and MLCad's Buffer Exchange feature (discussed later in this chapter).

NOTE *When Jake uses sub-models, he will use as many as makes sense for the design process. For the final build of the instructions, he combines some sub-models into larger components, which makes the output (not creation) easier for people to understand as instructions.*

Jake also saves each revision of his LDraw model as a separate file so that he can track his development process from start to finish. This yields a series of design sketches, and the entire project — from start to finish — is documented. He tests the instructions he is creating along the way (in some cases as many as seven or eight times) so he has a good idea of where the problems are. He also learns where he needs to highlight parts or sub-assemblies by using an MLCad Buffer Exchange.

To finish the instruction creation process, he asks friends, family, and other LEGO fans to try to build the final design. Because he's tested and revised the instructions himself along the way, there are usually very few bugs to fix at this point.

If you are building your model entirely in LDraw, without using physical LEGO bricks, Jake's techniques may not apply to you. We believe it is always a good idea to test your model out of physical LEGO bricks before publishing your building instructions. Remember, different people use different processes to create their physical LEGO creations and building instructions for them. Learn from others, and then develop the process that works the best for you.

Using Steps

As you already know, building instructions are made up of a series of construction steps, each step containing a few more parts than the previous one did. Believe it or not, a lot of thought needs to go in to how the construction process is organized. One way to learn how to make your own instructions is to take a look at those made for existing models. Check out instructions for a variety of models, from different themes and for different types of creations. BrickShelf is an excellent website that features archives of LEGO building instruction scans (http://library.brickshelf.com/scans/).

Finding a Logical Starting Point

Sometimes it is easy to know where to start building instructions for a model. For example, on a small car or truck, you naturally start at the bottom with the lower plates and axles. If you were instead designing a large Technic model or a spaceship, the starting point would not be so clear. Because most models are built "studs-up," try to find a place near the bottom of a model for the first step. It's a good idea to consider starting with a portion of the model that is fairly stable.

If you are building a complex model with many functional components, start with the core structure. Pick an area of the model that's solid, does not move, and that supports other moving components. For example, the Scumcraft (shown in Figure 12-2) begins with the fuselage.

Adding Parts

Take careful note of how many parts you add per step. When creating building instructions, you want to make it easy for the builder to follow along. Adding too many parts per step can make the building process difficult and even overwhelming to a younger builder. On the other hand, adding too few parts per step may annoy or tire someone trying to assemble your model, not to mention waste paper (if you are printing them).

You can get a good idea of how many parts to add per step by looking at official LEGO building instructions. While LEGO does a great job of making clear instructions, not all models add the same average number of parts per step. Small to medium sized models, such as the 6923 Particle Ionizer (http://library.brickshelf.com/scans/6000/6923/), add an average of four or five parts per step. On the other hand, some large models, such as the 8480 Space Shuttle (http://library.brickshelf.com/scans/8000/8480/) can add up to twenty or more parts per step!

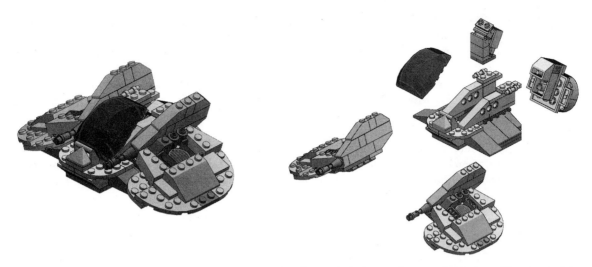

Figure 12-2: Jon Palmer's Scumcraft (left). Exploded view showing the Scumcraft's sub-models (right). The Scumcraft's core structure is the cockpit piece, where the wings, engine, and canopy attach via hinges.

NOTE   *If you're adding a lot of parts per step in your building instructions, it is a good idea to include a parts list image for each step. LEGO Technic models do this, as well as complex models like the 3451 Sopwith Camel. We will teach you how to make these in Chapter 15, when we discuss LPub.*

Common Sense

Of course, it's okay to add one or two parts in a step if those parts are of structural or functional importance to the model. Official LEGO building instructions do not follow a hard, fast rule requiring every single step to include a certain number of new parts. Use common sense and good judgment when deciding how many parts each step should contain. It might help to imagine your creation as if it were an official LEGO model; how would its instructions look if the LEGO Company created them? After some practice, and some time spent testing your instructions, it will become easier to determine where steps should go.

MLCad's Autosteps Feature

MLCad has a feature that you can use to insert steps into your model automatically. You can access the Autosteps feature by selecting **Edit > Add > Autosteps**. Figure 12-3 shows you the Autosteps dialog. This feature allows you to insert steps when the height (location of bricks in 3D space) changes as the model progresses. It also allows you to select an increment, and MLCad will insert a step every so many parts.

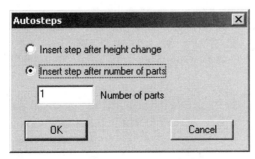

Figure 12-3: MLCad's Autosteps dialog.

Arranging Parts

It is also important to think about how you position parts when creating building instructions. As a rule, don't add a part that will hide another part you need to add in a future step. This can be difficult to avoid. When documenting the model, it is easy to do this accidentally and have to correct yourself later. Thanks to MLCad's robust editing capabilities, correcting a mistake such as this is fairly easy.

MLCad's Sort Feature

MLCad has a feature that can sort a model's parts by position (in 3D space), color, or part number. You can also limit its use only to currently selected parts, and choose whether to sort ascending (starting with the bottom-most part) or descending (starting with the uppermost part). To use this Sort feature, select **Edit > Sort**. You will be presented with the Sorting Parameters dialog, as seen in Figure 12-4.

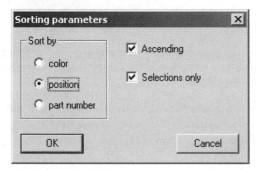

Figure 12-4: MLCad's Sorting Parameters dialog.

Part Connections

If you are creating instructions with the goal of having someone else build your model, consider the types of connections you use in your model's construction. This might not seem important to you as the model's creator, but it could be an issue to someone else who is experiencing your creation for the first time.

Number of Connection Points

When you are adding large parts to a model, keep in mind that some builders (especially younger ones) may have a hard time attaching too many studs at once. LEGO's master builders, the people who assemble the gigantic sculptures for theme parks and special events, sometimes use rubber mallets to pound down large parts. When building with your physical bricks, you may notice how difficult it is to attach one large plate to another, and how hard you have to press down to make sure all studs connect; keep this is mind if you would like younger children to be able to build your creation successfully. Young children are not as strong or agile with their hands as teens or adults, so putting pieces together that have many stud connections may be physically challenging. Also, depending on the age and disposition of the child, he or she may become frustrated when building your model and give up. This is definitely not what you want to happen!

Changing the View Angle

Sometimes building instructions call for you to rotate the model as you are working on it, to attach parts to an area that isn't visible in the current step. This seems commonplace when you are building with official LEGO instructions, but how do you get this to work within the LDraw system? Fortunately, MLCad has developed a feature called the "Rotation Step" to solve the problem.

MLCad's Rotation Steps

A Rotation Step is a feature unique to MLCad, which allows you to change the angle of your model when creating building instructions. You have likely seen this illustrated many times when building official LEGO models. For example, in one step you may be adding parts to the top of a model when the next step requires you to turn your model over and add parts to the bottom. The Rotation Step feature allows you to do just that; it changes the view of the model for a particular step or steps in your instructions. See Figure 12-5 for an example.

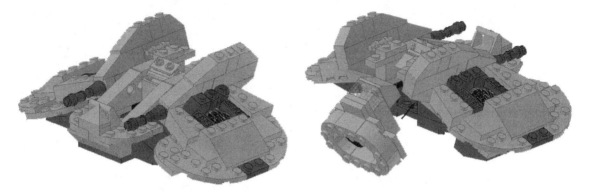

Figure 12-5: Rotation Steps are used to rotate the Scumcraft so the back is seen in front. This makes it possible to see the engine being attached in the proper position.

Three Types

There are three types of rotation steps: Relative, Absolute, and Additive. *Relative rotation steps* will rotate the model relative to the default view angle, which is the angle the model is displayed in the 3D view pane. *Absolute rotation steps* disregard the defaults and allow you to insert the exact angle on the 3D coordinate plane. *Additive rotation steps* rotate the model based on the angle the model is at when the model is rotated. Relative and Additive rotation steps seem similar, but are in fact different: The former adds to MLCad's *default* angles, the latter adds to the *current* angle.

Figure 12-6 shows you LDraw's 3D coordinate system. The image was rendered in POV-Ray by Lars C. Hassing, and shows the X, Y, and Z axes as well as the rotation angles (which coincide with the polar coordinate system, indicated by the words Latitude and Longitude). The three dots near the top represent POV-Ray lights. We will cover these lights in Chapters 14 and 15, and the Polar coordinate system in Chapter 14. For now, you can use the degree markings in this image to help you envision the angles for the Rotation Steps.

NOTE *As shown in the image, rotating on the X-axis will move the model along the circle that intersects Z and Y. Rotating on the Y-axis will turn the model along the circle that intersects X and Z. Rotating on the Z-axis will turn the model along the circle that intersects X and Y.*

Relative Rotation Steps

Relative rotation steps rotate a model relative to the default view angle. The L3 Globe shows the model in the default 3D view. This is the same view that you see in MLCad's 3D view pane. If you were to rotate the model 180 degrees on the Y-axis, the car would appear turned around, with its rear facing the direction the front faces in the above image.

Figure 12-6: Lars C. Hassing's L3 Globe image helps you see the angles of rotation when learning Rotation Steps.

Absolute Rotation Steps

Absolute rotation steps allow you to enter the exact angles as represented on the L3 Globe. Using an absolute rotation step of 0, 0, 0 shows the front view of the car. To get the 3D view seen in MLCad, you enter 30, 45, 0 — placing the camera at 30 degrees on the X-axis and 45 degrees on the Y-axis.

Additive Rotation Steps

Additive rotation steps rotate a model relative to the model's position in the last rotation step issued. Therefore, they *add* to the previous rotation angle. While this type of rotation step is nice for completeness, it is much more confusing than the other two types. For now, we recommend you stick with Relative or Absolute rotation steps. In addition, we tested the Additive rotation steps on MLCad 3.00, and the preview feature didn't work properly. Therefore, to set up an Additive rotation step, you need to create it and then test it in View mode. Doing this repeatedly isn't practical when there are two easier methods of applying rotation steps available.

Using Rotation Steps

To issue the Rotation Step command in MLCad, use the 🅢 icon.

In our example, we used the Relative rotation step. We have found this is the easiest type of rotation step to grasp because it's based on MLCad's default 3D view. After clicking the Rotation Step icon (or choosing **Edit > Add > Rotation Step . . .**), the Rotation Step dialog appears (see Figure 12-7). We decided to turn the Scumcraft so the rear faced us, so we entered **180** for the Y-axis. The Y-axis spins the model horizontally; therefore, entering 180 makes the rear of the model face forward. Pressing the **Preview** button will show you what the rotation step will look like when applied. You can also spin the model to the desired rotation by dragging the preview image, and MLCad will calculate the angles for you automatically.

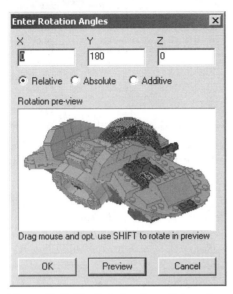

Figure 12-7: The Rotation Step dialog allows you to enter the rotation angles and see the effect in the preview pane. You can also rotate the model in the preview pane to have MLCad automatically calculate the rotation angles.

Placing Rotation Steps in the Proper Order

In order to get rotation steps to act the way you want them to, you need to make sure you position regular steps in the proper places as well. It's a tricky sequence, and it took a little experimenting on our end to figure it out too. You can see the proper sequence in Figure 12-8 on the next page, and the rotation step is highlighted in the model file window.

NOTE   *The example in Figure 12-8 shows the MLCad Buffer Exchanges used in the Scumcraft file. We will cover the Buffer Exchange feature in the "Using Exploded Views" section later in this chapter. We show the Scumcraft's rotation steps for consistency. For now, disregard the presence of the buffer exchanges in the example; they are independent of rotation steps.*

1. Before initiating a rotation step, insert a normal step. A rotation step only rotates the model; it does not mark the changing of a building step. Besides that, rotation steps *only* work in MLCad; outside of MLCad they may not function, so you want your model to have a regular step in place.

2. Once you have inserted your rotation step to rotate the view of the model, go ahead and add the parts you want in the current step. In our case, we added the file for the landing skids. You can see in Figure 12-8 that we did not call an LDraw part number, but a separate model file.

3. After adding your parts for this step, add a traditional step.

4. When you want to rotate the model back to its default or previous position, use the Rotation End Step by clicking ▧ on the toolbar, or choosing **Edit > Add > Rotation End Step**. This rotates your model back to the position it was in before the original rotation step.

Type	Color	Position	Rotation	Part nr.	Description
▢ PART	Black	-1195.350,-28.6...	-0.948,0.317,0.000 0.317,0.948,0.00...	scumcraft_s5.dat	Right Wing
S STEP	--	------	------	------	
◔ ROT-S...	--	------	0.000,180.000,0.000	------	START REL
▰ BUFEX...	--	------	------	------	STORE C
◊ PART	Black	-1116.350,-57.5...	1.000,-0.009,0.000 -0.004,-0.500,0.8...	scumcraft_s3.dat	Ghost
◊ PART	Black	-1136.470,-79.1...	0.000,0.000,-1.000 0.866,-0.500,0.00...	Arrow.dat	Ghost
◊ PART	Black	-1096.470,-79.1...	0.000,0.000,-1.000 0.866,-0.500,0.00...	Arrow.dat	Ghost
S STEP	--	------	------	------	
▰ BUFEX...	--	------	------	------	RETRIEVE C
▢ PART	Black	-1116.350,-90.4...	1.000,-0.009,0.000 -0.004,-0.500,0.8...	scumcraft_s3.dat	Engine
S STEP	--	------	------	------	
▨ ROT-E...	--	------	------	------	END
▢ PART	Trans-Bl...	-1118.350,-68.2...	-1.000,-0.009,0.000 0.008,0.910,0.41...	41883.dat	Windscreen 4 x 6 x 2 Canopy

Figure 12-8: These are the order of steps, rotation steps, rotation end steps, and parts used for the Scumcraft instructions in Figure 12-6. We will cover the Buffer Exchange and Ghost commands later in this chapter; disregard them for now.

NOTE *Rotation Steps are MLCad-specific commands. Because the LDraw file format is plain text, it is easy for individual software authors to extend it to meet their own needs. This is how innovation occurs in the LDraw community: One developer creates an extension to the file format, and gradually others adopt it. As of the printing of this book, Rotation Steps only work in MLCad and LPub. It's worth noting that your rotation step settings won't be interpreted outside of these two applications, and the LDraw-based program will display your model in the default position at every step.*

Using Sub-Assemblies

A very important part of creating logical, easy-to-follow building instructions is the proper use of sub-assemblies. The LEGO Company makes frequent use of them in their own instructions. Sometimes they incorporate *substeps* — highlighted boxes within a step, showing the assembly of a couple of parts before the parts are inserted into the main model. They also use another technique: creating entire steps for sub-assemblies, and inserting the complete sub-assembly into the main model in a later step.

Sub-Assemblies and Orientation

Sometimes a sub-assembly's studs are oriented in a different direction than those of the main assembly, or a hinge or a turntable part allows the sub-assembly to rotate on the main assembly. In this case, you definitely need to insert the sub-assembly in the instructions as a separate set of steps.

An example of this can be seen in the Scumcraft instructions. As we mentioned earlier, Mr. Palmer first builds the fuselage. He creates the wings as separate sub-models and adds the sub-models to the main model as whole pieces (see Figure 12-9). Our building instructions for the Scumcraft display this as a sub-assembly.

Official Instructions and Functional Building

You may notice that on some official LEGO building instructions, they do not completely follow our rules of functional building. We have made a point to introduce you to this concept so you can create models that are easy to manipulate in 3D, just as you would manipulate a physical LEGO model. You want your hinges and turntables to function easily so you can pose your model however you like. When you are assembling a physical LEGO model from building instructions, this is not necessary because hinges will work regardless of the order in which you attach them to the model. This does not the case with LDraw models because of the way the software handles sub-models. You need to be sure to group all of your functional components into separate sub-models if you want your hinged components to be easily usable in LDraw.

13

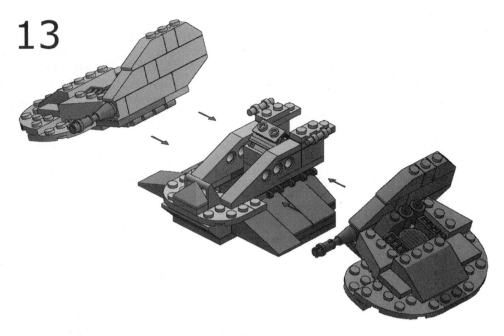

Figure 12-9: Scumcraft's wings being attached.

Using Exploded Views

Exploded views show a series of parts lined up, being inserted into place (see Figure 12-10). This is a common feature in official LEGO building instructions. We have all seen diagrams of exploded views in other applications — for example, when looking through an automotive repair manual. These manuals rely on exploded views to illustrate to the reader how to assemble and disassemble components of a car, and also to display all of the parts that exist (often with the manufacturer's part number and part name).

This technique is used in LEGO building instructions when a connection point is unclear, or if multiple parts are being stacked on top of each other in a single step. Exploded views can both eliminate ambiguity and save on the number of steps used in a building instruction manual.

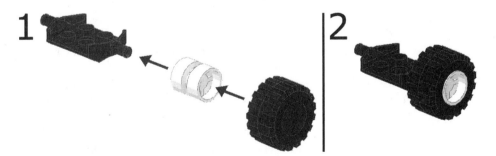

Figure 12-10: An exploded view shows the builder how to assemble parts like wheels.

Arrows

When creating a step with an exploded view, it's a good idea to use an arrow or some other indicator of how the parts are to be assembled. While we were writing this book, Willy Tschager sent us his arrow files to use as examples. Place the files **C:\LDraw\VLEGO\CH12\ arrow.dat, arrow_Ro.dat** and **arrow_Pt.dat** in your **C:\LDraw\PARTS** directory. You can use them in your model by entering the filename in the Select Part dialog, just as you would reference a sub-model. You can see the arrows in Figure 12-11.

NOTE *arrow_Ro.dat is the ¾ turn; arrow_Pt.dat is the arrow's point. These two files need to be aligned with each other by hand in MLCad. Use the four view panes and the different grid settings to ensure they are properly aligned with each other.*

Figure 12-11: Willy Tschager's arrow files help illustrate exploded view building instruction steps. arrow.dat (left) and arrow_Ro.dat aligned with arrow_Pt.dat (right).

One thing to take note of: The arrows are two-dimensional files. Depending on how they are rotated, you may not be able to see them in one or more of MLCad's view panes. A good idea is to rotate it so it is best visible from the 3D view. You saw in Figure 12-10 previously that the arrow has been rotated 45 degrees counter-clockwise along the X-axis, for better visibility.

MLCad's Buffer Exchange

Michael Lachmann has developed a very helpful feature for MLCad that lets you show parts in one location in one step, and clear them in the next. This feature, called "Buffer Exchange," was designed with building instructions in mind. It is used when you want to display a piece or sub-assembly being inserted into place in one step, and then show it completely in place in the following step. As you can imagine, this fits perfectly with creating exploded views for building instructions. It's difficult to describe this in words, so take a look at Figure 12-12 on the following page for an illustration.

How It Works

By inserting a Buffer Exchange STORE command, MLCad saves a *buffer*, or record of the model's parts, up to the line where the command is issued. The parts added in the step or steps immediately following the STORE command will be erased once the RETRIEVE command is issued. MLCad will then display the model as it was saved when STORE was issued. In Figure 12-12, a STORE command was issued before Step 11. After Step 11, a RETRIEVE command was entered to erase the mounting point and its arrows. With those parts erased, the mounting point could be added in its proper place unobstructed by the parts from the previous step.

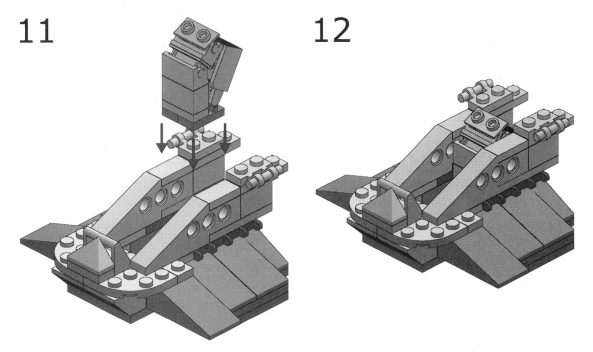

11   12

Figure 12-12: MLCad's Buffer Exchange essentially allows you to "scroll back" one or more steps, and keep building off of a previous step. This is perfect for presenting an exploded view for building instructions. Using it, you can place one element or sub-model in a temporary location, recall the saved buffer, and re-insert the part in its final location in the model.

Figure 12-12 displays the finished product after having been rendered in MegaPOV. You will see the same effect if you open the Scumcraft file from Chapter 8 in MLCad and look at it in View Mode, scrolling through step by step. However, if you take a look at the same file in Place mode, it will display all of the parts simultaneously. MLCad does this so you can see all of the parts you are editing, whereas when you view instructions, you are seeing the finished product. Figure 12-13 shows the difference between viewing the entire model in View Mode (left) and Place Mode (right).

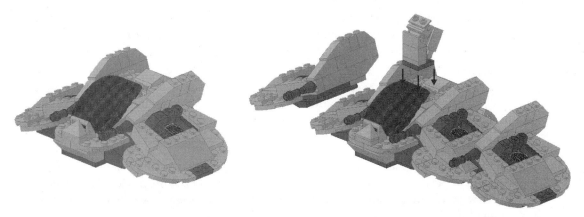

Figure 12-13: If you look at a model in View Mode (left), on the last step, it will not render parts intended to be erased by a buffer exchange. In Place Mode (right) you will see every part displayed.

NOTE *You might want to use MLCad's Hide feature to remove the Buffer Exchange parts from view while building. As you can see on the right side of Figure 12-13, the mounting point for the canopy that's above the model with arrows obscures part of the model, as well as the left wing. Get these out of the way using the Hide feature when you're building (　　 icon). MLCad doesn't save hidden parts when you save your file; therefore, hiding a part while building does not hide it if you view your model in another program.*

Using Buffer Exchange

Now that you've seen what Buffer Exchange can do, take a look at what it looks like inside the File View Window (see Figure 12-14).

Type	Color	Position	Rotation	Part nr.	Description
▭ PART	Light-Gray	-1116.470,-72.0...	1.000,0.000,0.000 0.000,1.000,0.000...	3048.DAT	Slope Brick 45 1 x 2 Triple
S STEP	--	------	------	------	
▪ BUFEX...	--	------			STORE A
♂ PART	Light-Gray	-1116.470,-152....	-1.000,0.000,0.000 0.000,1.000,0.00...	scumcraft_s2.ldr	Ghost
♂ PART	Black	-1136.470,-100....	0.707,0.000,0.707 0.000,1.000,0.000...	Arrow.dat	Ghost
♂ PART	Black	-1096.470,-100....	0.707,0.000,0.707 0.000,1.000,0.000...	Arrow.dat	Ghost
♂ PART	Black	-1096.470,-100....	0.707,0.000,0.707 0.000,1.000,0.000...	Arrow.dat	Ghost
S STEP	--	------	------	------	
▪ BUFEX...	--	------	------	------	RETRIEVE A
▭ PART	Light-Gray	-1116.470,-16.0...	-1.000,0.000,0.000 0.000,1.000,0.00...	scumcraft_s2.ldr	Mounting Point
S STEP	--	------	------	------	
▪ BUFEX...	--	------	------	------	STORE A
♂ PART	Black	-887.470,-28.98...	-0.951,-0.309,0.000 -0.309,0.951,0.0...	scumcraft_s4.ldr	Ghost

Figure 12-14: Buffer Exchange commands are used to achieve the results in Figure 12-12.

Since you already know that LDraw files are written in a sequence, it's easy to put together what is happening:

1. Before the Buffer Exchange command is issued, parts and steps are being added normally.

2. A Step separates the parts before the Buffer Exchange from the parts after it.

3. A Buffer Exchange STORE command is then called to store the current arrangement of parts in MLCad's memory. Note the Buffer Exchange line has "STORE A" at the description. A is the letter we assigned to the buffer; it can be given any letter designation between A and H.

4. More parts are inserted. These parts happen to be "Ghosted" (see explanation below). Scumcraft_s2.ldr is the mounting point sub-assembly, and the arrow.dats are Willy Tschager's arrows, which we mentioned earlier.

5. A Step separates the parts within the Buffer Exchange from the parts outside of it.

6. Finally, the Buffer Exchange RETRIEVE command is called. Note the description of the line reads "RETRIEVE A" instead of "STORE A." This erases the parts added in point number 4 from view, and restores the image as of the step that was issued in point number 2.

We recommend you take a look at the file **C:\LDraw\VLEGO\CH08\scumcraft.mpd** to see an example of Buffer Exchange used properly. Take a look at the model in Edit mode and View mode. In View Mode, be sure to scroll through the steps to see the Buffer Exchange in action.

Inserting a Buffer Exchange

To insert a Buffer Exchange, click 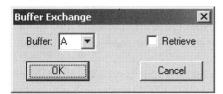 or choose **Edit > Add > Buffer Exchange** (see Figure 12-15). There are eight available buffers, each distinguished by letters A through H. Clicking **OK** will insert a STORE Buffer Exchange, whereas checking **Retrieve** before clicking OK will insert a RETRIEVE. We recommend that all of your STORE and RETRIEVE commands in a model use the same letter to identify the buffer exchange.

Figure 12-15: Buffer Exchange dialog.

Ghosting

You probably noticed that the parts inserted between the STORE and RETRIEVE Buffer Exchange commands are "Ghosted." When a part is ghosted, a white ghost icon appears on the left side of the line (instead of a green brick), and the word Ghost occupies the Description column. What does this mean? Ghosting lets you use the Buffer Exchange in sub-models without showing the unwanted exploded view parts from your steps. This feature does two things: only MLCad (and LPub, as of the time of this book's publication) will see the part; and the part is only visible if you are directly viewing the sub-model the part is in.

Just as with buffer exchange, the concept of ghosting is difficult to describe in words. To further illustrate this, we have included a small MPD file for your reference. Open the file **C:\LDraw\VLEGO\CH12\ghost.mpd**.

Take a look at Figure 12-16. MLCad should display the same image you see on the left hand side of the figure: a 4×8 plate with a 2×4 brick and a 1×2 brick on top of it. The two white bricks are actually a sub-model called "sub-model.ldr." Go ahead and select **sub-model.ldr** from the **Active Model** drop-down list above the File View Window.

The right hand side of Figure 12-16 is sub-model.ldr as seen from Place Mode. Recall from earlier in the chapter that in Place Mode, MLCad shows you all of the parts — including the ones within a buffer exchange. Take a look at the main model. Since the floating brick and arrows are ghosted, they're not displayed when view the main.ldr model.

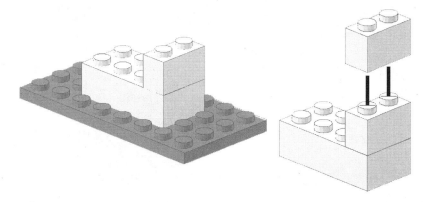

Figure 12-16: Ghosting lets you use the Buffer Exchange in sub-models without showing the unwanted exploded view parts from your steps.

Using Ghost

To Ghost a part, highlight it on the model file window and click the 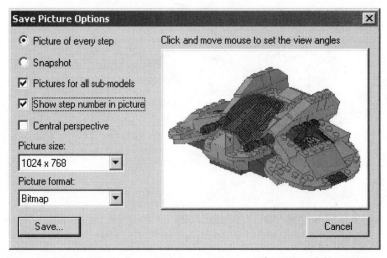 icon. To un-ghost a part, click .

Testing and Finalizing Your Building Instructions

Before releasing your building instructions to the public, you should spend some time refining and testing them. Since you are already familiar with the model you have created, ask someone else to build the model from the instructions you have made. Observe them or have them take note of where they struggle, and ask them to report to you where they found the instructions confusing.

Steve Barile finds it useful to test his instructions on a variety of different people. Because he is an adult and very familiar with LEGO parts, he has his wife (who is not a LEGO fan) test-build his models using his instructions. Aside from that, he invites the neighborhood kids over to help him test the instructions. By involving adults who are not LEGO fans and children in the testing process, he is able to pick out problems in his building instructions. From there, he can confidently make the necessary revisions to clarify the issues his testers find.

Publishing Instructions

Once you've tested your instructions and clarified any problems, you may want to publish your instructions. The examples we used in this chapter have been extensively post-processed using a combination of MLCad, LPub, L3PAO, MegaPOV, and an image editor. You will learn about L3PAO, LPub, and MegaPOV in the next four chapters. We discuss publishing and page layout techniques in Chapter 18.

Publishing Instructions in MLCad

If you are happy with MLCad output for your building instructions, you can use MLCad's Save Pictures feature that we discussed in Chapter 7 (**File > Save Picture(s) . . .**). You can see the Save Picture Dialog in Figure 12-17. If your instructions contain sub-models, make sure **Pictures for all sub-models** is selected. You probably want to include step numbers, so check **Show step number in picture** as well. Select your image size (in pixels) and image format.

When you click **OK**, you will be prompted with a directory to save the files to. The file name is by default the name of your model; MLCad will save your images as "filenameXX" where XX is your step number.

Figure 12-17: MLCad allows you to save pictures of your model's steps.

Showing Your Instructions to Others

Now that you know how to create effective building instructions using the LDraw tools, you can show your models to other LEGO fans online! You can use Jake McKee's Building Instruction Portal (www.bricksonthebrain.com/instructions/) to share your instructions, among other places.

The Building Instruction Portal sorts instructions by category: Castle, Space, Star Wars, Trains, and so on. The site accepts Building Instructions in any format: LDraw, Adobe Acrobat PDF, web pages with photographs, and more. It isn't a hosting site; you can't upload your files to have them stored on the site. It will provide links to your instruction files stored elsewhere on the Internet. There are many free services online you can host files from (and if you link directly to the files, you won't have to deal with annoying advertisements), or use your own service provider's web space if they allow you access.

The Building Instruction Portal is just one place online where you can share your instructions. You will learn about others when we discuss publishing your models online in Chapter 19, "LDraw and the Web: Viewing and Publishng Models Online."

Sample Instructions: Scumcraft

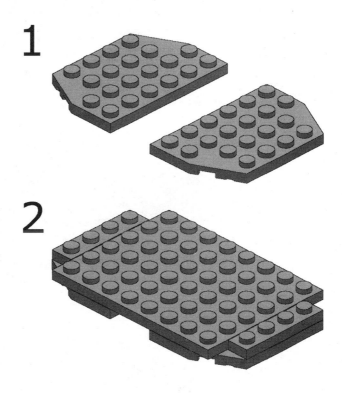

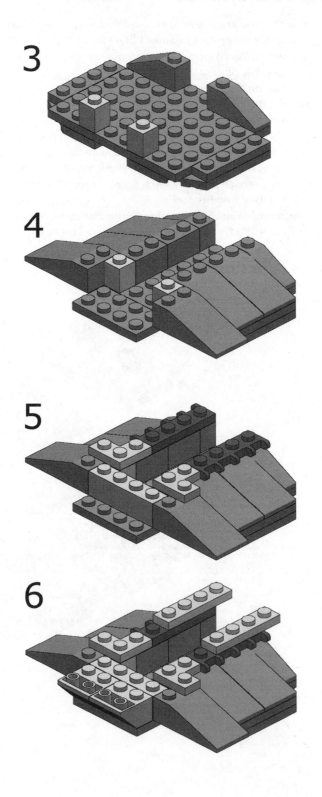

7

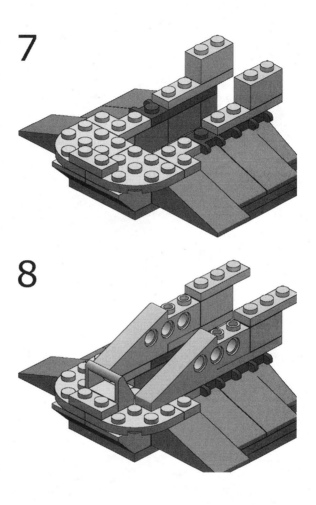

8

9

10

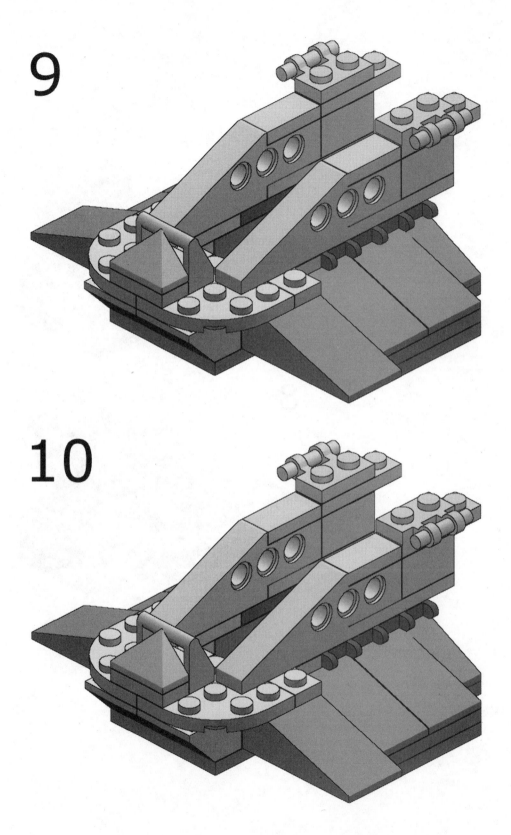

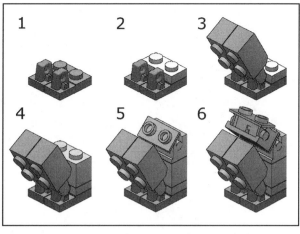

11

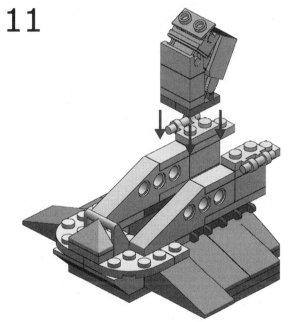

12

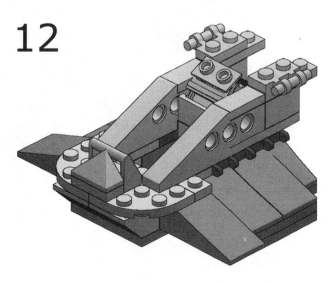

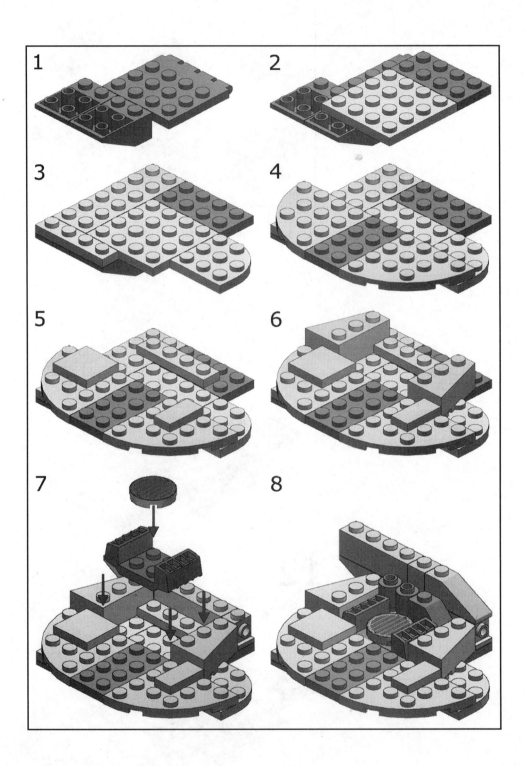

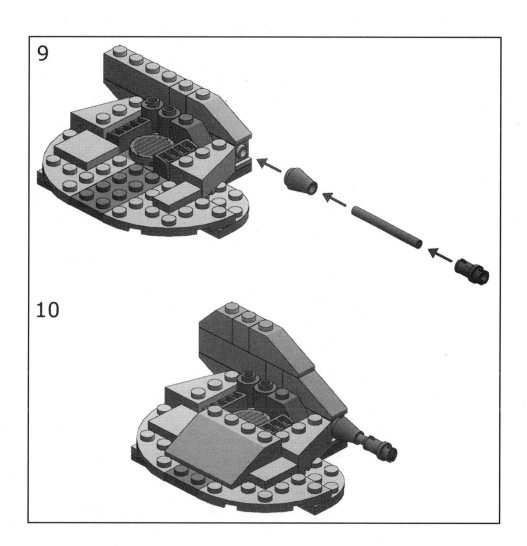

9

10

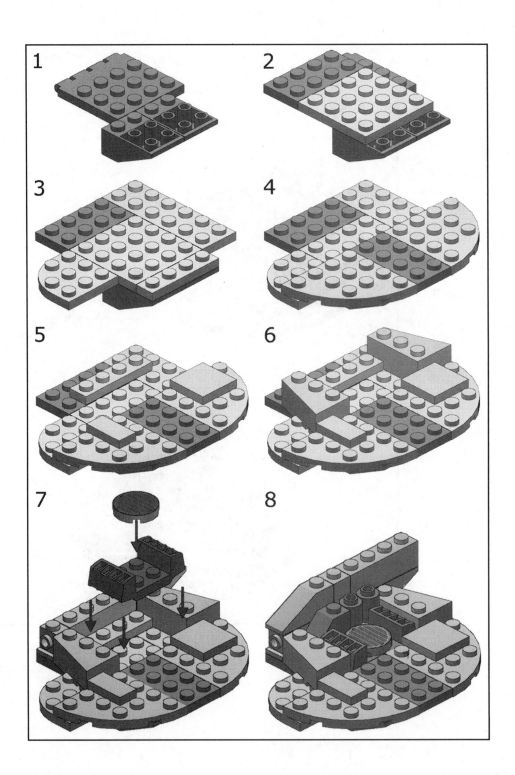

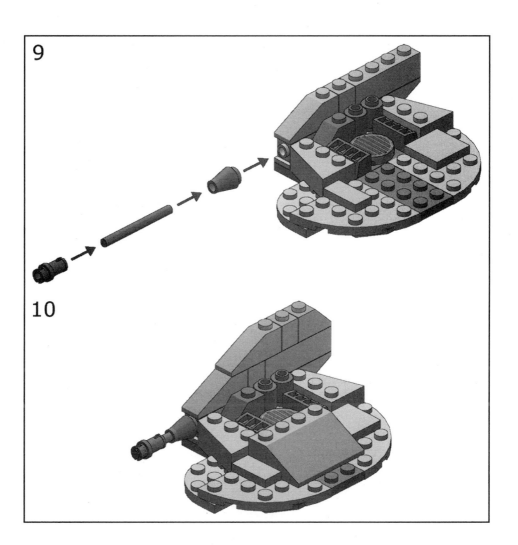

9

10

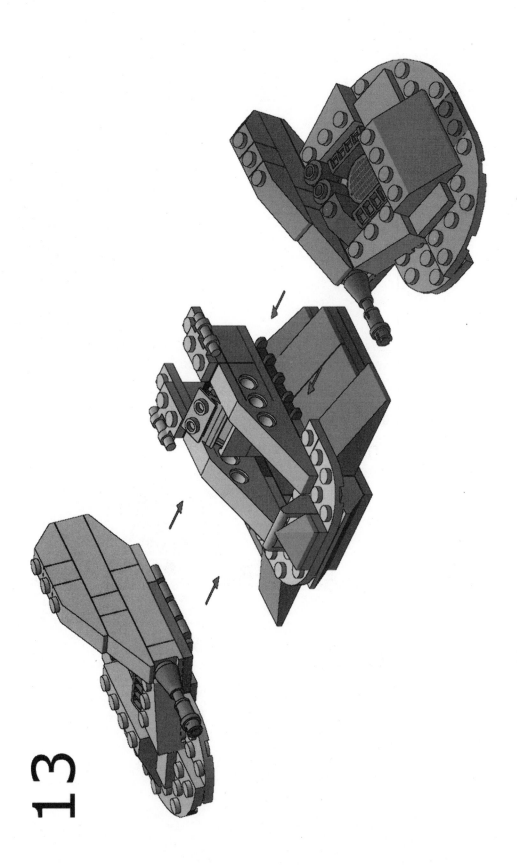

13

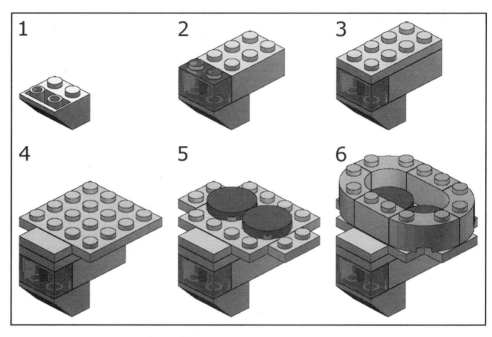

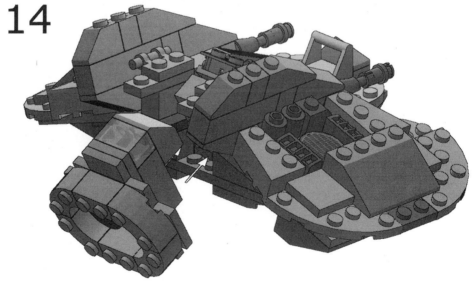

14

15

16

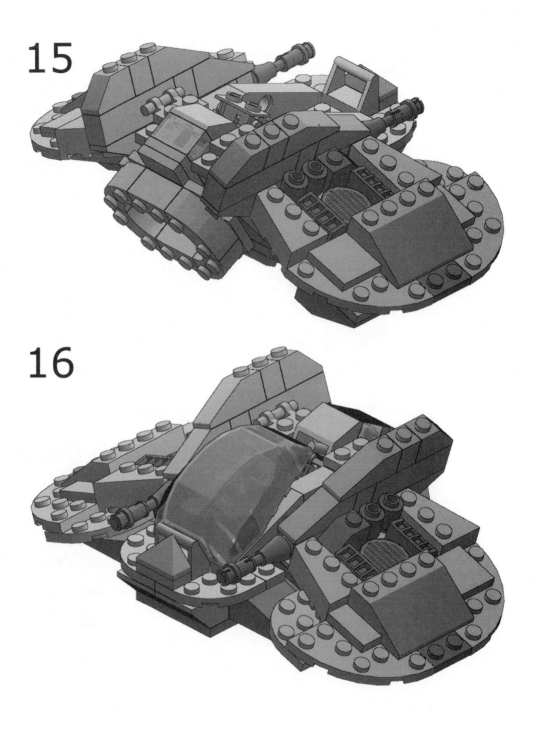

Summary

In this chapter, we discussed the techniques and ideas behind creating LDraw models for building instructions, citing examples. We introduced you to Steve Barile and Jake McKee, both pioneers in the realm of LDrawn building instructions, and shared some of their perspectives. This chapter also discussed advanced building instruction creation techniques such as MLCad's Rotation Steps and Buffer Exchange. Finally, we gave you tips on testing and publishing your LDraw instructions. The coming chapters will teach you more about high-quality rendering, including how to achieve nice MegaPOV renderings, such as the ones we used to create the Scumcraft instructions.

13

INTRODUCTION TO RAYTRACING SOFTWARE

This chapter provides a brief introduction to the free tools that are available for creating high-quality 3D images of your LDraw models. The four chapters that follow this one discuss the available tools in detail. The rendering processes we will teach you here require that you convert your LDraw file to another free 3D file format before actually generating an image. While this process may seem tedious, especially to those of you who are new to this concept, the rewards can be great. To see the potential, check out LDraw.org's Model of the Month at www.ldraw.org/community/contests/ or Jeroen de Haan's work at www.digitialbricks.nl.

Converting Options

The converting step is a very simple but necessary step to the raytracing process. It provides all of us with the ability to use POV-Ray and MegaPOV to create raytraced images and scenes of our models. Converters simply take an LDraw model and convert it to a file format that a 3D software package can understand. Some even go further than that and actually send commands to the 3D software package to render the image for you! We will teach you how to use L3PAO and LPub to convert LDraw files, both of which use L3P underneath. We explain this relationship in the following text.

L3P

The DOS file format converter, L3P, was created by Lars C. Hassing as a part of his L3 project (LEGO+LDraw+Lars). L3P converts an LDraw file to a POV-Ray file on-the-fly, without needing to use another library. However, it is capable of substituting its on-the-fly parts for Lutz Ulhmann's LGEO library of parts, which we will help you set up in the next chapter. This library, while it does not contain all of the current LDraw parts, produces higher quality results than L3P. To see the quality difference, take a look at Figure 13-1. You can see this image in color on the companion CD-ROM under \CH13\Figure 13-1.png.

LGEO Rendered Model   Non-LGEO Rendered Model

Figure 13-1: Using the LGEO library compared with straight L3P output.

The image rendered with the LGEO library is clearly a higher quality rendering than the one rendered without. LGEO parts have rounded edges and richer colors.

In order for users to benefit from the high quality LGEO library that Mr. Uhlmann developed, L3P provides a switch (-lgeo) that will cause L3P to use the available parts in the LGEO library before generating "on-the-fly" parts for the LDraw model.

L3P Add-on (L3PAO)

L3P Add-on (or L3PAO for short) is not really a converter itself, but a user interface for the DOS-based L3P. Remember, the original LDraw set of tools was created for DOS. For most basic computer users today, command line programs can be confusing and intimidating. One of the main problems with L3P in the command line was that the user needed to type every command correctly; if he or she did not, then L3P would generate an error and the user would have to type the entire line over again! For the LGEO image that is shown in Figure 13-2, the L3P command line would look like this:

```
"e:\LDraw\Apps\L3p\L3P.exe" "fig13-2.ldr" "E:\LDraw\MODELS" -cg45,45,0 -lg45,0,0 -lg30,120,0
-lg60,-120,0 -fg -bu -sw0.5 -q3 -lgeo -o
```

As you can imagine, this is quite a bit to type, especially when worried about mistakes. Jeff Boen solved this problem with L3PAO, a Windows interface that allows you to select your options visually, while it creates and runs the command line automatically.

L3PAO can also call POV-Ray and automatically render the converted file. While this may not seem like a big deal, it provides a way for users who are not familiar with POV-Ray at all to render images.

LPub

LPub is a groundbreaking piece of software created by Kevin Clague. LPub was designed to help create building instruction images for publishing in books. LPub automatically calls POV-Ray or MegaPOV to render the image for each instruction step. It also renders an image of each part needed to compile parts list images and a bill of materials for your model. LPub uses L3P to convert the LDraw file to POV-Ray format, just like L3PAO does. Before LPub was released, it would take countless hours editing files, rendering, and photo editing the building instructions to achieve what LPub now does in a single click. LPub can also create simple web pages of the building instructions. We teach you how to use LPub in Chapter 15, "LPub: Automate Building Instruction Renderings."

NOTE *We are only covering L3P/L3PAO and LPub for converting files from LDraw to POV-Ray format. There are other programs that do similar tasks, most notably the shareware 3DWin, capable of converting LDraw files to many professional 3D file formats. See 3DWin at www.tb-software.com.*

Raytracing Options: POV-Ray & MegaPOV

POV-Ray, short for Persistence of Vision Raytracer, is the raytracer of choice for the LDraw community. POV-Ray files are made up of a scene description language, essentially code that describes what the rendered scene will look like. We will first take you through L3PAO and LPub, so you can experience these results without learning the code. In Chapters 16, "POV-Ray," and 17, "MegaPOV," we teach you a few basic elements of POV-Ray's scene description language, and how to edit it inside of POV-Ray and MegaPOV. Take a look at Figure 13-2 for an example of POV-Ray's scene description language.

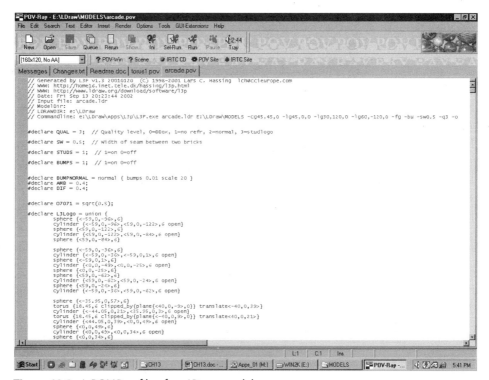

Figure 13-2: A POV-Ray file of an LDraw model.

MegaPOV vs. POV-Ray

MegaPOV is an unofficial build of POV-Ray, with a few more added features. The LDraw community has found these features useful for rendering building instructions, which is why we chose to discuss it in this book. We will talk about MegaPOV in detail in Chapter 17.

Summary

In this short chapter, we gave you a brief introduction to the various programs we will teach you in the next section. L3P converts LDraw files to POV-Ray format. L3PAO and LPub provide interfaces to L3P, and LPub can generate building instruction images automatically. POV-Ray and MegaPOV are used to render the high-quality images.

14

L3P AND L3P ADD-ON

In this chapter, we will introduce you to L3PAO, a Windows program that provides an interface for the DOS file format converter, L3P. L3P converts LDraw files to POV-Ray files so you can render high-quality images and scenes. You can also use MegaPOV, which we will discuss in Chapter 17, to do special things like putting edge lines on bricks to make them appear illustrated. We will more fully explore the wonders of POV-Ray and its unofficial version, MegaPOV, in later chapters. For now, we will teach you how to convert files and use L3PAO's various settings, or "switches."

The L3PAO Interface

To get started, open L3PAO.

Upon first launch, you will see a prompt to locate **L3P.exe**, the DOS file format converter L3PAO uses (see Figure 14-1 on the following page). Remember, L3PAO is a graphical user interface for the command-line DOS program L3P, which does the converting from LDraw to POV-Ray. L3PAO initially cannot locate L3P since it is only looking in its own directory.

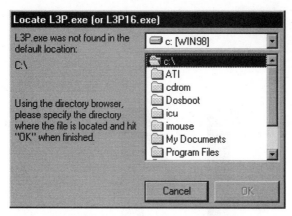

Figure 14-1: L3PAO looks for L3P upon first launch.

You will need to tell L3PAO where to look for L3P manually. Follow these steps:

1. In the drop-down menu on top, select the **C:** drive.
2. Locate the **LDraw** folder.
3. Select the Programs subfolder, than the L3P subfolder. Click **OK.**

A new window will appear, as shown in Figure 14-2, showing that L3PAO is looking for the LGEO library.

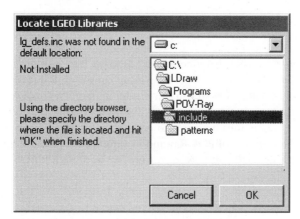

Figure 14-2: L3PAO looks for the LGEO library.

You will need to tell L3PAO where to look for the LGEO library manually. Follow these steps:

1. In the drop-down menu on top, select the **C:** drive.
2. Locate the **LDraw** folder.
3. Browse for the **Programs\POV-Ray\include** subfolder. Click **OK.**

A new window will appear, as shown in Figure 14-3, showing that L3PAO is looking for LEdit.exe.

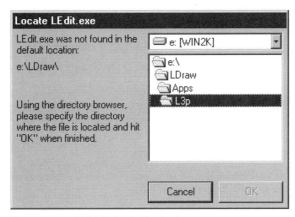

Figure 14-3: L3PAO looks for Ledit.exe.

You will need to tell L3PAO where to look for the Ledit.exe manually. Follow these steps:

1. In the drop-down menu on top, select the **C:** drive.
2. Locate the **LDraw** folder. Click **OK.**

Finally, you should see a window similar to Figure 14-4.

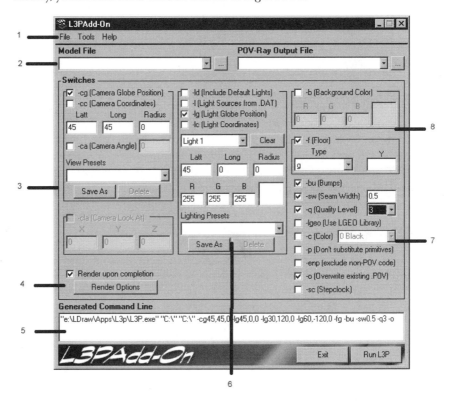

Figure 14-4: The L3PAO interface.

Here is a quick orientation of the main L3PAO window:

Menus (1): These are standard on all Windows applications; L3PAO is no different.

File Controls (2): This tells L3PAO which LDraw file to convert and where and what to call the output POV-Ray file.

Switches – Camera (3): This area provides the user with the ability to control the placement of the POV-Ray camera as well as where the POV-Ray camera should look. Several common camera placements (front, back, and so forth) are pre-defined and available via the drop-down menu.

POV-Ray Rendering Options (4): This provides specific POV-Ray instructions such as size, name and format type of the output rendered file.

L3P Command Window (5): This window shows what the command line instruction for L3P would look like if you were not using L3PAO. You cannot edit this widow.

Switches – Lights (6): This area provides the user with the ability to control the number and location of the POV-Ray lights.

Switches – Miscellaneous (7): This area provides the user with the ability to specify miscellaneous controls.

Switches – Floor and Background (8): This area provides the user with the ability to specify a background color and floor for the POV-Ray file.

Our First Rendering: No POV-Ray Knowledge Required

Using L3PAO, we can render an image in two steps! First we need to select an LDraw file.

Step 1: Select an LDraw File

At the top, under Model File, click on the the **...** button. Using the dialogue window that appears, open the file **C:\LDraw\VLEGO\CH14\airplane.ldr**. Click **OK**. Once you click OK, immediately to the right of the Model File menu you will see "airplane.pov" created under "POV-Ray Output File." This file will be created in the same directory where airplane.ldr is located. Here you can change the name and final location of the output file by clicking on the **...** button next to POV-Ray Output file. For now, though, leave this alone.

Step 2: Run L3P via L3PAO

To run L3P, click the **Run L3P** button located at the bottom right hand corner of the window. Soon afterward, a DOS window will open and close, then POV-Ray will run. If the POV-Ray Legal Notice window appears, click **OK** so that POV-Ray can proceed; now you can just sit back and enjoy the show while POV-Ray renders the LDraw model. Depending on the amount of memory and the speed of your PC this may take some time. In the end, POV-Ray will render (in color) what is shown in Figure 14-5. Once you are done looking at the image, close the render window and the POV-Ray application.

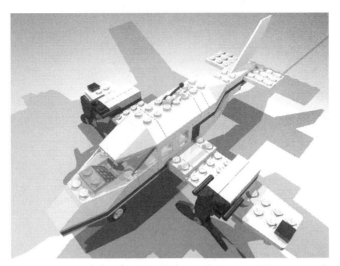

Figure 14-5: Our first render of an airplane. Model courtesy of Kevin Dixon.

See how easy that was! Of course, there are many things to be desired. The rendering, as you may have noticed, was very slow. The image has too many shadows, the camera location may not be were you wanted it, etc. We will teach you how to change these things later in this chapter, and teach you even more advanced techniques in Chapter 16, "POV-Ray." This is a quick introduction to the capabilities of POV-Ray, without having to learn any POV-Ray code.

The L3P Switches

There are four switch types for L3P: those that control the camera, the lights, the background, and the miscellaneous features.

The Camera Switches

There are four camera switches and two L3PAO features for the camera. The switches are: -cg (Camera Globe Position), -cc (Camera Coordinates), -ca (Camera Angle), and -cla (Camera Look_At). The L3PAO features are View Presets and Saving/Deleting Presets.

By default, L3P's automatic camera positioning will move the camera close to the model so that the rendered image is completely filled by the model. To prevent this from occurring, you must specify a radius in the -cg switch or use the -cc option.

Switch -cg: Camera Globe Position

L3P and L3PAO allow you to input your camera's position using either the Cartesian or the Polar coordinate systems. You can choose which system you prefer based on what you are more comfortable with. The Camera Globe Position uses the Polar coordinate system, whereas Camera Coordinates uses the Cartesian system. L3PAO chooses Camera Globe by default.

To use Polar Coordinates, think of the 3D world you are creating divided up like a globe with latitude and longitude lines. L3P places your model inside this "globe" with the center of the model located at the center of the globe. To place the camera, all you'll need to do is provide the Latitude, Longitude and Radius of the location you desire. Latitude is in the range from -90 degrees (south, along the positive Y-axis) to 90 degrees (north, along the negative Y-axis). 0 degrees is at the equator. Longitude is in the range from -180 to 180, where 0 degrees is along the negative Z-axis. 90 degrees (east) is then along the positive X-axis and -90 degrees (west) is along the negative X-axis. Figure 14-6 explains this concept visually.

Figure 14-6: The Cartesian and Polar coordinate systems. Image courtesy of Lars C. Hassing.

Switch -cc: Camera Coordinates

The Camera Coordinates switch uses the Cartesian coordinate system. To use this system, simply click on the **-cc** checkbox. Now you can define your camera location in terms of X, Y, and Z coordinates. Use Figure 14-6 as a guide for this coordinate system as well. Note that the arrows on each axis point in the positive direction. The units used are equivalent to LDraw Units, taught in Chapter 5.

NOTE *Using the Cartesian system, the floor is parallel to the XZ plane and Y is negative upwards. This means that -10 Y is 10 units above the floor. The Cartesian coordinates are the same ones that MLCad uses when you build your model.*

Switch -ca: Camera Angle

L3P uses a camera angle of 67.38 degrees by default. The Camera Angle setting is similar to the angle of a lens in real life photography. Smaller angles give a telephoto effect, zooming in on the subject and showing less perspective. Larger angles give a fisheye effect. We examine the Camera Angle more in depth in Chapter 16.

Switch -cla: Camera Look_At

This switch tells POV-Ray which point the camera is looking at in the Cartesian coordinate system. The only limitation here is that the camera location and the look_at location cannot be the same point.

L3PAO Camera Feature: Pre-defined Views

If you do not want to deal with having to input coordinates by hand, you can also use a set of pre-defined coordinates. Using the View Presets drop-down menu, L3PAO provides the user with some common camera locations. Choose a location and L3PAO will calculate the required position of the camera for you!

The predefined settings are absolute within the 3D world. Use MLCad's preview windows to determine which way is front, back, and so on for your model, and note that it may or may not coincide with these presets. By default, "Front" looks towards +Z. Therefore "Back" looks at -Z, "Left" looks at -X, "Right" looks at +X, "Above" looks at +Y (down), and "Below" looks at -Y. You can see a visual example of this in the L3 Globe image from Figure 14-6.

L3PAO Camera Feature: Saving/Deleting Camera Settings

You can save up to 20 user-defined camera locations by clicking on the **Save As** button located underneath the View Presets. This can come in handy if there is a particular camera angle that you always use. Likewise, you can delete any of the settings you saved using the delete button.

The Light Switches

There are four light switches and two L3PAO features for lights. The switches are: -ld (Include Default Lights), -l (Light source from .DAT), -lg (Light Globe Position) and, -lc (Light Coordinates). The L3PAO features are Lighting Presets and Saving/Deleting Presets.

Lights, just like the POV-Ray camera, are defined by either the Cartesian or Polar coordinate system. In addition to a location, lights require you to define a color. Color is determined by the light's RGB (Red, Green, Blue) value. The value of white is RGB = 255, 255, 255; while black is RGB 0, 0, 0; red is RGB = 255, 0, 0; and so on. If you are more comfortable with a color table, simply click on the box next to the RGB values for a standard window color box. By default, L3P will create one light source (color white) so that the rendered model is not completely in the dark.

L3P is only capable of producing one type of POV-Ray light (light_source). POV-Ray has additional light types (spotlight, area lights, and so on) that will be discussed further in Chapter 16. These additional types of lights must be manually hand-coded by the user, as L3PAO provides no interface with which to create them.

Switch -ld: Include Default Lights

This switch will override any custom lights you have placed, using L3P's default lights.

Switch -l: Light Source From .DAT

You can insert POV-Ray lights into a LDraw file when editing in MLCad. To do this, simply insert the file **light.dat** (called "POV-Ray Light Source") into your model where you want the light. This invisible part is represented in MLCad as a small "x" and will not show up in any renderings. If you have inserted light.dat parts in your model, select the **-l** switch in order to use them when you render.

Switch -lg: Light Global Position

L3PAO uses this switch by default (you don't need to check it). It uses the Polar coordinate system to place the lights. If you use a radius of 0, L3PAO will place the lights at the same radius where the camera is, effectively making the lights the same distance from the look_at point as the camera. If you use a negative radius, it will place the light source some percentage behind the camera. (For example, a radius of -10 will place the light source 10 percent further from the look_at point than the camera itself.)

Switch -lc: Light Coordinates

This switch is similar to -lg. The only difference is that the Cartesian coordinate system is used.

L3PAO Lights Feature: Lighting Presets

Similar to the camera presets, L3PAO also provides the user with some common light settings. Use the drop-down menu to view and select the presets.

L3PAO Lights Feature: Saving/Deleting Light Settings

You can save up to 20 user-defined lights by clicking on the **Save As** button located underneath the Lighting Presets. Likewise, you can delete any of the settings you saved using the DELETE button.

The Background Switches

There are two background switches. These switches are -b (Background Color) and -f (Floor). The default color for a background is Black, and the default floor color is Gray. The floor is placed directly below the lowest point in the LDraw model, so the model appears to be resting on it.

Switch -b: Background Color

To change the color of the background, use the RGB values or click on the color swatch on the right to use a color palette.

Switch -f: Floor

You can choose a Gray floor (Type g) or a red and white checkered floor (Type c). You can also change the floor's Y-value, moving it up or down in relation to the model. Remember to use a positive number to lower the floor, because +Y faces down in LDraw (see Figure 14-6).

Miscellaneous Switches

There are nine miscellaneous switches for L3P. These are -bu (bumps), -sw (Seam Width), -q (Quality level), -lgeo (use LGEO library), -c (Color), -p (don't substitute Primitives), -enp (Exclude Non-POV code), -o (Overwrite existing .POV), and -sc (Stepclock). By default, L3PAO activates the switches "bumps," "seam width" = 0.5, "quality level" = 3, and "overwrite existing .POV."

Switch -bu: Bumps

Physical LEGO bricks do not have smooth surfaces due to the mold injection process that is used to create them. Look closely at any brick and you will see a slightly uneven surface. This effects how the light is reflected off of a brick. Turn this switch on to have more realistic reflections on the sides of your parts.

Switch -sw: Seam Width

The default seam width, or distance between individual parts, is 0.5. The seam width feature adds another layer of realism to your model, simulating the slight gap between individual LEGO pieces in real life. You can select this switch to change the seam width value. A seam width value of 0 allows for no space between pieces.

Switch -q: Quality

L3P can render your model in four different quality levels. The lower quality levels are useful for rendering test images to set up the camera and lights, while saving the high-quality render (which takes longer) for the final image. By default, L3PAO sets the quality level at 2, the second highest level.

Level 0: Renders bounding boxes for parts rather than the part's actual shape.

Level 1: Parts are rendered by their normal shape but any reflection or refraction of light is not calculated.

Level 2: Parts are rendered normally, including reflection or refraction.

Level 3: This highest-quality setting also includes the word LEGO on each stud, like real LEGO parts.

See Figure 14-7 for a comparison of the four quality levels.

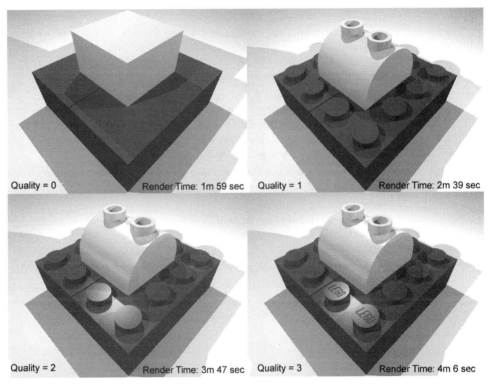

Figure 14-7: POV-Ray quality levels.

Introduction to Rendering Times

Several factors affect how long it takes POV-Ray to render an image: the size of the output image, the size and complexity of the model, and your system's specifications (processor speed, RAM, and so on). Because of the number of light calculations, transparent parts take significantly longer to render than solid or opaque parts. See Figure 14-8 on the next page for an example, and note the rendering times below each image. We explain factors for rendering time in greater depth in Chapter 16.

Rendering Time: 1hr 12min 28 sec   Rendering Time: 45 sec

PC Specifications: - AMD Atholon 800 Mhz, 512 MB RAM PC, 8MB Cache on Hard Drive

Figure 14-8: Transparent parts take significantly longer to render than opaque parts.

Switch -lgeo: Use LGEO Library

L3P can substitute LGEO parts for L3P parts where they exist. These parts are higher quality than L3P's parts, which are generated on the fly each time you run L3P via L3PAO.

NOTE   *For this to work, POV-Ray must also know where to locate the LGEO library. To do this you must edit the master POVRay.ini file. Follow these steps to edit the file:*
Step 1. Open POV-Ray.
*Step 2. Click on the **Tools** menu.*
*Step 3. Click on the menu item **Edit master POVRAY.INI** and a text file will open.*
Step 4. Scroll to the bottom of the POVRAY.INI file and on the last line insert the following:
Library_Path=<full path to the LGEO library>
*If you installed LDraw using the companion CD-ROM's installer, the full path will be C:\LDRAW\ Programs\POV-Ray\include\,so you should type: **Library_Path= C:\LDRAW\Programs\POV-Ray\ include**.*
Step 5. Save and close the text file; close POV-Ray.

Switch -c: Color

L3P has an issue with models that contain only one part. If you feed L3P an LDraw file with one part in it, L3P will override the part's color and make it gray. To get around this, use the Color switch and select your desired color from the drop-down list. This workaround only supports the original 16 LDraw colors (not transparent or dithered colors).

Switch -p: Don't Substitute Primitives

To increase the quality of an LDraw file rendered by POV-Ray, some parts (primitives) are automatically replaced by their POV-Ray equivalents. For example, a LEGO minifig head in LDraw is defined by a 16-edge polygon. L3P substitutes this 16-edge polygon with a true cylinder so that the rendered image is of higher quality, as shown in Figure 14-9. To deactivate this option, active the **-p** switch. (Note: Minifig heads with printed faces will still appear polygonal, except for the stud on top. This is because of the way the pattern is embedded in the part; a solution has yet to be devised for this problem.)

Switch -enp: Exclude Non-POV Code

When L3P converts an LDraw file, it includes the original LDraw file's lines as comments, along with all the comments within an LDraw file. If you don't want these comments in your POV file, select this option. We find these lines useful for navigating a very complicated POV-Ray file.

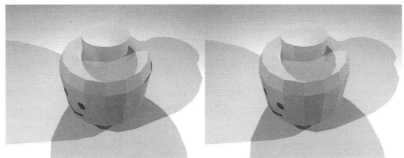

Quality Level 2 with POV-Ray Primitives   Quality Level 2 without POV-Ray Primitives

Figure 14-9: L3P will substitute primitives where it can, such as the stud on the minifig head in the left pane. The printed portion of the head remains unchanged because there's no solution for maintaining the integrity of the pattern on the smooth surface. The -p switch can turn off primitive substitution.

Switch -o: Overwrite Existing .POV

If you're overwriting a file you've already created, you need to use this switch to confirm the overwrite. Otherwise, L3P will not overwrite the file. This is to safeguard against accidentally overwriting a file. If you do not want to overwrite the file, you can always change the name of the output file via the field in the upper right hand corner of the L3PAO interface.

Switch -sc: Stepclock

By default, L3P will render an LDraw file as a single object or model. If the LDraw file has steps in it, you can use this switch to render each step when you render your model in POV-Ray. Refer to L3PAO's Render Options in the next section for more information on this feature. Though Stepclock is useful for models with simple steps, it does not provide nearly as much capability as LPub, and we strongly recommend using LPub to render building instructions. You can learn about how to generate instructions using LPub in the next chapter.

L3PAO's POV-Ray Interface

L3PAO allows you to control a few POV-Ray options from its interface without needing to edit the POV-Ray code L3P generates or configure POV-Ray's initialization file. This is an excellent feature that saves you from the tedious nature of POV-Ray's scene language, but at the same time limits your options. If you are comfortable with these options, and don't want to bother with POV-Ray, this is right for you.

L3PAO can render an image in POV-Ray automatically, as you learned at the beginning of this chapter. To access the POV-Ray Render Settings in L3PAO, you must select the checkbox **Render Upon Completion** in the lower left of the interface, then select the **Render Options** button immediately below the checkbox. Figure 14-10 shows the POV-Ray Render Settings dialog.

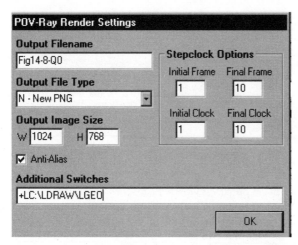

Figure 14-10: L3PAO allows you to configure specific POV-Ray render settings through its interface.

Image Options

You can select the filename you want for your final rendered image and change the output file type and image size (dimensions). If you don't specify these fields, L3P will use the name of the output POV file as your image name, a BMP image format, and an image size of 640×480.

Using the Stepclock

You can use the Stepclock (**-sc**) options in the Render Settings window provided you have checked the switch on the main L3PAO interface first. If you have not, these fields will appear grayed-out. In Figure 14-10, the model being used contains ten steps. To properly render all ten steps, fill in **1** and **10** for the Initial Frame and Final Frame values, respectively, and the same with Initial Clock and Final Clock. Because we recommend LPub for rendering building instructions, we are not covering this in depth. To find out more information on the Stepclock, see POV-Ray's help file, or Ahui Herrera's animation tutorials on LDraw.org at www.ldraw.org/reference/tutorials/.

Saving Your Settings

Finally, L3PAO allows you to save the rendering settings you've configured, so you don't have to reset all of the switches, lighting presets, and other items each time you want to run the program. To do this, in the main L3PAO window select **File > Save Scene**. You can open saved files by selecting **File > Open Scene**.

Exercises

Here are some exercises to get you accustomed to using L3PAO and manipulating the various settings. All of them use the LDraw file **C:\LDraw\VLEGO\CH14\space_scooter.mpd**. After rendering each exercise, remember to close all open windows of POV-Ray before moving on to the next.

NOTE   *Each time you have L3PAO automatically render an image, it will open a new POV-Ray window, even if you have rendered the file before. We recommend you close the POV-Ray window upon completion. Find the rendered images saved in the directory of your source file.*

Exercise 14.1: Camera Placement

If you render the exercise model using L3P's default values, you will get the image shown in Figure 14-11. To do this, check **Render Upon Completion** and then **Run L3P** in the bottom right corner.

Figure 14-11: Space Scooter rendered in POV-Ray.

Now let's render the model from the point of view of the minifig so we get an image similar to Figure 14-12.

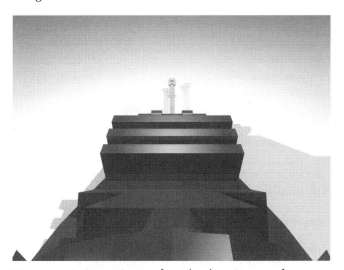

Figure 14-12: Space Scooter from the driver's point of view.

You can do this by simply changing the location and look_at position of the POV-Ray camera. Render the image using the following steps:

1. Place the camera in the general area of the minifig's eyes. To do this, use MLCad to locate the x, y, z coordinates of the minifig's eyes.

    (We used: x = 0, y = -96, z = 0.) You can find these by placing your cursor on the screen and watching the left-hand side of the status bar at the bottom of the MLCad window. It will display your cursor's X, Y, Z location.

2. Activate the **-cc** (Camera Coordinates) switch and provide the new camera location by entering the values you just found.

3. Activate the **-cla** (Camera Look_At) switch and provide a look_at location.

   (We made the minifig look straight and downward, at point [x = 0, y = 0, z = -130].)

4. Render the model in a lower quality level such as 1 and a small image size to test if you are correct. Once you are satisfied, render the model in the size and quality level you want. To change image sizes, select **Render Upon Completion**, and edit the Render Options.

Exercise 14.2: Using Colored Lights

Sometimes you want to use colored lights when lighting a model, most often when attempting to compile a scene. While we will discuss the basics of scene creation in Chapter 16, let's take a look at creating colored lights in the comfort of L3PAO. You can use colored lights to create an eerie effect, simulating the glow of candles in a dungeon, or the distant glow of stars in outer space. For our exercise, let's add a blue light directly on top of the model and a yellow light behind the model to see what effects this has on the final image. Use L3P's default values for the camera location. The easiest way to obtain the default values is to close and restart L3PAO. Follow these steps to render the image:

1. Use MLCad to locate the x, y, z coordinates for the location of the blue light.

   (We used: x = 0, y = -128, z = 20.)

2. Use MLCad to locate the x, y, z coordinates for the location of the yellow light.

   (We used: x = 120, y = -128, z = 10.)

3. Activate the **-lc** (Light Coordinates) switch and provide the coordinates for light 1. Change the color of the light to Blue (RGB: 0, 0, 255) using the RGB fields below the light's coordinates.

4. Use the drop-down menu under Lighting Coordinates to select **light 2**, and provide the coordinates for the second light you found in MLCad. You will have to reactivate the **-lc** switch for light 2 because L3PAO uses the -lg switch by default for all its lights. Change this light's color to Yellow (RGB: 255, 255, 0).

5. Render the scene. Select **Render Upon Completion**, and then Run L3P.

Notice that although you used blue and yellow lights, the light seen in your rendered image is green. Why? The blue and yellow lights blended together to create green light.

Exercise 14.3: Background & Floor

For this exercise, let's make the Space Scooter hover above the ground. To make it "hover," we need to lower the floor. See if you can get something similar to Figure 14-13 by following these steps:

1. Use your mouse and the status bar in MLCad to determine where a floor should be placed. (We used Y = 200.)

2. Select the Floor (**-f**) switch, and insert the floor's desired Y-value in the Y field.

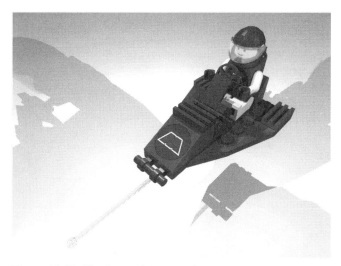

Figure 14-13: The Space Scooter is hovering.

Now let's change the background color. We chose purple, though you may want to choose another color.

3.   Activate the **-b** (Background Color) switch in L3PAO.

4.   Choose the color purple by entering its RGB value. We used: RGB = 128, 0, 255.

5.   Render the scene.

You might notice the default camera settings don't give a good angle to see the scooter hovering. Change the camera location by using the preset "right," and re-render the image. Your image should look like Figure 14-14.

Figure 14-14: The Space Scooter is hovering and the background color is purple.

Exercise 14.4: Using the Stepclock

For the last exercise, let's use the Stepclock to create basic instructions for the scooter. Note these are not the same quality or style of instructions we will teach you to create using LPub in the next chapter. However, we decided to include this exercise to teach you how to use the Stepclock feature. Follow the steps below to render the five images, seen in Figure 14-15. Close L3PAO and restart it to clear the settings before following the steps below.

Step 1   Step 2   Step 3

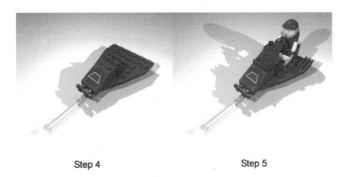

Step 4   Step 5

Figure 14-15: The Space Scooter instruction images.

1. Look at the file in MLCad and count the number of steps in the model (5).
2. Activate the **-sc** (Stepclock) switch in L3PAO.
3. Make sure **Render Upon Completion** is checked, then select **Render Options** to provide the frames and clock values.

   (We used: Initial Frame = 1; Final Frame = 5; Initial Clock = 1; Final Clock = 5.)
4. To see the effect of the Output File Name option, enter **SS_** in that field.

NOTE   *When POV-Ray is told to render several images in sequence, it will add numbers 1, 2, 3, and so on to the end of the output filename you provided. POV-Ray can only output file names with eight characters and three-character extensions, following the old DOS limitations.*

Summary

In this chapter, you learned how to convert LDraw files to POV-Ray files and render them all within the L3PAO interface. L3PAO runs the DOS application L3P underneath, and provides an interface for you to configure options so you do not have to manually write all of the DOS command. POV-Ray renders the high-quality image, not L3P or L3PAO.

You can configure many options that affect your output image right from the comfort of the L3PAO interface. Among these settings are camera and lighting controls, background color, floor, rendering quality, and more. You can also control some of the rendering settings when you tell L3PAO to automatically render the model in POV-Ray after converting the file format.

In short, L3PAO is an interface for a file format converter, which can also render the converted file and control limited POV-Ray settings.

15

LPUB: AUTOMATE BUILDING INSTRUCTION RENDERINGS

One of the hottest new programs in the world of LDraw tools is LPub. This program was designed to automatically render the building instruction steps that will be published in books about LEGO MINDSTORMS. Since its creation, LPub's author, Kevin Clague, has released the program to the LDraw community. In the months leading up to this book's publication, LPub has gained widespread popularity among LDraw users, and promises to become a key application in the future of the LDraw system.

In the previous chapter, we introduced you to L3PAO, the Windows interface to the L3P file format converter. We also introduced you to several concepts within POV-Ray, such as camera placement, light sources, background images, quality settings, and more. LPub can do all of the things L3PAO can do as an LDraw-specific user interface for L3P and POV-Ray. Furthermore, it handles automating building-instruction renders far better than L3P's Stepclock option, as illustrated in Figure 15-1 on the next page. LPub can automatically generate instructions of sub-models within an MPD file as well as recognize the advanced MLCad features Buffer Exchange and Rotation Step. It also provides an interface to control MegaPOV settings to create edge lines around renders for building instructions. LPub can also automatically generate web pages (HTML format) for your model. Wrapping all of these features into one package that automates the process makes LPub the clear choice for generating high-quality building instructions.

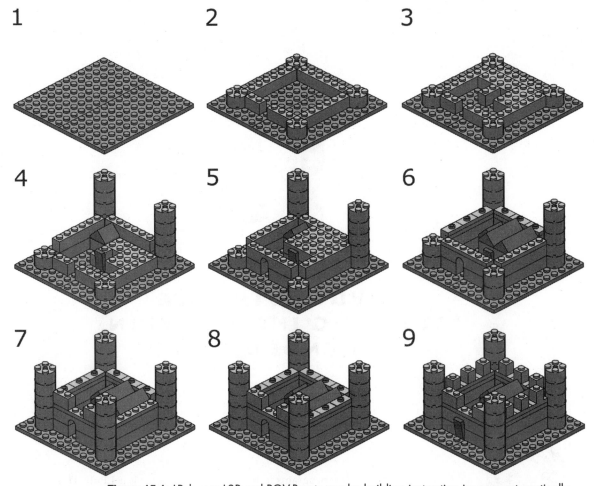

Figure 15-1: LPub uses L3P and POV-Ray to render building instruction images automatically.

NOTE *This chapter uses LPub 2.1.0.6. Check Kevin Clague's website for an updated version: http://www.users.qwest.net/~kclague/LPub/.*

Configuring LPub

The first time you use LPub, it will ask for the location of L3P (see Figure 15-2). It needs this information to function properly, because just like L3PAO, it uses L3P to convert the LDraw files to POV-Ray format.

Figure 15-2: LPub looking for the L3P.EXE file upon first launch.

After clicking **OK** in the dialog that is pictured in Figure 15-2, browse until you find L3P under the default install directory **C:\LDraw\Programs\L3P.exe**. Once you have done this and selected OK, LPub is ready to run.

The LPub Interface

Figure 15-3 shows the LPub interface. At first glance, the options folder tabs on the left-hand side of the window can look intimidating; after a few times through, they become much easier to understand and navigate. The blank space on the right hand side of the interface is a preview area; when POV-Ray generates your instruction images, LPub displays them there. You will see this happen when you run LPub for yourself.

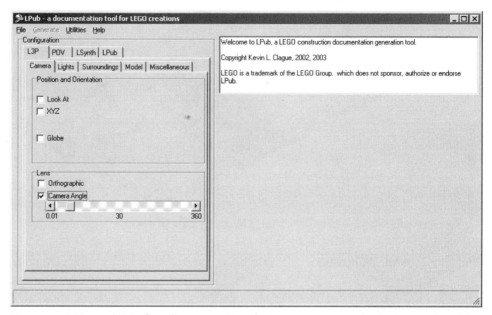

Figure 15-3: The LPub interface features options for L3P, POV-Ray, and LSynth, as well as LPub-specific settings.

LPub provides you with the same options as L3PAO, plus a few more. You can clearly see that the options are arranged much differently than they are in L3PAO. While the options are still grouped in similar categories, you need to look at many different "folder" panes to configure all of the necessary options.

Touring LPub

Before we discuss the many options for outputting building instruction images in LPub, why not take a tour? For the most part, we will use LPub's default settings to run a simple model through LPub. The model we will be using for this chapter is Chris Maddison's Microfig Scale Castle, as shown in Figure 15-4 on the following page.

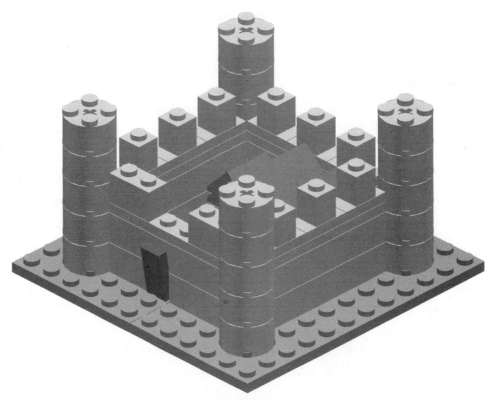

Figure 15-4: Chris Maddison's Microfig Scale Castle.

Microfig Scale?

Microfig Scale is a term coined by the online LEGO fan community. It simply means building to a scale smaller than minifig scale. This would include such models as the LEGO Company's Star Wars™ MINI sets, which are small versions of popular Star Wars vehicles. Some creations are even small enough to be considered "nanofig" or "picofig" scale! In Chris Maddison's creation, the castle's inhabitants would be much smaller than minifig size, so we call it "microfig scale."

Loading an LDraw File

Before you generate building instruction images of any LDraw model, you must load the desired LDraw file into LPub. To do this, select **File > Open LDraw File** from the menu. Browse for the desired file. In this case, we're looking for **C:\LDraw\VLEGO\ CH15\minicastle.ldr**. Browse for that file and load it into LPub.

Adjusting the Settings for Your First LPub Building Instructions

Now you should adjust a few settings in LPub before generating your first set of building instruction images. The LPub defaults are very well chosen, but one option selected by default makes renderings of smaller models like the Microfig Scale Castle less than ideal. This setting is the Minimum Distance. We will explain this in depth later in the chapter. It is very useful for multi-part models when you want to render instruction steps of multiple sub-models. However, for small models without sub-models, like the castle, this option makes the model render smaller in the image, so we need to disable it. To do so, follow these steps:

1. Select the **LPub** top-level folder tab.
2. Select the **Steps** second-level folder tab, inside of the LPub tab.
3. At the bottom, in the **Scale Controls** box, uncheck **Enable Minimum Distance**.

   You can leave the other the LPub options set to their default. Now you are ready to generate the first set of building instruction images for this model.

Running LPub

Now you are ready to generate building instruction images for the castle in LPub. Remember, you're taking this for a dry run before learning the settings, so don't think you need to be locked in to the look these images create.

   When you run LPub, the program will launch POV-Ray, similar to the way L3PAO can launch POV-Ray through its interface. When POV-Ray comes up, you may need to click **OK** manually on the legal notice (Figure 15-5) if you have not run POV-Ray once earlier in your Windows session.

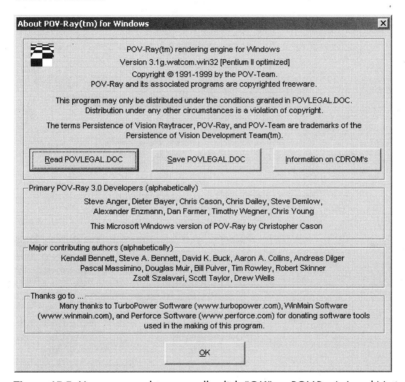

Figure 15-5: You may need to manually click "OK" on POV-Ray's Legal Notice for LPub to initiate rendering.

To generate building instruction images with LPub, select **Generate > Instruction Images** from the menu. LPub will create individual files for each step (as seen in LPub's white log window), launch POV-Ray, and POV-Ray will automatically render the images for you. When these programs are generating images, do not touch LPub or POV-Ray; you may interrupt the process, causing it to stop midway through. If you need/want to stop the generation of building instructions, you can select the **Cancel** button on LPub's "Progress Status" window.

Viewing the Results

When LPub is done generating images, POV-Ray will cease rendering. LPub will show a preview of the last-generated image, as you can see in Figure 15-6. In this case, it is the Bill of Materials. The Bill of Materials (if selected) is always the last image to be generated by LPub.

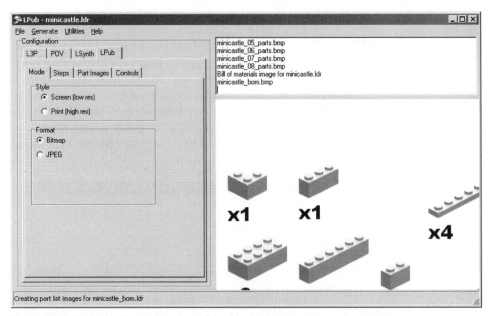

Figure 15-6: LPub displays a preview of the last image it generated.

LPub saves the images it generates in the same place the source LDraw file is located. In this case, look in **C:\LDraw\VLEGO\CH15**. In this folder, you should see the bitmap files LPub generates in this folder (see Figure 15-7).

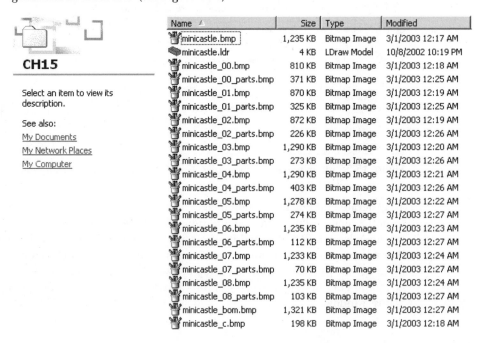

Figure 15-7: LPub places generated building instruction images in the same folder with the working LDraw file.

LPub uses a naming convention that is based on the name of your LDraw file (by default). Steps start off at 00 (versus 01, as you're used to seeing the first step numbered in instruction books). The file minicastle_01.bmp is the image for Step 1, the second step of your file (see Figure 15-8).

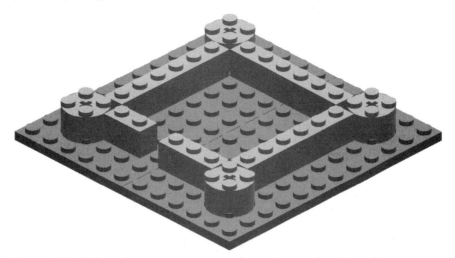

Figure 15-8: LPub renders a separate image for each step of an LDraw file.

The file minicastle_01_parts.bmp is the parts list for that step (see Figure 15-9). This image shows the parts needed to complete the step with the quantity of each part marked below. This is much like the building instructions for Technic sets, or more complicated models like the 3451 Sopwith Camel.

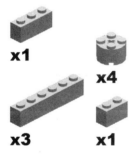

Figure 15-9: Parts List images such as minicastle_01_parts.bmp display the type and quantity of pieces needed for a particular step.

The file minicastle_bom.bmp is what LPub calls a "Bill of Materials" for the model. This contains an image of each part (in each color) needed for the model, complete with the total quantity needed as shown in Figure 15-10 on the next page. The Bill of Materials is a very useful image to generate for building instructions. Depending on the application and the complexity of the model you are working with, you may or may not want to include this image in the final instruction layout.

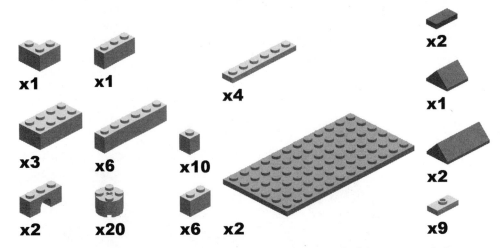

Figure 15-10: The Bill of Materials (minicastle_bom.bmp) shows all of the parts needed to create the model, as well as their quantities.

Finally, LPub generates two images of the completed assembly, as you'll recall were shown previously in Figure 15-4. One has the same dimensions as the instruction steps (default 800x600, customizable); the other is a smaller thumbnail. The full resolution image for the castle is minicastle.bmp, and the thumbnail is minicastle_c.bmp. The thumbnail image is used when you generate web pages for your building instructions.

Learning LPub's Interface and Options

Now that you have seen LPub's output for our simple model, the Microfig Scale Castle, let's move on to discuss LPub's interface and the various options it contains.

The top row of folder tabs in LPub represents broader categories. These allow you to configure options that are specific to different programs. All of the L3P options are available through the L3P tab. POV-Ray (and MegaPOV) options are available through the POV tab. The LSynth tab holds only one checkbox, to enable the synthesis of flexible elements on the fly while LPub is generating the LDraw files needed for the renders (we will describe the usefulness of this checkbox later in the chapter). Finally, the LPub tab allows you to configure options specific to LPub, such as image format, which images to generate (construction images, parts list images, and bill of materials; all of which we'll define later) and various other controls specific to LPub's brand of batch rendering.

The second row of folder tabs is specific to each top row. The L3P options have a different set of second-row folder tabs than the POV options do. These are in place to categorize the various options within each program's tab. For new users, the fact that these second-row tabs change dependent on the first-row tab selected can be a bit confusing. As you begin to use and learn LPub, working with this style of interface becomes easier and easier.

NOTE *Before you familiarize yourself with the specific options, perhaps a good exercise would be to play with selecting different first-row tabs and then browsing through second-row tabs. Observe the set of second-row tabs change as you select a different first row tab. This should accustom you to the interface a bit.*

L3P Options

First let's take a look at the L3P options in LPub. They are conveniently divided into Camera, Lights, Surroundings, Model, and Miscellaneous. Most of the settings you learned from L3PAO are found within this first-level folder tab. There are a few others in the POV-Ray tab; the rest of the settings are unique to LPub.

Camera

LPub's camera options give an interface for declaring the location and attributes of the camera in POV-Ray. A common element of LPub's interfaces is the checkbox; when you check the box next to a particular item, it displays the respective form fields. While unique, having to actively check a box can serve as a clear reminder of which options you have changed. In Figure 15-11, we show you all of the boxes checked, displaying all of the form fields. In practice, you may only have XYZ or Globe checked at once, not simultaneously. Checking the Camera Angle box displays a scrollbar. The number in the middle below the bar (in Figure 15-11 this number is 30) is the actual angle currently selected by the scrollbar. Also, you must select either Orthographic or Camera Angle, because they cannot be checked simultaneously. Orthographic removes all perspective, so it forces the camera angle (very close) to 0.

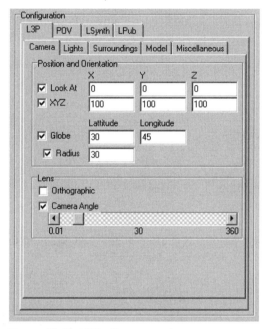

Figure 15-11: L3P's Camera options.

Lights

This tab contains the familiar L3PAO light switches. In Figure 15-12, we show the Custom Lights expanded by turning Light #1 on. The list box under Type in the Custom Lights section can allow a light to be set to either Off (no coordinates shown), Globe, or XYZ. You can select the color of the light by clicking on the color swatch below "Color."

Consistent with the checkbox system, Ambient, Diffuse, and Reflection need to be checked in order to display the dialog. Ambient controls the amount of ambient light present in the scene, Diffuse regulates the percentage of diffuse reflection, and Reflection controls how reflective the parts are. These settings are not available in L3PAO. Were you rendering an image or scene in L3PAO, you would need to manually edit the POV file to change these items. LPub allows you to change them through the interface, making knowledge of POV-code and searching through the POV file unnecessary.

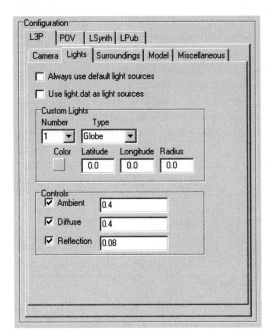

Figure 15-12: L3P's Lights options.

NOTE *To learn more about Ambient, Diffuse, and Reflection items, see POV-Ray's "Help on the POV-Ray Scene Language" document, under POV-Ray's Help menu. Search the index for "Ambient," "Diffuse Reflection Items," and "Using Reflection and Metallic."*

Surroundings

The Surroundings tab displays the familiar L3P settings for Background Color and Floor. These are the same as in L3PAO, as shown in Figure 15-13.

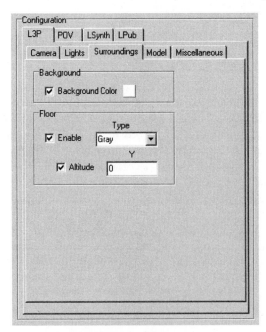

Figure 15-13: L3P's Surroundings options.

Model

The Model tab displays further L3P options that center on how the model is rendered. As you can see in Figure 15-14, these include Seam Width, Surface Bumps, Color (for single-part LDraw files, including color swatch), Render Quality (Box = Quality Level 0, Low = 1 Medium = 2, High = 3), and the optional LGEO library substitution.

NOTE *If you are using MegaPOV to render your model (we describe the MegaPOV settings that occur under the POV-Ray tab later), you may want to use a Seam Width of 1, to help the Find Edges process.*

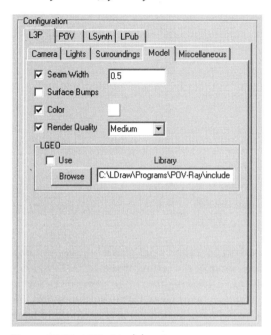

Figure 15-14: L3P's Model options.

Miscellaneous

The final tab in the L3P options is Miscellaneous, as shown in Figure 15-15. Here you can choose not to substitute primitives for POV-Ray equivalents (this was discussed in Chapter 14, and amounts to smoothing of cylinders and some other rounded objects). The latter two options make little sense for LPub, but were included for the sake of being complete. You can force L3P to exclude non-POV code (the extra comments it generates relating to the LDraw file). However, LPub intends for you to never need to interact with a POV file, so you would rarely need to use this option, if at all. Also, enabling Stepclock in LPub is a big no-no! Again, this is included for the sake of being complete, but since Stepclock and LPub perform similar functions, we do *not* recommend you use Stepclock in LPub.

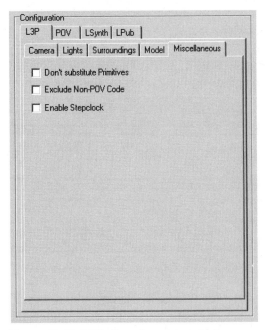

Figure 15-15: L3P's Miscellaneous options.

POV-Ray Options

The second first-level folder tab in LPub is "POV." This is where you can access all of the POV-Ray options. This contains some of the settings from L3PAO's Render Options window, and then some. Most notably, you can configure and use MegaPOV for rendering to create nice edge lines around the vertices of your parts, capturing a look very much like official LEGO building instructions. We will touch on these settings in this chapter, and discuss them more in depth in Chapter 17, "MegaPOV."

Rendering

The first tab of the POV-Ray settings includes options for selecting rendering quality. The top drop-down menu allows you to select which version or build of POV-Ray to render with. The default is POV-Ray 3.1. However, LPub also works with MegaPOV and newer versions of POV-Ray. In this book, we are covering basic usage of POV-Ray 3.1 and MegaPOV 0.7. Figure 15-16 reflects MegaPOV being the selected renderer. Later in this chapter, we will walk you through creating building instructions in LPub using MegaPOV.

The various Quality and Anti-Aliasing settings in this window allow you to change options you otherwise would need to configure directly in POV-Ray itself. For most applications, the defaults are acceptable. For anti-aliasing, you can edit the threshold, depth, and jitter. Since the defaults are OK for most applications, and since POV-Ray's help file covers them, we will not cover what they do in detail. You can access this information from POV-Ray itself by doing a Find (CTRL+F) in **Help > Help** on the POV-Ray Scene Language. For these options, remember to use the checkbox to access the form fields.

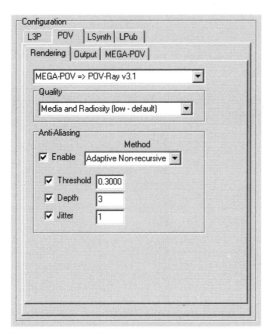

Figure 15-16: POV-Ray's Rendering options.

NOTE *We have provided a series of test renders on the CD-ROM to give examples of the different Quality settings at **C:\LDraw\VLEGO\CH15**.*

Output

This pane allows you to control various aspects of the images POV-Ray will output (see Figure 15-17). At the top, you can set the image dimensions. The default is 800×600. We recommend using common screen resolutions, which follow a 4:3 aspect ratio, such as 800×600, 1024×768, 1280×1024, 1600×1200, and so on. You can also select which image format you would like the final images in, choosing from BMP, JPEG, PNG, and Targa (compressed or uncompressed). Selecting **Specify File Name** allows you to give LPub a base filename for output images; the results will be numbered off of that. Finally, you can use Output Buffering, and specify the size of the buffer in kilobytes (for information on Output Buffering, see http://www.povray.org/documentation/view/128/, section 5.2.2.3.3).

NOTE *When entering custom image dimensions, remember that it must follow a 4:3 aspect ratio. POV-Ray's camera is set to 4:3 by default, and LPub allows no option to change this value. If you set custom image dimensions that are not 4:3, the image will render stretched or compressed.*

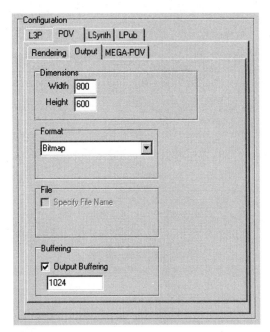

Figure 15-17: POV-Ray's Output options.

MegaPOV

The final POV options pane deals specifically with the **MegaPOV** (unofficial) build of POV-Ray (see Figure 15-18). These settings tell MegaPOV how to find edge lines in renderings (like the Scumcraft model's instructions in Chapter 12). These settings are explained in depth in Chapter 17. Later in this chapter, we will give you settings to use in an exercise.

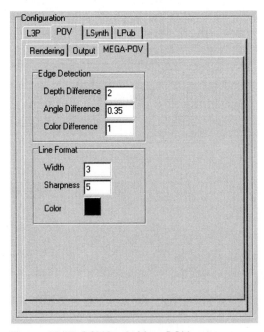

Figure 15-18: POV-Ray's MegaPOV options.

LSynth

LSynth gets its own top-level folder tab in LPub, with only one checkbox (see Figure 15-19). LPub is capable of using LSynth to synthesize flexible elements within models on the fly before generating building instructions. This is very useful because LSynth can create large LDraw files. After you are satisfied with the flow of flexible elements in your model, you can load the unsynthesized model into LPub. Checking **Enable Flexible Part Synthesis** will ensure your instructions, if LSynth commands are properly inserted in the model, will contain the flexible elements.

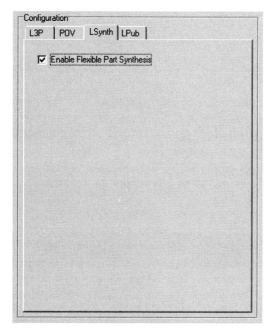

Figure 15-19: LPub can call LSynth to synthesize flexible elements on the fly before rendering building instructions.

LPub Options

Finally, LPub has a group of options that are specific to itself. These options cover image format, types of images it generates, and how it handles other LPub-specific features such as Previous Step Color Scaling, Parts List Images, and files it generates as a part of the rendering process.

Mode

These options control master settings within LPub (see Figure 15-20 on the following page). You can also control resolution and image format in the POV settings. However, these options can override the POV settings. When Style is set to "Screen," LPub uses the POV-Ray settings. When Style is "Print," LPub outputs images at 2048×1536 (note that when you change this switch, you can look in **POV > Output** and see these numbers have replaced the ones that previously occupied it). When you switch back to Screen resolution, the original settings are replaced. You can also override the image format you set in the POV-Ray options, and create JPEG images where you control the quality level. This is useful if you are creating building instructions for the Web, but is strongly discouraged for print because JPEG is a lossy format. It's important to note that the LPub > Mode settings are master settings that override or change settings in the POV-Ray menu.

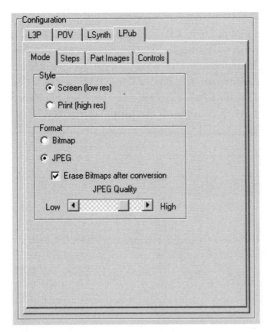

Figure 15-20: LPub's Mode options.

Steps

Here you can control how LPub handles steps and the types of images it generates. The first option, Previous Step Color Scaling, can be used to fade out all of the parts in previous steps. The scrollbar, as shown in Figure 15-21, allows you to set a percentage between the original part color and an arbitrary color (in this case, white). You can select the color you want LPub to fade the previous step's parts to by clicking on the white color swatch.

NOTE   *Kevin Clague, the author of LPub, prefers to fade out the parts from previous steps. This preference is evident in the books he has written. However, the authors of this book differ in building instruction philosophy and prefer to use the original part color for all parts, including those from previous steps. When you are using these tools, experiment with this setting to find your own preference; is it faded, original, or somewhere in-between? The scrollbar for this option lets you determine the degree of fading.*

The second section, Generate, controls which types of images LPub will generate when rendering LDraw files that contain steps (building instructions). Construction Images are the primary step images; they show the model being constructed. Part List Images are images that display the parts needed for each step, along with the quantity (as you saw previously in Figure 15-9). Bill of Materials is an image of all of the parts in a model, including quantities (as shown previously in Figure 15-10). Finally, Include Sub Assemblies allows you to decide whether or not LPub should render building instruction steps of each sub-model in an MPD file. If you render an MPD file in LPub, the output will use the sub-model name as the base name for the images, followed by the step number and "parts" or "bom" for parts lists and the bill of materials.

The third option, Enable Minimum Distance (Figure 15-21), is a very useful feature when rendering models with sub-models. Without a minimum distance defined, L3P fills the image space with the model. Previous to this feature, each sub-model would maintain a consistent scale, that is, size relative to the image resolution. However, different sub-models within the same model would have different scales dependent on their size, since L3P would fill the image space with the model. The Minimum Distance feature, if enabled, allows you to define a minimum distance the camera must be from *all* sub-models in a file, thus maintaining consistent scale across multiple sub-models. This is extremely important when relating to MegaPOV and edge lines, since the edge line settings are relative to image resolution, and not to the parts themselves. When you apply MegaPOV edge lines to a group of sub-models with inconsistent scale, some sub-models' parts will appear to have very thick lines, where others will appear thin. Enabling Minimum Distance forces a consistent scale, solving this problem.

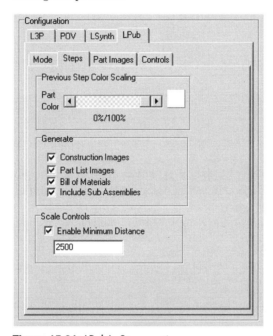

Figure 15-21: LPub's Steps options.

Part Images

LPub's Part Images options (Figure 15-22) allow you to set the camera position used to render the parts list images. The minimum distance ensures a consistent scale among all of the parts.

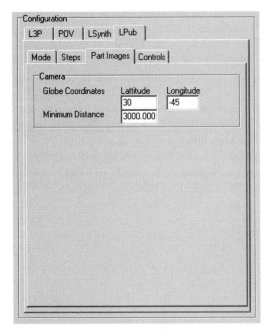

Figure 15-22: LPub's Part Images options.

Controls

Finally, LPub's Controls options (Figure 15-23) allow you to determine how LPub handles the files it generates as well as how it calls POV-Ray, crops images, handles configuration files, and so forth. If **Crop Images** is enabled, LPub will crop all of the white space from each construction step image and leave no margins around the rendering.

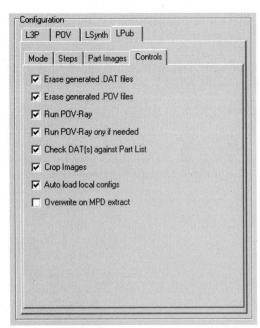

Figure 15-23: LPub's Controls options.

Putting LPub to Use: Two Sample Models

Now that you have become familiar with LPub's interface and some of its capabilities, we will walk you through setting up LPub and generating building instruction images and web pages for a couple of different models: a simple model, the Castle; and a model with sub-models, the Scumcraft.

Configuring LPub

Before generating building instructions with LPub, you should spend some time configuring the options to your preferences. We will walk you through our preferred method of creating instructions to give you an example. When you are making your own instructions, you may choose to use other settings. Experiment with different options to determine your preferred options for generating building instructions. Let's look at each program LPub uses separately and make our changes.

Configuring L3P Settings

We want to make the following changes to the L3P options to create better building instructions: use Orthographic Lens, turn off Reflections, and ensure render quality is set to Medium. Here are the benefits of each:

Orthographic Lens: Orthographic lens eliminates perspective from your renderings (Figure 15-24, on the left). When you use perspective (like the castle on the right hand side of the image), your model has a single vanishing point in the image. If you intend to lay multiple instruction steps on one page, or have an arrangement where multiple step images can be viewed at once, using perspective creates an awkward appearance. The human eye is used to seeing a single vanishing point in front, or two to either side, very far apart from each other. Having two or more vanishing points in the center of the field of view is unnatural and uncomfortable to look at. This is a distraction to viewers of building instructions. See an introduction to basic 2-dimensional perspective at http://drawsketch.about.com/library/weekly/aa021603b.htm. To turn on Orthographic lens, select **Orthographic** at the bottom of the **L3P > Camera** options pane.

Figure 15-24: Microfig Scale Castle rendered with orthographic camera (left). Without orthographic camera, using the default camera angle (right).

No Reflections: To better mimic official LEGO building instructions, turn reflections off. When reflections are turned on, you can see reflections of surrounding parts on the surface of the model, as shown, for example, in Figures 15-4 and 15-8, respectively. To turn off reflections, select the **Reflections** checkbox at the bottom of the **L3P > Lights** pane. Change the value in the form field to 0.

Medium Quality: The last L3P option you need to make sure of is the quality level. Use Medium Quality (Quality 2) when rendering building instructions. Rendering at Quality 3 (using the LEGO logo on the studs) does not look good when using the MegaPOV edge lines we will introduce you to later in the chapter. To ensure the quality level is set to Medium, go to **L3P > Model**, check **Render Quality**, and select **Medium** from the drop-down list.

Configuring POV Settings

For this group of settings, we want to use MegaPOV as the renderer, and configure the MegaPOV edge detection and line format settings. You will learn about MegaPOV in depth in Chapter 17, but for now follow the instructions to configure these settings.

Setting MegaPOV as the Renderer: To set MegaPOV as the renderer, select **POV > Rendering**. Next, select **MEGA-POV => POV-Ray v3.1** from the list. This replaces POV-Ray (the official build) with MegaPOV (the unofficial build capable of the desired edge-lines).

Configuring MegaPOV's Edge Detection and Line Format Settings: To configure the edge detection and line format settings, select the **MEGA-POV** second-level folder tab. Use these settings:

> Depth Difference: **2**
> Angle Difference: **0.35**
> Color Difference: **1**
> Width: **1.3**
> Sharpness: **5**
> Color: **Black**

Configuring LPub Settings

Finally, we must configure a few LPub-specific settings to achieve our preferred style of building instructions. First, eliminate Previous Step Color Scaling. Next, check which images to generate, and make a decision on the Minimum Distance.

Eliminating Previous Step Color Scaling: We want to remove Previous Step Color Scaling so all parts display their proper colors. To do this, go to **LPub > Steps**, and move the Previous Step Color Scaling scrollbar to the far left. Below the scrollbar should read **100%/0%**, signifying the Part Color shows through 100 percent, and 0 percent of the selected color to the right is visible. If you like, you can set the scrollbar to any place in-between and produce various degrees of fading on the previous steps' parts.

Check Which Images to Generate: Keep tabs on which images you are telling LPub to generate. You can select these the same options pane under **Generate**. All images are checked by default. Decide whether you want LPub to generate Parts List Images and a Bill of Materials. We recommend you generate all of these.

Minimum Distance: Remember that the Minimum Distance feature is best used on models that contain sub-models, to ensure consistent scaling across the board. On simple models such as the Castle, it is not needed. Make sure it is unchecked. However, when rendering models with sub-models, it is a good idea to check this box, especially when using MegaPOV's Find Edges.

Saving Your Configuration Settings

LPub allows you to save your configuration settings three different ways. First, you can save your configuration as an LPub configuration file, which you can load in a separate LPub session by opening it. Second, you can save a "Local Config," which becomes the default configuration used for a specific directory. Whenever you load an LDraw file from that directory, LPub uses the settings from your Local Config. Thirdly, you can save a "Global Config." This re-writes LPub's default settings and becomes LPub's new default settings.

Saving LPub Configuration Settings

You can save any of these three types of config files via the File menu. To save a Global or a Local config, all you need to do is select **Save Global Config** or **Save Local Config**, and LPub saves it automatically. To save a configuration file, you must specify the filename by clicking **File > Save Config As**. This enables you to save a configuration file that is neither a global nor a local config (therefore LPub does not load it automatically; instead you open it manually via the File menu).

For our purposes, perhaps it is a good idea to save a Local Config. First, make sure **minicastle.ldr** is open (so LPub saves the config file to the proper directory), then select **File > Save Local Config**.

Restoring Your Presets

If you want to restore your saved global or local config settings without closing and re-opening LPub, you can do so via the File menu. Simply select **File > Open Global Config** or **File > Open Local Config**, and LPub will restore the options to the ones saved in the respective file.

Generating Instructions for a Simple Model: Microfig Scale Castle

Now that you have checked your LPub settings, you are ready to generate building instructions. We had you quickly generate instructions for the castle near the beginning of this chapter, to give you an idea of what LPub can do. Now let's generate the castle again using our optimized settings, so you can see the difference. You'll recall the Scumcraft instructions we displayed in Chapter 12. This is how you create instructions that look like that. In fact, after you generate the castle, we will walk you through generating the Scumcraft instructions, and teach you how to handle special conditions when you are creating instructions for a model with sub-models. For now, let's generate the castle again with the new settings.

Generating the Instruction Images

Ensure that the file minicastle.ldr is loaded in LPub. We have saved our configuration file to **C:\LDraw\VLEGO\CH15\minicastle.lpb**, which you can retrieve by selecting **File > Open Config File**.

To render the instruction images, select **Generate > Instruction Images**. Sit back for a little while for LPub to generate the images. You can see an example of what LPub will generate in Figure 15-25.

NOTE *MegaPOV will not render the edge lines on the first pass, but on the second. The image will appear to hang at 99 percent complete. When it does this, do not cancel the operations within LPub (unless you want to render these images some other time instead); the program has not hung. After some delay, MegaPOV will draw the edge lines and move on to rendering the next image. The time it takes to post-process edge lines is dependent upon the resolution that you are rendering at. Be advised: When rendering at print resolution, it will take a considerable amount of time to generate the edge lines for each image.*

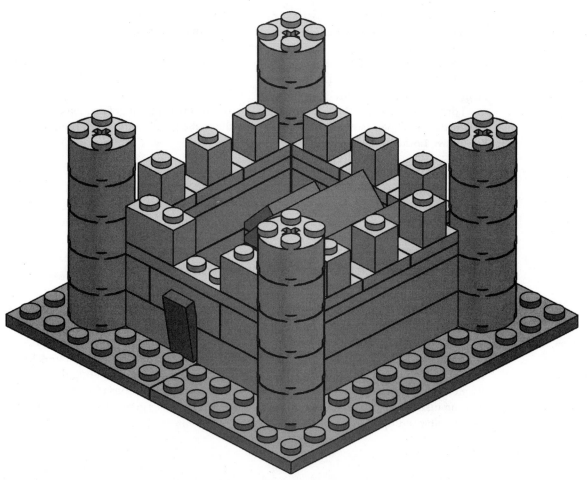

Figure 15-25: Chris Maddison's Microfig Scale Castle.

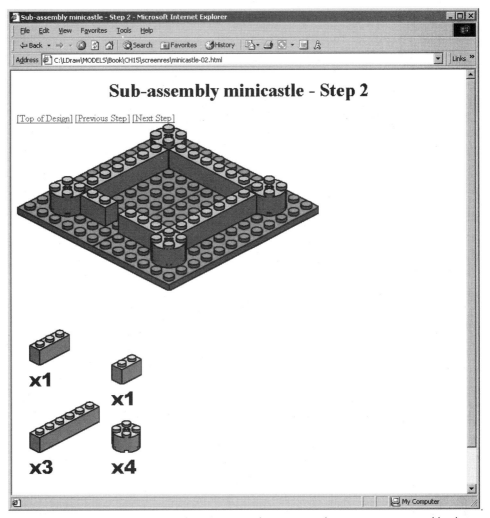

Figure 15-26: LPub can create simple "Screen Web Pages," with one page per step, like this one.

Generating Web Pages

After LPub has finished rendering the instruction images, you can generate web pages for your model. LPub currently has two styles of pages that it can generate: "Screen Web Pages" (see Figure 15-26), which creates a single, simple page for each step including navigation; and a "Web Page Table" (see Figure 15-27), which creates a single page with a table showing all generated images, which link to individual pages displaying only the selected image full-size. These pages do not contain navigation. The Web Page Table option was designed for publishers who need to assemble and print building instructions in books.

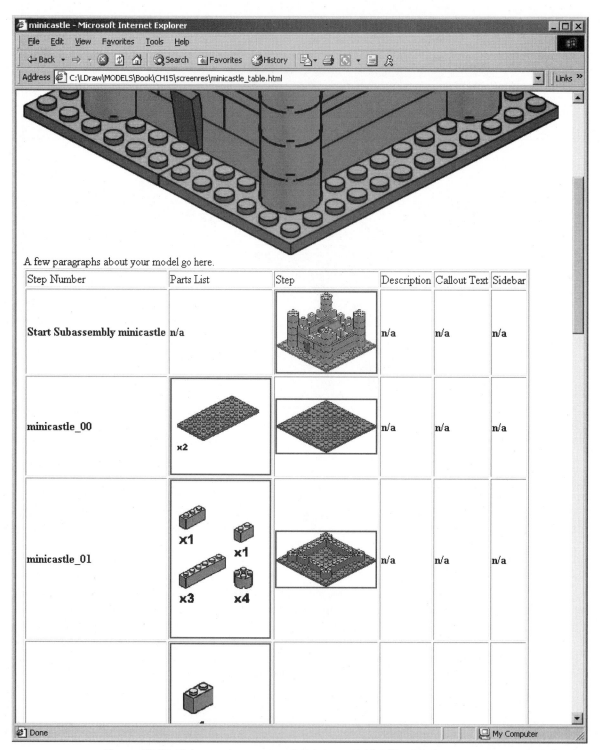

Figure 15-27: LPub can also create a "Web Page Table" like this, laying every image out for you.

The two forms of web pages in LPub are pre-defined. As of this version of the software, you cannot modify the template LPub uses to generate them. A possible future feature for LPub is in fact user-defined templates for generating web pages. At that time, you will be able to plug in your own website design and generate pages with the LPub instructions automatically! For now, though, the simple LPub pages will have to suffice.

To see how this works, generate a set of "Screen Web Pages" for the castle. Choose **Generate > Screen Web Pages**. LPub will automatically create the pages in the same directory the currently loaded LDraw file is located in. When done, LPub will launch your default web browser and open the first page. Click **Next Step** to navigate between the different pages for each step of the model.

Generating Instructions for a Model with Sub-Models: Scumcraft

There are a few more issues to consider when generating a model with sub-models. Let's walk through those with the Scumcraft.

Avoiding Duplicate Parts in Parts List Images with Models that Use Buffer Exchange

Models that use MLCad's Buffer Exchange pose a problem to LPub's parts list images: They contain duplicate entries of the same part. These duplicate entries exist because the part must be placed once in the exploded view location, and a second time in the final location. In order for LPub not to display the same part in two different parts list images, you need to essentially "comment out" the duplicate parts by adding meta-commands before and after the targeted parts. LPub has a set of meta-commands specifically dedicated to this. Before such meta-commands, LPub users would experience erroneous parts list images that displayed the same part in a model in two different parts list images.

To learn how to use these meta-commands, open up the Scumcraft model in MLCad. We will show you how to insert these commands into the file yourself in preparation for generating the building instructions via LPub. The Scumcraft model can be found in **C:\LDraw\VLEGO\CH08\scumcraft.mpd**. After opening the Scumcraft:

1. First, make sure "Draw to Selection" is enabled, by clicking the ![icon] icon on the toolbar. This makes it easier to know what you have selected while looking for the proper locations to insert the meta-commands.

2. Next, find the first **RETRIEVE A** statement (see Figure 15-28). Insert the comment (using the **ab|** toolbar icon) **PLIST BEGIN IGN**, which expanded means "Parts List Begin Ignore," or "ignore the parts between this statement and the end statement."

Type	Color	Position	Rotation	Part nr.	Description
◻PART	Light-Gray	-1116.470,-72.0...	1.000,0.000,0.000 0.000,1.000,0.000...	3048.DAT	Slope Brick 45 1 x 2 Triple
S STEP	--	-------	-------	-------	
BUFEX...	--	-------	-------	-------	STORE A
◔PART	Light-Gray	-1116.470,-152...	-1.000,0.000,0.000 0.000,1.000,0.000...	scumcraft_s2.ldr	Ghost
◔PART	Black	-1136.470,-100....	0.707,0.000,0.707 0.000,1.000,0.000...	arrow.dat	Ghost
◔PART	Black	-1096.470,-100....	0.707,0.000,0.707 0.000,1.000,0.000...	arrow.dat	Ghost
◔PART	Black	-1096.470,-100....	0.707,0.000,0.707 0.000,1.000,0.000...	arrow.dat	Ghost
S STEP	--	-------	-------	-------	
BUFEX...	--	-------	-------	-------	RETRIEVE A
◻PART	Light-Gray	-1116.470,-16.0...	-1.000,0.000,0.000 0.000,1.000,0.00...	scumcraft_s2.ldr	Mounting Point
S STEP	--	-------	-------	-------	
BUFEX...	--	-------	-------	-------	STORE A
◔PART	Black	-887.470,-28.98...	-0.951,-0.309,0.000 -0.309,0.951,0.0...	scumcraft_s4.ldr	Ghost
◔PART	Black	-1345.350,-28.6...	-0.949,0.317,0.000 0.317,0.949,0.00...	scumcraft_s5.ldr	Ghost

Figure 15-28: Select the first RETRIEVE A statement in the Scumcraft's primary sub-model.

3. Finally, select the next line, the reference to the Mounting Point (**scumcraft_s2.ldr**) sub-model. Insert another comment at this point, **PLIST END**. This ends the section of commented-out parts. These lines should now look like the lines in Figure 15-29.

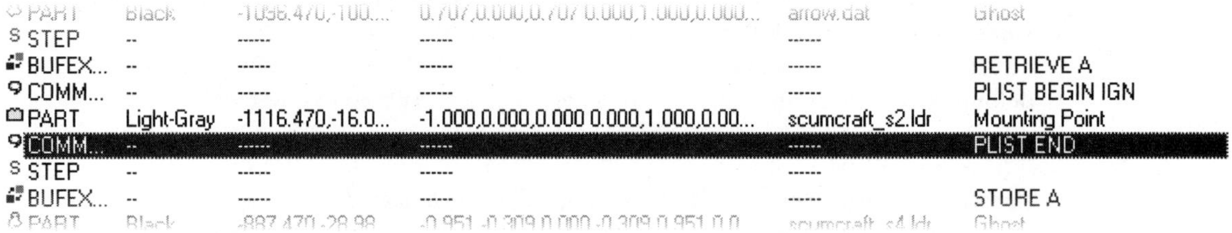

⊘ PART	Black	-1036.470,-100...	0.707,0.000,0.707 0.000,1.000,0.000...	arrow.dat	Ghost
S STEP	--	------	------	------	
⊿ BUFEX...	--	------	------	------	RETRIEVE A
⊙ COMM...	--	------	------	------	PLIST BEGIN IGN
⊡ PART	Light-Gray	-1116.470,-16.0...	-1.000,0.000,0.000 0.000,1.000,0.00...	scumcraft_s2.ldr	Mounting Point
⊙ COMM...	--	------	------	------	PLIST END
S STEP	--	------	------	------	
⊿ BUFEX...	--	------	------	------	STORE A
⊙ PART	Black	-887.470,-28.98	-0.951,-0.309,0.000,-0.309,0.951,0.0	scumcraft_s4.ldr	Ghost

Figure 15-29: You can insert meta-commands into an LDraw file to eliminate duplicate parts in LPub Parts List Images.

Now you have learned how to add the necessary meta-commands. The Scumcraft has several points that need these comments. After each RETRIEVE A statement, the exploded parts are restated, and these parts need to be commented out. To get a feel for this, go through the rest of the Scumcraft file. There are two more RETRIEVE A statements in the main sub-model, and several in each of the l-wing.ldr and r-wing.ldr sub-models. Find these and replace them before proceeding.

NOTE *If you don't want to take the time to do this, the file C:\LDraw\VLEGO\CH15\scumcraft.mpd has all of the meta-commands added.*

Multi-Piece Parts and Part List Images

LPub also has a meta-command to help you combine multi-piece parts. Some parts in the LDraw library, such as the 4×4 turntable (part numbers 3403 and 3404), shock absorbers, pneumatic pistons, and so on are each one LEGO "part" made up of different moveable (or non-moveable) components. Each molded component is modeled as a separate part in the LDraw library. When you place these parts in your model and generate Part List Images, LPub will read them as two separate parts (with separate entries in the Part List Image). In order for LPub to consider these as one part (and therefore render it properly in the Part List Image and Bill of Materials), you can use the **PLIST BEGIN SUB** meta-command. To use this command, simply group the parts, like this:

```
0 PLIST BEGIN SUB turntable.ldr
1 7 0 0 0 1 0 0 0 1 0 0 0 1 3403.DAT
1 7 0 0 0 1 0 0 0 1 0 0 0 1 3404.DAT
0 PLIST END
```

The turntable.ldr argument within the **BEGIN** command tells LPub to treat the encapsulated parts as a sub-model "turntable.ldr." The **END** command ends the grouping.

Buffer Exchange and Previous Step Color Scaling

The Scumcraft uses Buffer Exchange for many steps. Using Previous Step Color Scaling where previous steps' parts are "grayed" or faded out (a technique we the authors don't prefer, but you may) on a model that uses buffer exchange poses an issue. Since you have to restate the exploded part in the current step after retrieving the buffer, that part will not appear faded out as if it were inserted in a previous step. If you want that part to be faded out according to your Previous Step Color Scaling options, surround it with these meta-commands:

```
0 BI BEGIN GREYED
```

and

```
0 BI END
```

The parts in between these meta-commands will appear faded according to your Previous Step Color Scaling settings. If you are using our configuration file to render this model in LPub, you do not need to add these meta-commands, since we are not using Previous Step Color Scaling.

Undesirable Sub-models

The Scumcraft is a unique case in that it uses unofficial LDraw parts. These parts are included in the MPD file as sub-models, so anyone opening it can see all of the parts in the file, regardless of whether or not they have all of these unofficial parts on their computer. Because we have set up the file in this manner, when you render the images in LPub, it will render some unnecessary images. These extra images are not sub-models of the Scumcraft, but the unofficial parts. LPub does not allow you to select which sub-models you want to render; it only gives you the option to render all sub-models or render no sub-models. Be aware that you won't need some of these images to create the Scumcraft's building instructions.

Clearing the Parts List Cache

LPub saves a cache of each parts list image, that is, each individual part used to create the parts list images. LPub will reuse the cached files in the parts list images rather than re-render the part. If you change *any* settings that could affect the parts preview images, you should flush your cache before rendering building instructions. If you do not, your parts list images may display some parts rendered differently than others. To do this, select **Utilities > Erase Part Image Caches**. Go ahead and do this now to clear the cache of parts generated when rendering the castle.

Generating the Instruction Images

Open the Scumcraft file in LPub. You can find a totally prepared Scumcraft file under **C:\LDraw\VLEGO\CH15\scumcraft.mpd**. Also, you can use the Castle's config file, available under **C:\LDraw\VLEGO\CH15\minicastle.lpb**.

To render the instruction images for the Scumcraft, select **Generate > Instruction Images**, as shown in Figure 15-30. This will take considerably longer to render than the castle because there are many more steps and sub-models.

Organizing Your Images

When you generate instructions for a complex model with many sub-models, like the Scumcraft, you will be left with a lot of images. Note that all of the images pertaining to a particular sub-model will have that sub-model's name as a prefix. The numbering system is as discussed earlier in the chapter. Take a look at the files LPub generates for the Scumcraft to get a better understanding of this.

LPub is only capable of generating the images via POV-Ray and MegaPOV, plus creating simple web pages. When you finish using LPub, you are left with a batch of raw images. These need to be assembled into a nice page layout to complete the process of creating building instructions. We will introduce you to some techniques for doing this in Chapter 18, "Post-Processing Your Building Instructions."

Figure 15-30: Jon Palmer's Scumcraft.

Generating Single Images in LPub

LPub can also create single images in POV-Ray, much like L3PAO. To generate an image of a complete model, select **Generate > Complete Assembly**. We have not focused on this feature since we prefer L3PAO for single-image rendering. We do find it useful, however, to render a single image to check LPub settings before rendering an entire set of building instruction images.

Summary

In this chapter, you learned how to create building instructions using LPub. We taught you about LPub's many options and how to optimize them to create attractive instruction images. LPub is a program that uses POV-Ray or MegaPOV as a rendering engine, and batch-processes the rendering of many different images to form each step. In the next two chapters, you will learn about the technical details of POV-Ray and MegaPOV, which are followed by chapters on how to assemble and present the models you have rendered in an attractive package.

16

POV-RAY

In this chapter, we will teach you several techniques for editing the POV-Ray scene language code. With this knowledge, you will be well on your way to learning how to tweak your own POV files. This chapter covers advanced techniques, which are not for the faint of heart. However, once you have engaged and familiarized yourself with the material, you will be able to create some stunning results using POV-Ray's features. With some experimentation, you can be on your way to creating scenes worthy of winning LDraw.org's popular Scene of the Month competition.

NOTE *The Scene of the Month competition can be found at http://www.ldraw.org/community/contests/.*

In the previous chapter, we introduced you to Kevin Clague's LPub. You learned to harness the power of POV-Ray without needing to learn POV-Ray's language. This chapter introduces you to the nuts and bolts of the POV-Ray raytracer. By the end of this chapter, you should have a cursory understanding of POV-Ray's scene language approach to 3D modeling, its syntax, camera and lighting techniques, and other features that make POV-Ray a powerful, free choice for 3D rendering. By learning POV-Ray's language, you will be able to take your renderings to another level. L3P only converts the information directly pertaining to your LDraw model. When you use POV-Ray, you can control variables outside of your LDraw model, which can be used to create a total environment for your model to exist within.

POV-Ray's User Interface

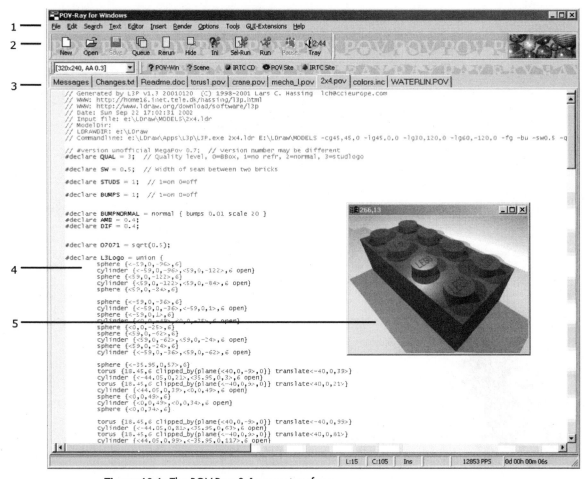

Figure 16-1: The POV-Ray 3.1g user interface.

Here is a quick orientation of the POV-Ray interface:

Menus (1): POV-Ray's standard Windows menus.

Toolbar Icons (2): Commonly used commands that can be found in the menus also have icons for easier access.

File Tabs (3): These tabs allow you to toggle between all currently open POV files.

Workspace (4): You can view and edit the POV-Ray scene language code in this window.

Render Window (5): This window appears when POV is rendering a file. When POV-Ray is not rendering, you can only see the code; you cannot "preview" a model to be rendered.

Rendering Inside POV-Ray

Throughout this chapter, we'll instruct you to render files that have been included with this book's CD-ROM and should have been installed on your hard drive along with the LDraw software. Until now, you've rendered via L3PAO or LPub, and have not had to use POV-Ray's interface to render. Here, we'll quickly walk you through the process of setting the image size and rendering an image.

We've created a POV file for you to use as an example, which we will refer to several more times throughout the chapter. Open POV-Ray, and open the file **C:\LDraw\VLEGO\CH16\2x4.pov**. This is a simple 2×4 brick. We chose it because it can provide a good example, while at the same time it's simple enough that it takes very little time to render.

Setting the Image Size

You need to select the size of the image POV-Ray will render. To set the image size, use the drop-down list on the left-hand side, above the workspace (see Figure 16-2).

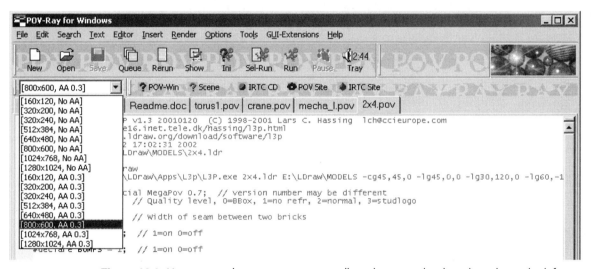

Figure 16-2: You can set the size your image will render at via the drop-down list at the left.

We recommend you use an image size of **800×600, No AA** when rendering for this chapter and rendering test scenes. It is an average size, not very large, yet big enough for you to distinguish details.

Anti-Aliasing

The "AA" and "No AA" suffixes on the image sizes in the drop-down list refer to anti-aliasing, a process computers use to smooth edges of contrast lines. If you have anti-aliasing turned on ("AA"), POV-Ray will take longer to render the image. If you turn it off ("No AA"), POV-Ray will not take nearly as long to render, but elements within your scene will appear to have jagged edges. Turn anti-aliasing on for finished-grade renders, and off for test renders. See Figure 16-3 on the next page for an example of a rendering without anti-aliasing (left) and with (right).

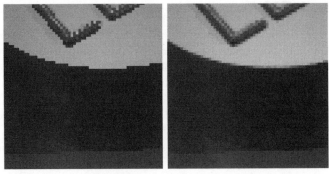

Figure 16-3: Left: 2x4 Brick rendered using Quality 3 with anti-aliasing off (rendering time: 19 seconds); right: 2x4 Brick rendered using Quality 3 with anti-aliasing on (rendering time: 25 seconds). Zoom: 500%.

Changing the Quality Level

In L3PAO, you change the quality level of a POV file via the interface. In POV-Ray, you need to change a piece of code to do the same thing. When L3P converts an LDraw file, it places the quality declare statement immediately below the block of comments at the top of the POV file. If you take a look in the 2×4.pov file, you should see this code on line 10:

```
#declare QUAL = 3;  // Quality level, 0=BBox, 1=no refr, 2=normal, 3=studlogo
```

NOTE *To find the line number where your cursor is, use the status bar at the bottom of the POV-Ray window. To the right of center is a field that says L: <number>. Move your cursor up and down and you will see this change. This is the line number.*

The comment later on the same line gives you some brief information on what each quality level means. Remember that quality 2 renders without LEGO on the stud, where quality 3 includes the markings. Quality level 3 takes the longest to render. If you are rendering a test image, consider stepping down the quality level as appropriate.

Rendering

To render an image in POV-Ray, simply press the **Run** button on the toolbar. Go ahead and try this now with the **2×4.pov** file.

What If It Won't Render?

If you hit Run and the model doesn't render, either something is wrong with your code, or LPub has reconfigured the initialization settings. The Messages window will display a list of code errors, including line numbers. You can view the messages window by selecting it from the file tabs (it's always the first tab on the right). Take note of the later section "POV-Ray Syntax" to be sure you have used the proper syntax in any code you have edited. Also, note the section below, "WARNING: LPub Residue."

Using and Customizing Initialization Files

As your POV-Ray knowledge grows, you may want to set custom image sizes and use the (step)clock, which require editing the initialization files. POV-Ray reads initialization files (.INI) before rendering to tell it which settings to use. When you selected the Run button a few seconds ago, it used one of these files, although you probably didn't realize it. These initialization files are plain-text files that you can read and edit yourself. They tell POV-Ray settings like which image size to render at, which sizes to offer as options in the drop-down, and where to look for include files (which we will introduce you to later in this chapter). POV-Ray uses two main initialization files, QUICKRES.INI and POVRAY.INI.

NOTE *L3P and L3PAO refer to POV-Ray's* clock *feature as the "stepclock." This references the LDraw step feature, which L3P employed to create a POV file that would generate building instruction images.*

QUICKRES.INI

QUICKRES.INI tells POV-Ray which image sizes to offer as options from the drop-down list. Go ahead and open it from the folder where you installed POV-Ray. It's located in the subfolder **renderer\QUICKRES.INI**. We have POV-Ray installed under **C:\LDraw\Programs\POV-Ray**, per the recommendations in Chapter 2. We will reference this folder as the POV-Ray install folder throughout the rest of the chapter. You can copy and paste an entry to create custom image sizes, as we have done below:

```
[1600x1200, AA 0.3]
Width=1600
Height=1200
Antialias=On
Antialias_Threshold=0.3
```

POVRAY.INI

POVRAY.INI, which is stored in the same directory as QUICKRES.INI, maintains the last used image size, and tells POV-Ray where to look for include files when rendering. If you open this file, you will see at the bottom the Library_Path points to the POV-Ray include\ directory. You can set additional locations to check for include files by adding additional lines, like this:

```
Library_Path=C:\LDraw\Programs\POV-Ray\INCLUDE
Library_Path=C:\LDraw\MyPovScenes
```

You can include up to twenty instances of the Library_Path command. DO NOT include a \ or / at the end of the path, or it will not function properly.

Advanced Feature: Setting a Custom Initialization File

If you ever use POV-Ray's clock feature (called Stepclock in L3PAO and L3P) to create building instructions or to animate a model or scene, you must use a custom initialization file. To set a custom file, click the **INI** icon on the Toolbar. This will open POV-Ray's render settings interface. Use the **Browse** button to locate the file of your choice. Once you have selected a file, exit by selecting **Set, but do not render**. POV-Ray will use the .ini file you provide for all renderings until you set a different file via this same method. If you select **Default**, POV-Ray will revert to using the QUICKERS.INI file.

CAUTION *This is an advanced feature. We don't recommend you set a separate initialization file unless you know what you are doing. For more information, read the POV-Ray Scene Language help file.*

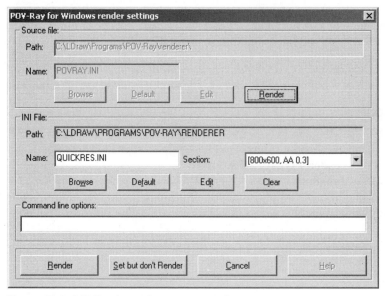

Figure 16-4: POV-Ray's render settings window.

You can get started using custom initialization files by using the one we created, located under **C:\LDraw\VLEGO\CH16\steps.ini**.

WARNING: LPub Residue

In the last chapter, we taught you how to render building instruction images automatically via LPub. When LPub calls POV-Ray or MegaPOV to render images for it, it leaves a little something behind when it's done. In the previous section, we taught you how to set a custom initialization file. An additional feature, the command line options, is also available via the dialog shown in Figure 16-4.

If you want to render directly in POV-Ray after using LPub, you need to eliminate the command line options LPub automatically sets. First, open the render settings window via the **INI** icon on the toolbar. Next, find LPub's command line options in the command line field, and delete them. Select **Set but don't Render** to close the window.

Getting Oriented with the POV-Ray Language

This book was written to teach LEGO fans about the LDraw system of tools. Therefore, we will only discuss issues that relate to rendering converted LDraw models. Our discussion will barely scratch the surface of what POV-Ray is capable of doing. If you want to learn how to use this software to its fullest potential, we recommend you read the tutorials found in POV-Ray's Help menu. You can also explore the official POV-Ray website at http://www.povray.org for documentation and help via their online newsgroups.

NOTE *In this book, we will be using POV-Ray 3.1g for Windows. The POV-Team has released version 3.5. However, at the time of this writing, there remain outstanding compatibility issues between L3P and POV 3.5. While it is possible to render a converted file in POV 3.5, we recommend this only for advanced users, and even then, with reservation.*

The Key Elements

There are four basic elements of POV-Ray files that you as an LDraw user need to be aware of: objects, floor/background, camera, and lighting. In this sense, you can think of using POV-Ray like taking a picture. We will introduce you to each of these elements and discuss how they are used.

When L3P converts your model to POV-Ray format, it arranges the elements in this order:

1. Objects
2. Floor and Background
3. Camera
4. Lights

The Camera is by far the element you will need to modify most frequently, followed by lights. We'll also teach you to create and modify a floor plane. In general, you don't want to touch the code that makes up the subject (your model), a long and incredibly complex section of code that defines every element of every part converted. This is why L3P/L3PAO does the work for you!

Preparing Yourself Mentally

One thing that is important to realize is how far you can get by modifying, or "hacking," existing code. Try not to become overwhelmed, but look at the code snippets we will introduce you to throughout this chapter and you should start to gain a familiarity with what they are doing. Chances are you won't have to write much code from scratch unless you are doing advanced work. POV-Ray's scene language is not impossible to use, no matter how foreign it might look at first glance.

POV-Ray Syntax

The syntax for most POV-Ray code is as follows:

```
Keyword { functions of keyword }
```

The braces, "{" and "}," are essential in order for POV-Ray to know what information to attribute to each keyword. It uses the braces to section off code that modifies one keyword from code that modifies another. When you read a POV file in POV-Ray, it highlights keywords in purple to help you distinguish between them and other elements of the POV code.

If you are using a #declare statement, or something similar (you can see these at the top of a converted POV file), your line should end in a semicolon. See the example below:

```
#declare QUAL = 3;
```

Using Comments

We've already briefly mentioned comments earlier in the chapter. Comments are an important part of the POV-Ray file. When L3P converts an LDraw file, it includes many comments that are useful information for you, the user. You can use these to get your way around the scene language without having fully studied each element.

Throughout this chapter, we'll ask you on several occasions to "comment something out." When we say this, we mean place comment markings in front of or around the specific piece of code. Just as in the LDraw format, when a line or a block of code is distinguished as a comment, POV-Ray does not parse (or read) it when rendering a scene.

When you are working on your own, you may want to use comments to write notes to yourself about what you are doing in a certain area, so you can come back to a file and work on it later, or so you can send it to a friend to have him or her take a look at what you are doing.

Comment Syntax

Comments in POV-Ray take two forms. A single-line comment is preceded by two forward slashes, like this:

```
// This one line is a comment
```

Or, they are encapsulated by a slash and an asterisk, like this:

```
/* All of the code
on these three lines
is commented out */
```

Remember:

```
This line is not a comment, since it has no comment marks.
// This line is a comment
This one is not. If you don't comment out this line, POV-Ray will give you an error when you try to render something.
```

Using POV-Ray's Camera

A basic camera statement in POV-Ray is defined by the following syntax:

```
Camera{ location <x,y,z> look_at <x,y,z> }
```

In an L3P-converted file, the camera statement is usually a bit more complex. It will look like this:

```
camera {
    #declare PCT = 0; // Percentage further away
    #declare STEREO = 0; // Normal view
    location vaxis_rotate(<40,-40,-20> + PCT/100.0*<39.8995,-56.4264,-39.8995>,          <-
2251.39,-3183.94,2251.39>,STEREO)
    sky    -y
    right   -4/3*x
    look_at <10,-5, 25> // calculated
    angle  67.3801
    rotate  <0,1e-5,0> // Prevent gap between adjacent quads
    //orthographic
}
```

This code can look a bit scary, but there is nothing to fear. You only need to configure the camera location, the camera look_at position, and the lens angle. These three elements are clearly displayed in the code above.

Camera Location

The camera location may look complicated, but for our purposes, you only need to be familiar with the first portion of the code.

```
location vaxis_rotate(<40,-40,-20> + PCT/100.0*<39.8995,-56.4264,-39.8995>,        <-
2251.39,-3183.94,2251.39>,STEREO)
```

In the code above, the location of the camera is <40, -40, -20>. The rest of the line is used by POV-Ray to further define the location of the camera. These values have been calculated by L3P for the model in question and you do not need to be concerned with it. You could even erase the rest of the location code and rewrite it as follows, and still get the same results in your rendering.

```
location <40,-40,-20>
```

You can try this out for yourself if you like. In POV-Ray, open the file **C:\LDraw\VLEGO\CH16\2×4.pov**. To render the file, click the **Run** button on the Toolbar. Your rendering should produce an image like the one seen in Figure 16-5.

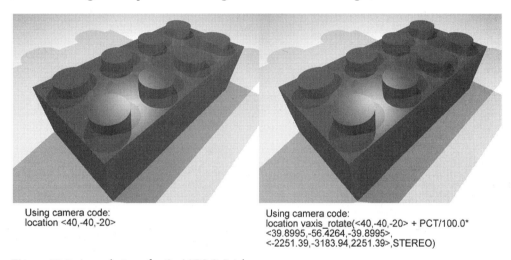

Using camera code:
location <40,-40,-20>

Using camera code:
location vaxis_rotate(<40,-40,-20> + PCT/100.0*
<39.8995,-56.4264,-39.8995>,
<-2251.39,-3183.94,2251.39>,STEREO)

Figure 16-5: A rendering of a 2x4 LEGO Brick.

To see for yourself that the extra camera code is irrelevant, find the lines that read:

```
location vaxis_rotate(<39.6387,-39.3198,-20.2106> + PCT/100.0*<39.8995,-56.4264,-39.8995>,
<-2251.39,-3183.94,2251.39>,STEREO)
```

(This code is located on lines 189 and 190).

Comment these lines out by placing two slashes // at the beginning of each line. By doing this, POV-Ray will ignore the code on that line. After you have commented out both of the lines, type a new line below them with the simpler camera location code we mentioned earlier.

```
location <40, -40, -20>
```

Now re-render the file by clicking **Run** on the toolbar. Do you see a difference?

Now that we have illustrated there is no difference in the rendering, you don't have to bother with the long-winded camera code. The point to remember here is that you only need to be concerned with camera location, look_at position, and angle of the lens. The rest is unnecessary for our purposes.

Lens Angle and Camera Look_At

These other camera elements are straightforward in the code L3P generates, with no potentially confusing additional code. The values in the POV code are the same values you input in the L3PAO interface.

Using POV-Ray's Lights

POV-Ray has several different types of lights. L3P/L3PAO can only create one type of POV-Ray light, the Pointlight. POV-Ray also has spotlights, cylinder lights, area lights, and ambient lights. These have different properties, and can be used for various purposes. Here we will teach you a bit about POV-Ray lighting code, and introduce you to each type of light.

Let's start by re-rendering the 2×4 brick file. This time, locate the three light sources and comment them out (they are located on lines 199 through 211, immediately below the camera code). Figure 16-6 shows you the image you will get when you render the file with the lights commented out.

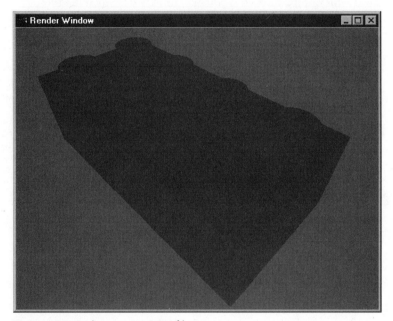

Figure 16-6: Lights out in a POV file.

In Figure 16-6, you can barely make out the general shape of the model. But, if you turned off all the lights, why can you see still see the model? You would think the image would be completely dark. In POV-Ray, the ambient light is turned on by default, but you can disable it. This light, white by default, is used to create the shadows where the light sources are not present. In the case of Figure 16-6, since no lights are present, the ambient light makes everything a shadow and that is why we can still see the model.

Pointlights

The most basic light in POV-Ray is the Pointlight. Pointlights illuminate everything in the scene equally, no matter how close or far away they are from the light's location in the 3D world. Pointlights are defined by two parameters: a location and a color. The syntax for a pointlight is:

```
light_source { <x,y,z>, color rgb <r,g,b> }
```

Note that r, g, and b above in brackets represent the numbers assigned to create the color. For white, use **<1, 1, 1>**.

POV-Ray defines its RGB (red, green, blue) values differently than most graphics programs do. Instead of allowing you to enter a value 0-255, it only accepts values 0-1. This means that if you have an RGB value, you will need to convert it to POV-Ray's system in order for POV-Ray to be able to render it. To convert a color from the standard 1-255 RGB scale to POV-Ray's scale, divide each separate value by 255. (Example: 255/255=1, 0/255=0, 128/255=0.5, etc.)

Using this system, rgb <1, 1, 1> is white light, and black is <0, 0, 0>. You can combine the colors to get new colors using basic color mixing theory. Purple would be rgb <1, 0, 1> because red and blue make purple.

NOTE *It is possible to use RGB values greater than 1 and less than 0. Doing this will create strange effects on your colors in a rendering. If you increase an RGB value to greater than 1, the color will appear more saturated than any other color in the scene (for a cool neon green effect, try using* rgb <0, 2, 0>*. Likewise if you use a negative number, such as* rgb <0, -1, 0>*, it will wash out all of the green in the scene.*

Spotlights

The spotlight is ideal for focusing the attention of the viewer on a specific area of an image, just like a physical spotlight focuses an audience's attention on a specific place on a stage. In addition to providing the two basic parameters needed for a pointlight, the spotlight adds three additional parameters: radius, falloff, and tightness. Here is some typical spotlight code:

```
light_source { <100,0,0> color rgb <0,2,0>
spotlight radius 15 falloff 20 tightness 10  point_at <0, 0, 0> }
```

Figure 16-7 is an illustration of how the different values relate to a spotlight. You can apply this illustration when you are determining the values for your own spotlights.

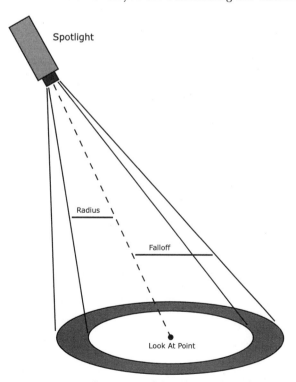

Figure 16-7: An illustration of the components of a Spotlight in POV-Ray (a similar image is shown in the help file).

radius and falloff are angles; the point_at point is a location in 3D space.

In addition to the elements we show in the illustration, there is a value called tightness. tightness is used to determine how hard or soft the edge of the falloff circle is, and determines the brightness of the light within the radius. The default value for tightness is 10. Lower values will make the spotlight brighter, but wider and with sharper edges. Higher values will dim the spotlight and make it narrower with softer edges. This is similar to a real spotlight: If you make the iris very small, the light radius will be small, and the light will be very intense within that radius; if you open the iris wide, the radius will spread also, and the falloff edge will be soft.

NOTE *For good graphical references on tightness and the other elements, see the POV-Ray Scene Language help file section on Spotlights.*

In Figure 16-8 you can compare the difference between a pointlight and a spotlight.

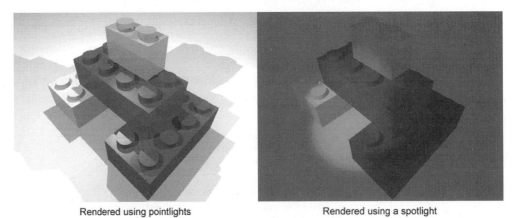

Rendered using pointlights   Rendered using a spotlight

Figure 16-8: Comparing pointlights with a spotlight.

Cylinder Lights

When you're working with spotlights, the further an object is from the light's source, the greater the radius must be to illuminate the object. Spotlights in POV-Ray diffuse over distances, just like real spotlights do. A cylinder light has a fixed radius no matter how far away an object is from its source. These can be useful for rendering laser beams, or even Star Wars™ Lightsabers™. The Lightsaber is cylindrical in shape, and in order to get the entire saber to glow, you need to use a cylinder light. The syntax for a cylinder light is exactly the same as the one for a spotlight, except instead of using the keyword "spotlight," you use the keyword **cylinder**, like this:

```
light_source { <100, 0, 0> color rgb <0, 2, 0>
cylinder radius 15 falloff 20 tightness 10  point_at <0, 0, 0> }
```

Area Lights

If your goal is to create truly realistic renderings, you will likely want to use area lights. Look at Figure 16-9 and note the difference between the two images. The image on the right, rendered with area lights, has much softer, more realistic shadows.

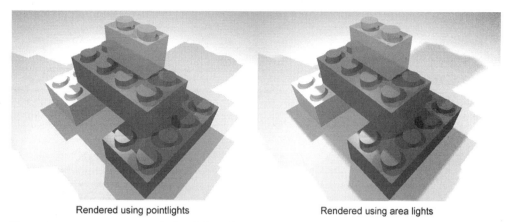

Rendered using pointlights   Rendered using area lights

Figure 16-9: Comparing pointlights (left) with area lights (right).

Pointlights provide direct illumination only. If an object is not in light, it is in shadow. The light from a `pointlight` doesn't diffuse or reflect like in real life. This causes sharp shadows in renderings. Pointlights are only close to realistic if you are rendering a space scene, where there is no atmosphere and little ambient light. In natural environments, or even indoors, light is affected by the surroundings. The ambient light present is diffuse and produces soft shadows. To simulate this in POV-Ray you can use the area light. Here is an example of area light syntax:

```
light_source { <0,0,0>, rgb <0,0,1>
area_light <5,0,0>, <0,0,5> 2, 5
adaptive 1 jitter }
```

See Figure 16-10 for a visual explanation of the area light code, and Table 16-1 for a breakdown of each element.

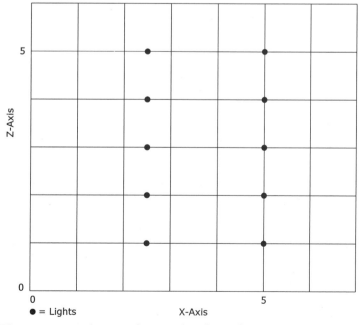

Figure 16-10: A diagram showing the effects of the above area light code.

Table 16-1: Area Light Code

Code	Explanation
`light_source { <0,0,0>, rgb <0,0,1>`	This is the beginning of the light source declaration. The coordinates (in the first set of brackets) signify the location in 3D space; the RGB value displayed here will render as blue.
`area_light <5,0,0>, <0,0,5> 2, 5`	`area_light` text tells POV-Ray this is an area light we are declaring. The coordinates in the brackets define the outer limits of the area lights. A pair of numbers follows the coordinates, which, in this example, are 2 and 5. The first number tells POV how many rows of lights to place; the second indicates how many lights are placed on each line.
`adaptive 1 jitter }`	`adaptive #` tells POV-Ray to only use lights that are necessary to determine the color of a specific pixel in question during the rendering process. A higher adaptive number gives more realistic shadows, but also adds to the rendering time. (Note: If you don't use the adaptive keyword, POV-Ray will use every light in the array to determine the color of every pixel. This can be extremely time consuming!) `jitter` tells POV-Ray to jitter, or vibrate, the lights a little bit during rendering. This creates softer shadows in the rendering.

Ambient Lights

We briefly mentioned ambient light earlier in the chapter. Ambient light affects everything in a POV-Ray scene. By default, the ambient light is white. In order to set the ambient light to a color other than white, you can add the following code to your POV-Ray file.

```
global_settings { ambient_light rgb <r,g,b> }
```

Most images don't need an ambient light color other than white, so this isn't used very often. Perhaps you would want to change this if you were rendering a scene of a LEGO space model landing on the planet Mars. Mars' surface would reflect red light, so you could use a red ambient light for your scene.

Special Lighting Conditions

POV also allows you to add a few special conditions to your lights. These are conditions that are declared in addition to your light source type; any type of light can have these conditions added to it. You can make a light shadowless, set it to fade off after a certain distance, and determine the rate of falloff. These three special light keywords are `shadowless`, `fade_distance`, and `fade_power`.

shadowless

The shadowless condition, as its name implies, provides illumination to an area without creating shadows. See Figure 16-11 for an example.

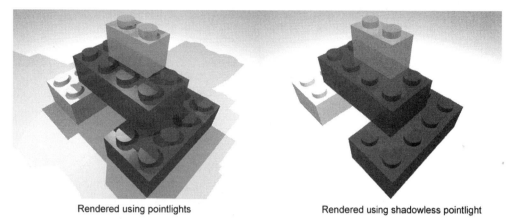

Rendered using pointlights   Rendered using shadowless pointlight

Figure 16-11: Left: Image rendered with a regular pointlight; Right: The same image using a pointlight with the shadowless condition enabled.

Shadowless lights help create clearer building instructions. LPub even uses shadowless lights by default. To use a shadowless light, simply insert the keyword **shadowless** at the end of your light code prior to closing your braces }. Here is an example of a shadowless pointlight:

```
light_source { <-10,-46,36>, color rgb <1,1,1> shadowless }
```

fade_distance and fade_power

These special lighting conditions provide a way to achieve yet another level of realism in a rendering. In real life, light fades over distance. An object 100 feet away from a light source should receive much less light than an object only 10 feet away. To make this happen in POV-Ray, you need to use fade_distance and fade_power.

To use these two conditions, place the keywords at the end of any light source code, before closing the braces }. fade_distance sets the distance to which a light illuminates the object. fade_power is an exponential value, which is used to determine the actual rate of illumination over the distance. If you use these, it will take longer to render your image, since POV-Ray needs to make more calculations when they are included. If your goal is a truly realistic looking image, these keywords are a must. You can see the POV-Ray Scene Language help file (find the topic "Light Fading") for more information.

Using POV-Ray Objects

This section is brief, simply because you will not need to seriously interact with POV-Ray's objects if you are creating a scene out of all LDraw parts converted from L3P. We will introduce you to some simple POV-Ray object code, and will teach you how to edit a plane (the object used for creating a ground/floor surface in scenes). As always, if you are interested in learning more about POV-Ray's Scene Language, we encourage you to read the help file.

To see what object code looks like in POV-Ray, see line 158 of the 2×4.pov file that you should have opened earlier (it's located at **C:\LDraw\VLEGO\CH16\2×4.pov** on your hard drive). The code looks like this:

```
#declare _2x4_dot_ldr = object {
// Untitled
// Name: 2x4.ldr
// Author: Ahui Herrera
// Unofficial Model
```

```
// ROTATION CENTER 0 0 0 1 "Custom"
// ROTATION CONFIG 0 0
    object {
        _3001_dot_dat
        matrix <1-SW/80,0,0,0,1-SW/28,0,0,0,1-SW/40,0,SW/2.8,0>
        matrix <1,0,0,0,1,0,0,0,1,-10,0,20>
        #if (version >= 3.1) material #else texture #end { Color1 }
    }
//
}
```

This is the code that defines the 2×4 brick, converted from the LDraw file via L3P. At the beginning of the code snippet, you will see a #declare statement. That statement names the code contained within the braces {} _2x4_dot_ldr. You will later see a statement calling the object in line 168:

```
object { _2x4_dot_ldr #if (version >= 3.1) material #else texture #end { Color1 }}
```

2×4.ldr was the name of the converted file. If you convert an LDraw file with multiple sub-models, each sub-model will be contained in one of these #declare statements.

Why does it take POV-Ray all of these lines of code to create a single 2×4 brick? A 2×4 brick, as simple as it appears to an average LEGO builder, is certainly not a basic geometric shape. If you scroll upwards, you will find the object 3001_dot_dat, the 2×4 brick represented in POV code. This block contains triangles, cylinders, and spheres. All of these basics shapes are used to create components of the 2×4 brick. Coding a model with multiple pieces by hand would be nightmarish at best for most users. Fortunately, you don't have to worry about writing this code yourself; L3P does it for you.

The Plane Object

Let's take a closer look at the "plane" object. If you are hand-editing a converted POV-Ray file, this is most likely the only object you will need to manually edit. If you ask it to, L3P will generate a plane for the floor in the POV file it writes. In the file **2×4.pov**, look at line 171. You should see the following code:

```
// Floor:
object {
    plane { y, 24 hollow }
    texture {
        pigment { color rgb <0.8,0.8,0.8> }
        finish { ambient 0.4 diffuse 0.4 }
    }
}
```

You can change the color of the floor by adjusting the RGB values in line 175. If you want to move the location of the floor, change the numeric values in line 173. Remember, take a look at the L3Globe image (Figure 14-6) to get your bearings. Positive-Y is down, and negative-Y is up.

NOTE   *If you are looking for specific objects in your POV file, note that they are well commented. When L3P converts an LDraw file, it comments the Floor, Background, Camera, and Lights. Also, recall from earlier that each LDraw part is named in a #declare statement.*

For more information on planes or other POV objects, see the POV-Ray help file on your computer.

Tips for Generating Scenes

Many people who get into POV-Ray do so in order to create scenes. A scene is more than a rendering of a simple model; it is a model placed within an environment. That environment can be entirely LEGO (therefore other LDraw parts), or it can be made up of other non-LEGO POV-Ray objects. In this section, we'll show you some tips for creating scenes using pre-made POV-Ray files, which you can insert and arrange within your file at will.

This is an advanced topic, but it is not impossible to learn. We recommend you read this section, then try the exercises at the end of the chapter. This section will introduce you to the concepts necessary to generate scenes on your own. With some determination, you can walk away from this chapter with the general knowledge you need to create your own scenes. Add in a bit of your own creativity and ingenuity, and you'll be well equipped to create your own stunning POV-Ray scenes!

Include Files

The first concept we would like to introduce you to is include files. This is a fundamental concept in the POV-Ray scene language, which originated from markup languages and programming. Include files are files that contain code you may want to use over and over again across multiple POV files, or they could contain code you aren't going to edit, which you wish you could just "collapse" into a single line. These files can contain other LDraw models, other POV-Ray objects, color definitions, camera code, or anything else. How does an include file work? You can issue a single #include statement in your file, which references the separate include file. When POV-Ray renders your scene, it will include all of the code in the separate file as if it were written in the file it is rendering. In your main file, you only see one line, but to POV-Ray it can represent hundreds or even thousands of lines of code. LDraw files work in the same manner — they reference (or "include") parts that are in separate LDraw files.

POV-Ray comes with standard include files for many objects. These include files can be found in the **\POV-Ray\include** folder. In addition, many people in the POV-Ray community have created other include files for any object imaginable (mountains, oceans, trees, people, outer space, and so forth). You can search for others' include files on the Web. We have even "included" on the CD-ROM some popular packages of POV-Ray include files for you to play with.

You will learn to use include files more and more as you learn about POV-Ray. Here we will introduce you to colors.inc, one of the files distributed with POV-Ray. We will also show you an example of two include file packages distributed by POV-Ray user Chris Colefax: a Galaxy Creator for creating space scenes, and a City Creator for creating backgrounds with city elements, such as buildings to create a skyline.

Using Include Files

As the name suggests, you need to "include" the colors.inc in your POV file before you can use its contents. This is done by entering a statement in the POV file like the one below:

```
#include "colors.inc"
```

Most #include statements are entered at or near the top of a POV file.

If you only enter the name of the file, POV-Ray will only look in its own "include" sub-folder (and other library paths you may have defined in POVRAY.INI). If your desired include file is stored somewhere else on your hard drive, you must enter its full path.

colors.inc

POV-Ray's colors.inc file contains 128 pre-set color definitions. You can use the color names conveniently in place of an RGB value. This means two things: You don't have to memorize the RGB values of a specific color and manually enter them across multiple elements, and you can also change the value of a color in one location, and apply the changes across every element that uses that color. You can find this file under **C:\LDraw\Programs\POV-Ray\include\colors.inc** on your hard drive. Browse the file to see the available colors.

Using colors.inc

Here is an example of using colors.inc in POV code. Note that instead of an RGB value, the code snippet below uses the color name "Gold." You can use any named color from colors.inc if you issue the previously mentioned include statement at the beginning of your file (it is important to include colors.inc at the top, so it is called before you use a color name).

```
// Gold Light
#include "colors.inc"
light_source {<-10, -20, 04> color Gold }
```

Galaxy Creator

Chris Colefax has assembled a group of POV-Ray objects that create an easy way to make custom space scenes. Before we can use Galaxy Creator, we need to install it because it does not come with the official POV-Ray files. On the companion CD-ROM, locate the directory **\POV Add-Ons\Galaxy Creator**. Select all of the folder's eleven files and place them in **C:\LDraw\Programs\POV-Ray\include**. We will walk you through the basics of using these files in Exercise 16.4.

Locate the file **C:\LDraw\VLEGO\CH16\Space.pov** on your hard drive. If you open the file and render it, you should see an image similar to Figure 16-12.

Figure 16-12: An example using the galaxy include file. Planet Earth model courtesy of the POV-Team.

This is just one example of the many possibilities of the Galaxy Creator. With these files, it's easy to place your own models against a starfield background with beautiful glowing nebulae like the one above. Later in the chapter, we will show you how to create your own backdrop using the Galaxy Creator. To learn more about Galaxy Creator, we recommend that you read the **Galaxy.htm** file in your POV-Ray include directory to learn about Galaxy Creator's various possibilities.

City Creator

Chris Colefax also created a city generator. The images in Figure 16-13 were created with his example POV files, included in the city creator package. Again, in order to use the files, you need to copy them from the companion CD-ROM into the **C:\LDraw\Programs\POV-Ray\include** directory. You can find the necessary files on the CD-ROM under **\POV Add-Ons\City Creator**.

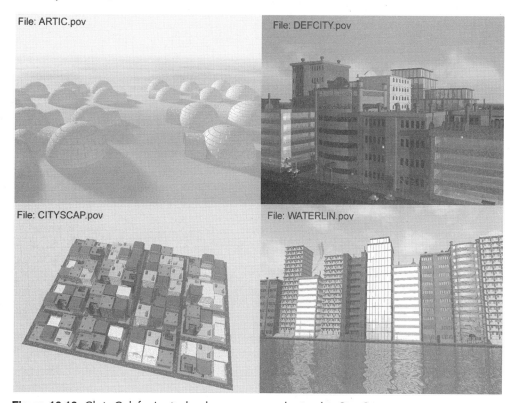

Figure 16-13: Chris Colefax's city landscapes, created using his City Creator.

With these pre-made backdrops, all that is left for you to do is insert your own models and manipulate the camera to find the best angle.

We have included more of Chris Colefax's include file packages on the companion CD-ROM under the folder **\POV Add-Ons**. We won't be talking about them in the book, but they are there for you to explore on your own. He has done a wonderful job providing a user manual for each of his applications, and we recommend you read each readme file fully before using the files in your projects.

Search the Web for POV-Ray Elements
With a little searching on the Internet, you can find other peoples' POV files for download. If you find a POV-Ray rendering online that contains elements you like, such as a particular image of sky or water, download the file and experiment with the objects contained inside of it!

Skies

POV-Ray provides several types of skies that you can cut-and-paste into your own code. You can find these files on your hard drive under C:\LDraw\Programs\POV-Ray\ scenes\textures\pigments\skies. You can open any of the sky files and render them to see what they look like. See some examples in Figure 16-14. Paste the sky code into your own POV-files and it will render with your own scenes.

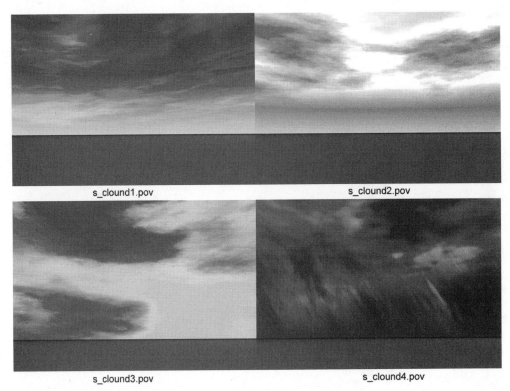

s_cloud1.pov   s_cloud2.pov

s_cloud3.pov   s_cloud4.pov

Figure 16-14: Various sky renderings provided with POV-Ray.

Using Pre-Made Scenes

Before you dive in and attempt to create your own scenes from scratch, look around the Internet — especially the POV-Ray community. You will find that almost any scene that you want has been created by someone else already. When you find a scene you like, try contacting the creator and ask to borrow his or her POV file for your own work. Most people are more than happy to help you out. Browse the companion CD-ROM for some other snippets of POV code that might come in handy for your own projects.

Generating Scenes Automatically

Forester and Terragen are two programs that you can use to automatically generate scene code. To use them, you simply define the desired parameters in their respective graphical interfaces. While Terragen doesn't write or export POV files, Forester can import Terragen files and export POV-Ray files. By using these programs in conjunction with each other, you can generate POV code for landscapes to go with your LEGO scenes. Note that both of these programs are shareware; therefore they have a limited array of features available in the downloadable version.

We did not include these programs in the installation process you used in Chapter 2. To use them, you must install them on your computer yourself. You will find folders for both programs on the CD-ROM under **\POV Add-Ons**. These folders contain the installation files for the programs. This next section briefly describes each program and shows an example of a simple scene we created. If you would like to follow along, install the programs. However, they are not necessary for the exercises at the end of the chapter, so you may want to read this section and install them later.

NOTE *To open a .zip file, you can use a program such as WinZip, available at www.winzip.com.*

Terragen

You can use Terragen to create landscapes in a matter of seconds. The majority of the commands are accessible through a series of graphical windows. Figure 16-15 is a screen shot.

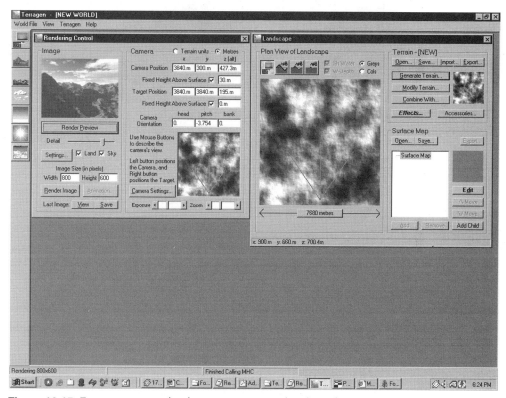

Figure 16-15: Terragen creates landscapes via its graphical interface.

Terragen can create landscapes including mountains, hills, bodies of water, and sky elements. Figure 16-16 is one example of what Terragen can create with a few clicks of the mouse.

Figure 16-16: A landscape created in Terragen.

The only drawback to this software is that it cannot save a POV-Ray file of the landscapes it creates. But you can save your landscape as a Terragen file (*.ter), import the file in Forester, and use it to export to POV-Ray.

NOTE   *You can visit Terragen's website at http://www.planetside.co.uk/terragen/.*

Forester

You can use Forester to populate the landscape you created in Terragen with things like trees, bodies of water, cabins, and more. Forester can import Terragen's landscape files. You can also use Forester to export files into the POV-Ray format. See Figure 16-17 on the next page for an illustration of what Forester can do.

The image in Figure 16-14 was created without hand-editing any POV-Ray code. Using the Forester application we placed the trees, the lake, and the cabin, and Forester generated all of the POV-Ray code for us! Forester comes with a variety of objects (trees, grass, plants, planes, homes, and so on) that you can place in a scene. See Figure 16-18 on the next page for a screen shot of Forester.

Figure 16-17: A forest scene created by Forester, rendered in POV-Ray.

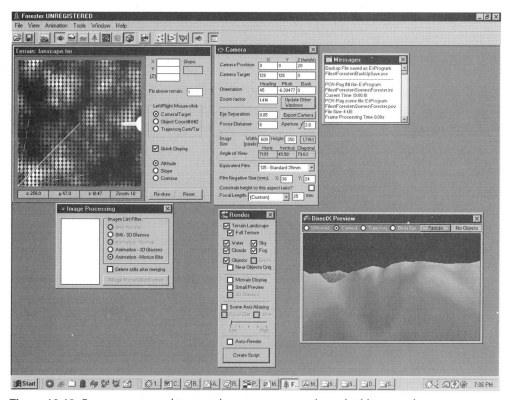

Figure 16-18: Forester creates objects such as trees, grass, plants, buildings, and more.

NOTE   *You can visit Forester's website at http://freespace.virgin.net/stephen.dartnall/.*

Correcting Scale Issues Between Terragen/Forester and L3P

If you cut and paste a scene generated using Terragen/Forester and place it in an L3P-converted POV-Ray file, you will notice the LEGO models are significantly out of proportion with the surrounding scene. See Figure 16-19 for a visual explanation of this phenomenon.

Figure 16-19: L3P-generated objects, such as the bottom of this airplane model, are significantly larger than Terragen/Forester generated objects. This requires correction via scaling.

L3P and Terragen/Forester use different scales to create objects. When you paste Forester-generated POV code into an L3P-generated file, you will end up with a miniature scene and seemingly giant LEGO models. You can fix this using the POV-Ray keyword scale inside the LEGO objects.

Scaling Code

In Figure 16-20, the following code was placed inside the plane object:

```
object { _plane _dot_ldr
    scale .02 translate <85, 10, 183> rotate <1,0,1>
#if (version >= 3.1) material #else texture #end { Color7 } }
```

Figure 16-20: You can scale down objects to bring them more into proportion with the surrounding scene.

The scale keyword shrinks or enlarges an object depending on the value you provide it. A scale of 1 maintains an object's size. To make an object smaller, enter a percentage as a decimal number, or to enlarge, enter a number greater than 1.

Learning Terragen and Forester

Both Terragen and Forester have excellent user manuals and step-by-step tutorials. You can search online for other users of these programs and ask for support. Both programs have strong communities of users who are willing to answer new users' questions. We recommend visiting these programs' respective websites, listed earlier in this section.

Exercises

The following sets of exercises have been designed to introduce you to some of the tasks you'll perform most frequently.

Exercise 16.1: Using the Camera

We have talked quite a bit about the camera location and look_at position, but have only briefly mentioned the camera lens angle. In L3PAO, the switch is called -ca (camera angle), and POV uses the keyword angle to define the lens angle. The default value provided by L3P is 67.3801. To get a feel for using the camera angle, let's see what happens as we increase or decrease this value.

1. In L3PAO, open the file **C:\LDraw\VLEGO\CH16\airplane.ldr**. Set the output directory to **C:\LDraw\MODELS** (you can set this path by clicking on the ellipsis **...** button to the right of the POV-Ray Output File box). Check the Floor switch (**-f**) and ensure the floor drop-down box says **g** to create a gray floor. Render the file through L3PAO by selecting **Render upon completion**.

2. POV generated a BMP file when it rendered this model. Find the file in the output directory you selected, and rename it to **airplane_angle_67.bmp**. This will allow you to render the file again and not overwrite the original rendering, which you will need to compare and contrast between attempts.

3. In POV-Ray, open the **airplane.pov** file you converted. Locate the angle keyword in the camera code (line 6839 in our file).

4. Change the angle from 67.3801 to **20**. Save the file and render it by clicking **Run** on the Toolbar.

5. Once it is finished rendering, rename the new airplane.bmp file to **airplane_angle_20.bmp**.

6. Now, go back to the camera code and change the angle value from 20 to **100** and render it a third time.

7. Once it is finished rendering, rename the latest airplane.bmp file to **airplane_angle_100.bmp**.

You should now have three images as shown in Figure 16-21.

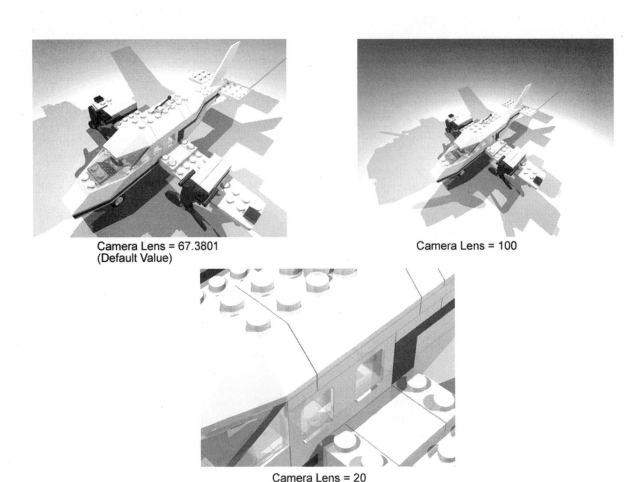

Camera Lens = 67.3801
(Default Value)

Camera Lens = 100

Camera Lens = 20

Figure 16-21: The airplane model rendered with various camera lens angles. Model courtesy of Kevin Dixon.

When you lower the value of the camera angle, you narrow the angle of the camera's field of view, and you get less perspective. This is similar to real photography. If you use a telephoto lens, you can zoom in on objects. If you increase the angle you get a wider, more panoramic image. But if you increase the angle too much, you get the fishbowl effect.

Exercise 16.2: Using Spotlights and Pointlights

You can use colored lights to change the mood of your image drastically. In this exercise, you will see how various colored lights can change the mood of a rendering.

1. Use the **airplane.pov** file from the previous exercise. Change the angle value back to **67** (you can leave out the decimal points L3P inserted originally).

2. Find the lighting code immediately below the camera code. Comment out the existing, L3P-generated light sources by enclosing them in block comments with a forward slash and an asterisk (`/* commented code */`). You should do this for lines 6845-6856.

3. Add a violet spotlight, pointing at the airplane's canopy. First, be sure to include colors.inc at the top of your file (directly above the `#declare QUAL` statement), by inserting the following:

```
#include "colors.inc"
```

Next, insert the code below immediately below the three pointlight statements you commented out in Step 2:

```
light_source { <-45,-300, -82> color Violet
    spotlight radius 15 falloff 20 tightness 10 point_at <-45, -52, -82> }
```

4. Save and render the image.

This attempt rendered a bit dark. We want to focus the light on the canopy, but still have the rest of the plane illuminated. To do that, let's brighten the color of the spotlight and add a couple shadowless pointlights to illuminate the rest of the model. We make them shadowless lights so their shadows don't interfere with the spotlight's, creating the feeling of ambient light.

1. Change the color of the spotlight to a brighter color. We used "LightSteelBlue." You can open up the file **C:\LDraw\Programs\POV-Ray\include\colors.inc** to browse for other colors, if you like.

2. Add two gray colored pointlights — place one on each side of the airplane model. The pointlight to the right should illuminate with half the value of the spotlight, and the pointlight to the left should illuminate three-fourths the value of the spotlight. You can use the percentage grays included in colors.inc to accomplish this task (see the file for more details). See our code below:

```
light_source { <-45,-300, -82> color LightSteelBlue
    spotlight radius 25 falloff 30 tightness 10 point_at <-45, -52, -82> }
// Left Side Light using ¾ (75%) the brightness of the spotlight
light_source { <-300,-150,60> color Gray75 shadowless }
// Right Side light using ½ (50%) the brightness of the spotlight
light_source {<300,-150,60> color Gray50 shadowless }
```

Render the image to view the results.

You can experiment with various lighting techniques to achieve the goal you desire. When you start making your own scenes, be patient, try things out, and see what works and what doesn't. With some experience, you will soon gain a feel for lighting in the POV-Ray scene language.

Exercise 16.3: Creating Your First Scene

Let's create a background scene for the airplane model. This will be your very first simple scene. For this exercise, we will use Chris Colefax's City Creator. You can see what will be the end result in Figure 16-22 on the following page.

Figure 16-22: We will teach you to create this simple scene with Kevin Dixon's airplane model.

Let's get started.

1. Make sure the airplane.pov file is open from the previous exercise.

2. In order to use the City Creator files, copy all of the contents of the CD-ROM directory **\POV Add-Ons\City Creator** to your hard drive under **C:\LDraw\Programs\POV-Ray\include**.

3. Open the file **C:\LDraw\Programs\POV-Ray\include\WATERLIN.POV**. Copy its contents and paste them into the bottom of airplane.pov. Save the file as **scene1.pov**.

4. Comment out the original camera statement L3P provided, immediately above the light source code you commented out in the previous exercise. Also, comment out the spotlight and pointlights you added in the last exercise. The WATERLIN.POV file includes its own camera and light statements.

5. Save the file and render it.

When you render the image, you should notice the plane's tire is very large and renders behind the skyline. We noted earlier in the chapter that there will be scale and orientation issues when using objects from include files. Now you need to correct for this by using the keywords scale and translate. Translate moves an object in 3D space. Find the object airplane_dot_ldr and insert the following rotate, translate, and scale code inside the last closing brace as shown:

```
object { airplane_dot_ldr
#if (version >= 3.1) material #else texture #end { Color7 }
    translate <X,Y,Z> rotate <X,Y,Z> scale ?? }
```

Figure 16-23 shows a series of test renders along with the values we used to create the final rendering you saw in Figure 16-22 earlier. Try some test renderings yourself, using the values we show in Figure 16-23 as a guide. We recommend you change the values of one keyword at a time to see the effects each one has on the image. If you need help, look at the file called **\CH16\plane_n_sky.pov**.

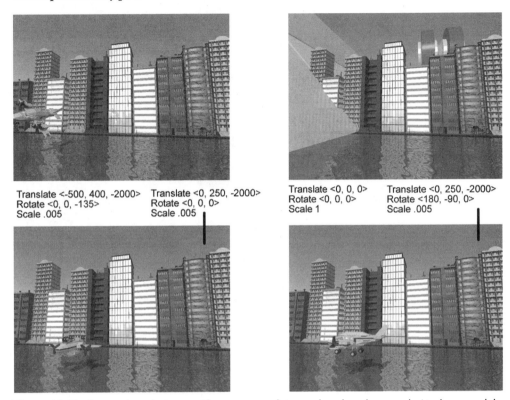

Translate <-500, 400, -2000>
Rotate <0, 0, -135>
Scale .005

Translate <0, 250, -2000>
Rotate <0, 0, 0>
Scale .005

Translate <0, 0, 0>
Rotate <0, 0, 0>
Scale 1

Translate <0, 250, -2000>
Rotate <180, -90, 0>
Scale .005

Figure 16-23: Our trial renderings, with rotate, translate, and scale values used. Airplane model courtesy of Kevin Dixon.

NOTE *When you are test rendering an image, it is best to change only one value at a time. This way, you isolate each setting and get a better feel for exactly what you are changing. Remember to turn down the quality, turn off anti-aliasing, and consider a smaller image size for test renders.*

Exercise 16.4: Creating a Space Backdrop

If you're a space fan, you will enjoy these next two exercises. In this exercise, we will walk you through making a backdrop for the space scene in the next exercise. We're going to use elements of Chris Colefax's Galaxy Creator. For a peek at what it will look like, see Figure 16-24 on the following page. The file **C:\LDraw\VLEGO\CH16\galaxy01.bmp** shows the scene in color.

Figure 16-24: A galaxy created using Chris Colefax's Galaxy Creator.

1. In order to use the Galaxy Creator files, copy all of the contents of the CD-ROM directory **\POV Add-Ons\Galaxy Creator** to your hard drive under **C:\LDraw\Programs\POV-Ray\include**.

2. Open the file **C:\LDraw\VLEGO\CH16\galaxy01.pov**. This time, we have defined most of the scene for you, minus a couple of star clusters. Render the file if you would like a quick look. Let's add the star clusters now.

3. In galaxy01.pov, insert the following code before the camera statement.

```
// LARGE STAR (Green Star)
  #declare galaxy_object_name = "Star1"
  #declare galaxy_colour1 = <1, -.2, 0>;
  #declare galaxy_object_scale = .75;
  #declare galaxy_object_position = <-10,-55,0>;
  #include "GALAXY.OBJ"
```

4. Save the file and render. Notice a new star cluster appears in the upper right-hand corner. We recommend you play with the galaxy_object_position values (one at a time) to get a feel for how those numbers effect the object's position.

5. Now let's place the second star cluster. Insert the code below. Once you have done this, save your file as **galaxy02.pov**, and re-render it.

```
// Large star (Red Star)
  #declare galaxy_object_name = "Star4"
  #declare galaxy_colour1 = <-.5,2,.75>;
  #declare galaxy_object_scale = 1.5;
  #declare galaxy_object_position = <35,-50,0>;
  #include "GALAXY.OBJ"
```

Make sure you have saved the file. You have just finished creating the galaxy backdrop. If you like, take the time to get used to the values for galaxy_object_scale and galaxy_object_position. Play with them one at a time, and do small test renders to see the results. When you are creating space backdrops on your own, be sure to refer to Chris Colefax's Galaxy Creator help file, which you moved to **C:\LDraw\Programs\POV-Ray\include\Galaxy.htm**.

Exercise 16.5: Creating a Space Scene

In this exercise, we will add the planet and the starfighter file to get the image shown in Figure 16-25.

Figure 16-25: A space scene, featuring Daniel Jassim's starfighter.

Let's get started creating the scene.

1. Open the file **C:\LDraw\VLEGO\CH16\Space.pov**. Select and copy lines 6 through 49. This is the code to create the planet.

2. Paste the copied code from Step 1 into **galaxy02.pov**, immediately before the camera code.

3. Save the file as **space_scene.pov**.

4. In L3PAO, convert the file **C:\LDraw\VLEGO\CH16\starfighter.mpd** using the default settings and Quality Level 2. Save the file as **C:\LDraw\MODELS\starfighter.pov**. *Do not render the file upon completion; simply run L3P.*

5. Open the POV file you just created. Select and copy lines 5 through 5718.

NOTE *This is a lot of code to select! For an easier method than simply using the mouse, place your cursor at the beginning of line 5. Hold down the SHIFT key, and while doing that press **Page Down** repeatedly until you are in the range of line 5718. With the SHIFT key still depressed, use the arrow keys to navigate single lines until your cursor is at line 5718.*

6. Paste the copied code from Step 5 into space_scene.pov, at the very top of the file.

7. Save and render the image.

You will see a small starfighter and a dim planet in your scene. To adjust the starfighter, use the keywords translate, rotate and scale inside of the object code for starfighter_dot_mpd. This time, you are scaling up, so your scale number should be greater than 1. Adding a pointlight at <0, 0, 0> will help illuminate the planet. This time, find and enter the code yourself; there are examples of what you need throughout the rest of this chapter.

No cheating this time! We want you to exercise the skills you've learned and experiment with these values for yourself *before* you look at our version. When you've done this, take a look at **C:\LDraw\VLEGO\CH16\galaxy03.pov** and compare it to your own scene.

Congratulations!

Exercise 16.6: Using Fog

For fun, let's play with the POV-Ray fog effect. You can find examples of the fog keyword's syntax in the POV-Ray Help Manual (**Help > Help** on the POV-Ray Scene Language, Help Topics, "Fog"). Before getting started with this exercise, take some time to read about the fog keyword in the help file. Familiarize yourself with the topics below.

Read in the Fog Topic in the POV-Ray Scene Language Help File:

- A Constant Fog
- Setting the Minimum Translucency
- Creating a Filtering Fog
- Adding Some Turbulence to the Fog
- Using Ground Fog
- Using Multiple Layers of Fog

Now that you are familiar with the above topics, let's have some fun. We will walk you through adding fog to the model **C:\LDraw\VLEGO\CH16\fog_of_war.mpd** to create the image seen on the right in Figure 16-26. If you like, you can have a look at our color image in the Chapter 16 folder.

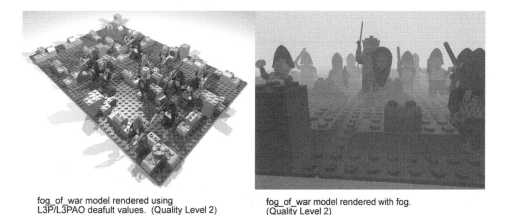

fog_of_war model rendered using L3P/L3PAO deafult values. (Quality Level 2)

fog_of_war model rendered with fog. (Quality Level 2)

Figure 16-26: Fog of War model with and without fog. Model by Ahui Herrera.

Follow these steps to insert the fog:

1. In L3PAO, convert the file **C:\LDraw\VLEGO\CH16\fog_of_war.mpd**, using the default settings. Save the file as **C:\LDraw\MODELS\fog_of_war.mpd**. When you render this image, it should look like the image on the left in Figure 16-26.

2. Adjust the camera location and look_at values to zoom in on the image.

    We used location <-60, -100, -240> and look_at <-90, -10, 210>.

3. Add the fog using the "Using Multiple Layers of Fog" code as your template. See our code below.

```
fog {
  distance 700 color rgbt <0.2, 0.5, 0.2, .08>
  fog_type 2 fog_offset -1 fog_alt 1
  turbulence 0.5turb_depth 0.2 }
 fog {
  distance 500 color rgbf <0.5, 0.1, 0.3, 0.4>
  fog_type 2 fog_offset 0 fog_alt 4
  turbulence 0.6 turb_depth 0.2 }
 fog {
  distance 900 color rgbt <0.6, 0.8, 0.6, 0.7>
  fog_type 2 fog_offset -.5 fog_alt -104
  turbulence 2 turb_depth 0.3 }
```

Render the image and compare it to our color image stored on your hard drive. Ours should have a bit more green in it than yours, and does not have shadows. To fix this, use the keyword shadowless on all your light sources, as described earlier in the "Special Lighting Conditions" section. Next, make two of the three lights in the file use the color green to achieve a greenish tint.

Render your scene. It should look similar to the image on the CD-ROM.

NOTE   *Save your fog_of_war.pov file. We will be using it again in the next chapter.*

Summary

In this chapter, you learned some of the key elements of the POV-Ray scene language. We taught you the features that are important for LDraw users desiring to take their renderings to the next level. Coming away from this chapter, you should be comfortable manipulating many aspects of the POV-Ray language.

Remember that you can control the camera, lighting, and certain elements of the model via L3P/L3PAO. When using POV-Ray itself, you only need to focus on the background and scene elements. You rarely will need to write POV-Ray code from scratch, since POV-Ray is distributed with excellent documentation and its online community has plenty of resources available for your use.

Have patience as you continue to learn how to use POV-Ray. In the end, you will be able to wow your friends with your ability to create these amazing rendered LEGO scenes.

17

MEGAPOV

MegaPOV is an unofficial "build" of POV-Ray that has been modified to include some new features. In early 2002, some members of the LDraw community discovered that this program could be used to create renderings with edge lines on virtual LEGO bricks, which are ideal for building instructions. We've already introduced you to MegaPOV's features during Chapter 15's discussion on LPub. In this chapter, we'll spend some time introducing you to the nuts and bolts of MegaPOV by discussing two features that differentiate it from the official build of POV-Ray.

MegaPOV contains all of the features POV-Ray 3.1 does, and then some. It allows you to use new types of light sources, objects, and more. We will focus on two post-processing techniques MegaPOV employs: find_edges and Focal Blur. As you already know from Chapter 15, find_edges generates edge lines similar to the ones you see in your real LEGO building instructions. Focal Blur simulates the focal properties of a physical camera lens — it allows you to create a depth-of-field, so some objects are in focus, and others are not.

If you are interested in learning about additional MegaPOV features, we encourage you to refer to its help manual. The help manual is composed of zipped HTML files, which are downloadable online at ftp://ftp.povray.org/pub/povray/Unofficial/MegaPOV/htmldocs07.zip. Since our book focuses on creating virtual LEGO models and building instructions, we won't be dealing with these features. If you are interested, you can have a look at examples of what MegaPOV can do on the official MegaPOV website, www.nathan.kopp.com/patched.htm.

In this chapter, we teach you MegaPOV version 0.7. MegaPOV 1.0 is available; however, just like POV-Ray 3.5, there are still some outstanding compatibility issues when it comes to LDraw-based tools such as L3P and LPub.

Rendering in MegaPOV

Since MegaPOV is essentially a souped-up version of POV-Ray, it uses the same file extensions the official version does: .pov and .ini. There is no fundamental difference between MegaPOV and POV-Ray, save MegaPOV's additional features.

Editing the Version Statement

When you are using MegaPOV, an unofficial version of POV-Ray, you need to use a #version statement at the top of your POV file. This is needed to deal with compatibility issues between different versions of POV-Ray. Usually, the line looks like this:

```
#version 3.1; // version number may be different
```

But, when you are creating files to be rendered in MegaPOV, you need to manually change this statement. The proper statement for a MegaPOV file is:

```
#version unofficial MegaPov 0.7; // version number may be different
```

Note that the "// version number may be different" part is a comment, so it has no effect on the rest of the line.

NOTE *L3P does not include a #version statement in converted files, since it only expects the files to be rendered by official releases of POV-Ray.*

find_edges

MegaPOV's find_edges feature is used to accentuate the edge lines on parts (or their vertices). It applies the edge lines in post-processing, which simply means that after your image has been completely rendered, MegaPOV adds in the edge lines. You can see an example of this in Figure 17-1.

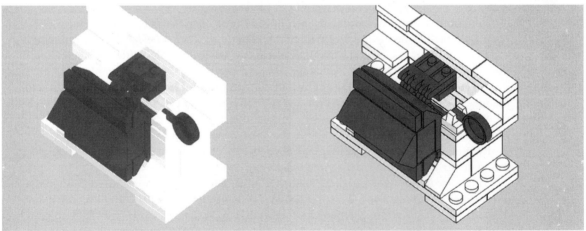

File: pizzaoven.bmp
Processed pizza oven image

File: pizzaovenPP.bmp
Post-Processed pizza oven image
using find_edges

Figure 17-1: Left: POV-Ray rendering of Marisela Herrera's Pizza Oven. Right: MegaPOV rendering using find_edges.

Go ahead and open MegaPOV. You will quickly notice that MegaPOV looks exactly the same as POV-Ray! The only difference between MegaPOV and POV-Ray is that MegaPOV has been patched to allow for new features. Open the file **C:\LDraw\VLEGO\CH17\ pizzaoven.pov**. Now, render the file at a low resolution such as **640×480, AA**. Pay special attention to the process.

Initially, the image will render exactly how you would expect it to in POV-Ray. However, when it is finished rendering, it applies the find_edges effect. When MegaPOV saves the rendered image, it actually saves two files: pizzaoven.bmp and pizzaovenPP.bmp. The pizzaovenPP.bmp file is the post-processed image with the edge lines applied. Pizza-oven.bmp is the initial rendering, with no post-processing applied.

Using find_edges

Take a look at the find_edges code below. This code needs to be encapsulated inside the global_settings code, a block of settings that are applied to the whole scene. In the pizzaoven.pov file, it starts on line 2118. It should look like this:

```
global_settings {
  post_process {
      find_edges {
          2,     //depth difference required for line
          0.35, //angle difference required for line
          0.2,  //color difference required for line
          1.2,  // 2.0 default line width
          20,   // 1.4 default line sharpness
          rgb <0,0,0> //color of line
      }
  }
}
```

MegaPOV uses three factors to determine where to draw edge lines. It calculates the difference in depth between objects, the difference in the angles on objects, and the difference in color (shading) between objects as rendered. The above code is a sample that we have found particularly effective in creating clean lines. However, depending on your model and especially on the resolution you are rendering at, you may need to tweak these settings.

We encourage you to experiment with the settings above to better learn exactly how they work. Change one number at a time to see what each setting does. If you don't play with one number at a time, you won't be able to totally isolate the exact effect each factor has.

Recall our introduction to MegaPOV's edge lines feature in Chapter 15. The line thickness is independent of the image size/resolution you render at. So, if you render a small image, the edge lines will be one size. If you render a much larger image but do not change the line thickness, the lines will appear thinner *relative to* the parts in the rendering. Actually, the lines don't change size at all — they remain a constant thickness (in pixels). However, since when you increase the image size, you also increase the size of the rendered model, the edge line thickness appears to change relative to the model.

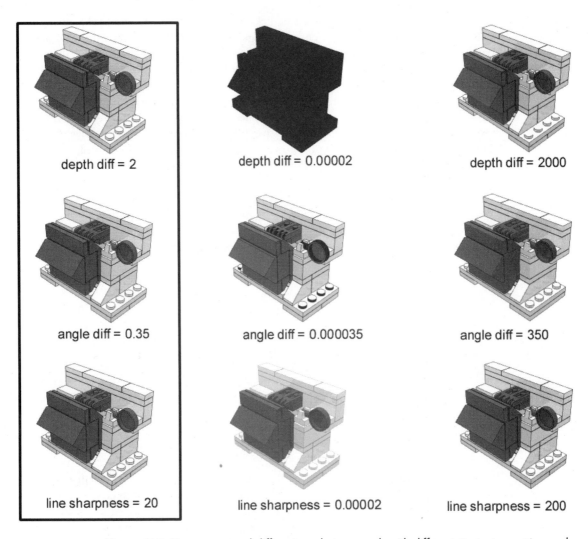

Figure 17-2: Here are several different renderings, each with different `find_edges` settings values. Model by Marisela Herrera.

Turning Off Reflections Manually

In Chapter 15, we discussed LPub's ability to turn off reflections in POV-Ray. This is an important effect to consider when rendering using `find_edges`. If you turn off reflections, you don't get mirrored images of parts reflected on other parts, resulting in a much smoother-looking image.

To turn off reflections manually, you need to locate the color definitions for the file. These appear below the #declare statements in L3P-generated files. A color definition statement looks like this:

```
#ifndef (Color7)
#declare Color7 = #if (version >= 3.1) material { #end texture {
    pigment { rgb <0.682353,0.682353,0.682353> }
    finish { ambient AMB diffuse DIF }
    #if (QUAL > 1)
        finish { phong 0.5 phong_size 40 reflection 0.08 }
        #if (BUMPS) normal { BUMPNORMAL } #end
```

```
        #end
} #if (version >= 3.1) } #end
#end
```

You can see the reflection setting on the sixth line, towards the left. To turn off reflections, simply change this value to 0. Doing this will turn the reflections off for that particular color. Repeat this through all similar blocks of code (they will look identical, minus the color number and RGB values) to turn reflections off in the entire model.

NOTE *Sometimes there are additional color definitions elsewhere in the file. To ensure that you have found them all, do a find (**Search > Find** or **CTRL+F**) for the string "reflection," and ensure all reflection settings are set to 0.*

Using Shadowless Light

If you turn off reflections, you will also want to use shadowless lights. Recall from the previous chapter how to do this. Simply add a "shadowless" statement to the end of each block of light source code, like this:

```
light_source {
    <10,-105.899,-149.899>  // Latitude,Longitude,Radius: 45,0,183.706
    color rgb <1,1,1>
    shadowless
}
```

Using Orthographic Camera

Another setting we mentioned in Chapter 15 was the use of orthographic camera. L3P does not handle orthographic camera when it converts files (LPub handles this on its own). Orthographic camera eliminates perspective from the image, similar to the way official LEGO building instructions are illustrated. You may want to use this feature in MegaPOV when you are hand-editing your own files.

To turn off orthographic camera, remove the comment marks in front of the //orthographic statement at the bottom of the camera code.

Compensate for the Camera Position

When you turn on orthographic camera in an L3P-converted POV file, you may find your model does not fit entirely inside the render window (and therefore, parts outside of the window are not rendered). To correct for this, we have found that you can typically adjust the camera's location and look_at points about -50 on the Y-axis. To maintain the same isometric view, you must change both of these values. We will walk you through that in Exercise 17.1.

When you are using LPub and select orthographic camera, you do not experience this problem. LPub contains a built-in workaround for this glitch.

focal_blur

MegaPOV's focal_blur feature blurs all or part of an image, just like you would see part of a real photograph in focus and another part out of focus. It creates a depth of field in your image, and blurs all that is rendered outside that depth. The depth of field in numbers is the distance between the closest point to and the furthest point away from the camera where the image is in focus. If any part of the subject extends outside of the depth of field, that is, in front of it or behind it, it will appear blurred. Figure 17-3 provides an illustration.

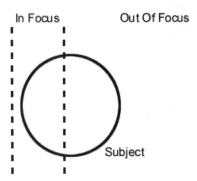

Camera

Figure 17-3: Depth of field illustrated. Inside the dotted lines, the image appears in focus; outside of it, the image appears blurred, or out of focus.

In photography, the depth of field is associated with the aperture of the lens. Aperture is the number given to the size of the opening of a lens, which regulates the amount of light allowed through to the film or CCD. A simple rule to remember: the smaller the aperture, the larger the depth of field.

In MegaPOV, there is no aperture setting for the lens of the camera. Instead, we define the size of the depth of field in numbers that correspond with distance. The code used for `focal_blur` is also inside of the `global_settings`, where `find_edges` also belongs:

```
global_settings {
post_process {
focal_blur {
                20, //field_start
25, //field_depth
6, // max_blur_radius
0 // keep_aa
}
}
}
```

The focal blur settings define the various elements that affect the blurring process. You can define where (in the 3D coordinate system) the field (area in focus) starts via `field_start`. `field_depth` indicates how deep the field is in LDraw units (LDU). The `blur_radius` affects how much blurring will occur outside of the field. Finally, `keep_aa` tells MegaPOV how much to blur anti-aliased pixels (1 will blur them 100 percent, 0 will blur them 0 percent).

NOTE *Nathan Kopp, MegaPOV's author, says that if your background is mostly a constant color, a* keep_aa *value of 1 works well. However, if your image contains a lot of color variety and sharp color changes in the background, you should use a lower* keep_aa *value.*

You can see what the focal blur looks like compared to an un-blurred rendering in Figure 17-4.

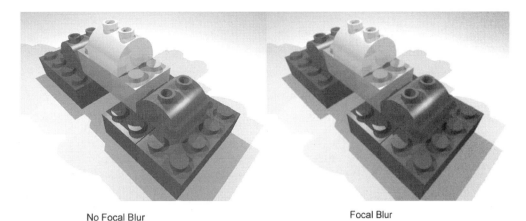

No Focal Blur   Focal Blur

Figure 17-4: Focal blur versus no focal blur on a model.

Exercises

The following two exercises have been designed to give you a feel for using find_edges and focal_blur.

Exercise 17.1: Rendering A Model Using find_edges

Here we'll walk you through the process of applying find_edges code to a model.

1. Open L3PAO. You need to convert the file **C:\LDraw\VLEGO\CH17\spa.ldr**. Set the background color to white and use quality level 2; the LEGO markings on the studs look bad using the find_edges filter. Be sure to set the directory you want L3PAO to save your file in (we recommend **C:\LDraw\MODELS**). Ensure that **Render Upon Completion** is *not* checked. Once this is all configured, Run L3P to convert the file.

2. Open the newly created file **spa.pov** in MegaPOV.

3. Be sure to change the #version statement at the top of the file to:

```
#version unofficial MegaPov 0.7;
```

4. Turn off reflections manually in the code as we discussed earlier. Set all of the reflection values in the blocks of color definitions to 0. If you need to, refer to the earlier section **Turning Off Reflections Manually**.

5. Set the lights to "shadowless" by adding the word shadowless inside of each block of light source code, before the closing brackets, as illustrated earlier in the "Using Shadowless Light" section.

6. Turn on orthographic camera by un-commenting the //orthographic statement at the bottom of the camera code. To compensate for the camera position, add **-50** to the Y-value of both the camera location and look_at values.

7. At the bottom of the file, insert this find_edges code:

```
global_settings {
  post_process {
      find_edges {
         2,    //depth difference required for line
         0.35, //angle difference required for line
         0.2,  //color difference required for line
         2,    // 2.0 default line width
```

```
        20,  // 1.4 default line sharpness
        rgb 0 //color of line
      }
   }
}
```

8. Save the file and render.
9. To familiarize yourself with the find edges code, spend some time tweaking individual values, or rendering the file at different image sizes. We optimized the sample `find_edges` code for a 1024×768 image. Try other image sizes and manipulate the `find_edges` settings to find the optimum numbers for image size you choose.

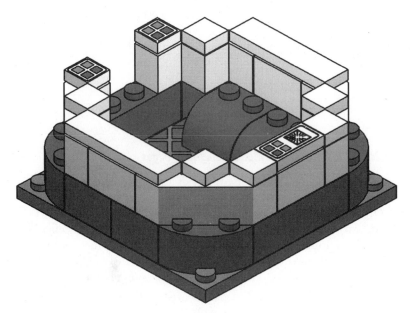

Figure 17-5: Ahui Herrera's Spa, rendered in MegaPOV with `find_edges` applied.

Exercise 17.2: Using Focal Blur

Now let's add the focal blur to the Fog of War file you created in Exercise 16.6.

1. Open the **fog_of_war.pov** file in MegaPOV.
2. Add the code **#version unofficial MegaPov 0.7;** to the top of the file.
3. Add the following code to the very bottom of the POV file:

```
global_settings {
post_process {
focal_blur {
                10, //field_start
35, //field_depth
6, // max_blur_radius
0 // keep_aa
}
}
}
```

4. Save and render the model.

5. After you have rendered the file, take some time and tweak the settings to play with the focal blur feature, noting what each setting means by referring to the earlier section "focal_blur."

See our *Fog of War* image with focal blur in Figure 17-6.

Figure 17-6: *Fog of War* with focal blur. Scene by Ahui Herrera.

Summary

In this chapter, you learned how to hand-edit two of MegaPOV's features. To learn more about using MegaPOV, its help manual (referenced in the introduction to this chapter) can be a valuable resource. The LDraw community only uses a few of MegaPOV's features. Discovering how to use this tool fully is a matter of having patience and experimenting with the code one line at a time.

18

POST-PROCESSING YOUR BUILDING INSTRUCTIONS

After learning how to generate building instruction images automatically using LPub, and gaining a better understanding of what lies behind POV-Ray and MegaPOV, you should be ready to post-process your building instruction images. For those of you not familiar with the term *post-processing*, it means simply to process something after it's been through primary production, or in our case, the act of arranging our building instruction images into a page layout and numbering them.

For this chapter, we rendered building instructions of the Mobile Crane, a minifig scale model by author Tim Courtney (see Figure 18-1 on the following page). Its design is in the spirit of Jennifer Clark's DEMAG AC-50-1, a part of which we showed you as an example in Chapter 1. We have included the raw print-resolution bitmap images for this model on the CD-ROM under **CH18****crane-raw-printres**\. (You did not copy these images to the hard drive when you used the installation application in Chapter 2.) Now you'll have an opportunity to see and experiment with the images on your own.

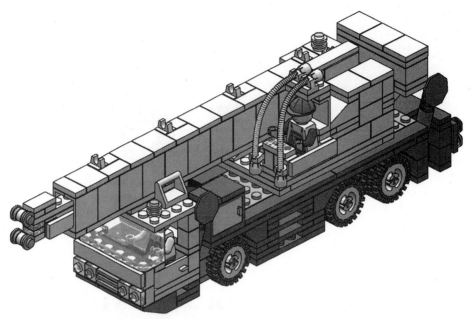

Figure 18-1: We created building instructions for Tim Courtney's Mobile Crane. (We discuss layouts in this chapter, and provide the full instructions in Chapter 20.)

In this chapter, we will describe some of the techniques we used to post-process the raw LPub images. You can see the final building instructions in Chapter 20, "Building Instructions: Mobile Crane." Remember, the focus of this book is creating virtual LEGO models, so we are not going to go in depth on other software tutorials and teach you everything you need to know to post-process instructions. After teaching you the objectives and processes for post-processing images, we will show you three examples of completed building instructions; one example is an official LEGO set, the other two are fan-created. You can use these examples as references in addition to the instructions that we created for Chapter 20.

Objectives of Post-Processing

By post-processing the raw instruction images LPub generates, you can create well laid out sets of building instructions to reach your target audience (the person you want to build your model) effectively. Remember that LPub does not generate refined, well laid out pages; it only creates the source images used to create these pages. It is up to you, the creator of building instructions, to work in an image editor or a page layout program (or both) to arrange these raw images in a logical, easy to follow, and attractive manner. You'll need to make the finishing touches to the instructions manually.

Create a Well Laid Out, Logical Set of Building Instructions

Remember that the goal of building instructions is to describe clearly to a builder how to construct a model. Consider your audience not only when you optimize a model for instructions (as we taught you in Chapter 12), but also when you lay out the final pages. Laying out raw images and pages in an orderly, clean fashion helps the reader as they move from step to step, and as they look at the parts list images to determine which pieces to gather for the current step. See Figure 18-2 for an example of the instruction layout we created for the book.

NOTE *We decided to omit parts list images in our example, since the model was simple enough not to need them. Adding the images would have complicated the layout process. Parts list images are more appropriate for complicated Technic models than for basic constructions.*

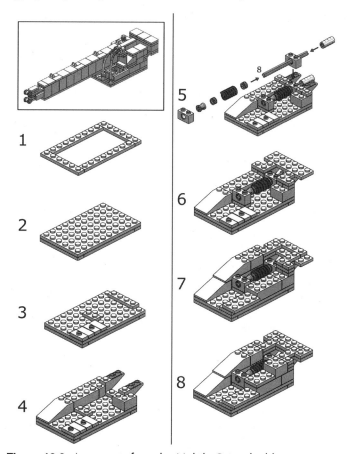

Figure 18-2: An excerpt from the Mobile Crane building instructions in Chapter 20.

Reach Your Target Audience Effectively

Finally, in everything you do with building instructions, you need to consider your target audience. All of your time and hard work assembling instructions is worthless if your audience doesn't understand how to build the model. Know your audience and tailor your instructions to them and their understanding.

Even though LEGO is primarily a kids' toy, there are many people who could use your instructions to build one of your creations. They could be your friends, students, teachers, kids, or adult fans of LEGO (AFOLs). Each type of person has a different level of understanding of LEGO bricks. The same model may be easy to build for one person, but difficult for someone else. Keep this in mind throughout the process of designing the model, documenting it, optimizing for instructions, and laying out rendered images in a page layout.

At BricksWest 2003 (www.brickswest.com), right next door to LEGOLAND California, Steve Barile gave a presentation entitled "Beyond LDraw: The Making of a Train Instruction Book." He discussed with the audience how he created his Freight Train Instruction Book from start to finish. Steve has allowed us to include his presentation on this book's CD-ROM. You will find an Adobe Acrobat file under **C:\LDraw\VLEGO\CH18\ Making_Train_Instruction_Book_SBarile_BW03.pdf** (also available in Microsoft PowerPoint format in the same directory).

Processes

Now we will describe some of the processes that you can use to create building- instruction page layouts. We won't go into detail here because the instructions involve techniques and knowledge outside the scope of this book. We will attempt, however, to point you to online resources where you can learn non-LDraw techniques such as image editing.

Selecting a Software Program

Before you get to work on your instructions, you should select a program to use to post-process your images. Post-processing your instruction images requires non-LDraw-based software. Some people use image editors such as Gimp, Paint Shop Pro, or Adobe Photoshop, while others have made use of Microsoft Word and Microsoft PowerPoint. Perhaps the ideal solution is using professional vector-based page layout programs such as Adobe Illustrator or Quark XPress; however, these programs are very expensive.

Image Editors

These image editing programs edit bitmap-style images. They are the same programs used to edit images for the Web. Learning how to use them properly can aid you in creating finished building instructions.

NOTE *When we say "bitmap-style" image, we don't only mean images with the extension BMP. A BMP file is a bitmap-style image, but not all bitmap-style images are BMP files. These bitmap images simply mean images comprised of individual pixels. Conversely, a vector image is an image file that is defined by mathematical lines and curves, so it is infinitely scalable. Bitmap images are also defined by their "resolution," a term you should be familiar with by now from reading this book. Bitmap images lose quality as you expand them beyond the dimensions of their pixels, so they are not infinitely scalable.*

The Gimp

The Gimp is a freeware image editing program, fostered by the open-source software community. Information on the Gimp can be found at http://www.gimp.org. To learn how to use the Gimp, check out http://manual.gimp.org.

Paint Shop Pro — http://www.paintshoppro.com

Paint Shop Pro is an affordably priced image editing program, which can produce professional results. You can purchase Paint Shop Pro at http://www.paintshoppro.com. To learn how to use Paint Shop Pro, check out the Learning Center on the Paint Shop Pro Users Group website at http://www.pspug.org

Adobe Photoshop

Adobe Photoshop is an industry standard professional image editor. This program has a hefty price tag, but is very powerful. You can purchase Photoshop via http://www.adobe.com/products/photoshop/. To learn how to use Photoshop we recommend searching the Internet with the terms "Photoshop tutorials." There are many excellent tutorials readily available for you to use.

Microsoft PowerPoint and Word

Some people who create building instructions use Microsoft PowerPoint or Microsoft Word to lay out their instructions. These two programs are included in the Microsoft Office suite. Their advantage is that the additional elements — text, shapes, and so on — are vector instead of bitmap. This means you can scale them. We will talk more about PowerPoint and Word when we show you the instruction examples at the end of the chapter.

Professional Page Layout Software

If you want to take this even further, beyond even the vector capabilities of Microsoft PowerPoint or Word, you can look at professional vector graphics or page layout programs such as Adobe Illustrator, Adobe InDesign or Quark XPress. Be aware though: These are professional tools with professional price tags and professional interfaces. The learning curve is high for novice users; however, for those of you who have the desire to learn and master these tools, you can produce incredible results in your LDrawn building instruction layouts as well as in other types of page layouts (and publishing in general).

Making Your Selection

Whatever your choice of page layout options, choose something that you're comfortable using. If you want a challenge, learn something that will grow your skills and give you better results in the end. We (the authors) use popular image editing software because that's what we are used to. Note that for the inexperienced, however, they aren't always the most efficient choice.

Preparing Your Images for Layout

Before you can start arranging your images into page layouts, you should prepare the renderings you have created. First, make certain that your renderings have turned out the way you want them. Next, find and organize your rendered images and prepare to lay out pages.

Debugging Instruction Images

LPub offers great automation, but sometimes you will realize that not everything rendered correctly. Before you lay out pages, you will want to correct any issues you have with the renderings. Take a look at every image LPub created for your model. The easiest way to do this is through a "1." Recall that in Chapter 15, we taught you how LPub can create two different types of basic web pages automatically. The web page table (see Figure 18-3) lays out clickable thumbnails of each image LPub has created for the current model, minus the Bill of Materials. To create this web page table via LPub, select **Generate > Web Page Table** from the menu.

A few things you can check for in the images:

MegaPOV Edge Lines: See to it that you are satisfied with how MegaPOV handles finding the edges, if you are using that feature. Tweak the MegaPOV settings in LPub to reflect this. You can also play with the Seam Width, since this affects how well MegaPOV can find edge lines. Adjust Seam Width settings via the **L3P > Model** folder in LPub. For our final rendering of the crane, we used a seam width of **1.0**.

Stray Parts and Arrows: If your model relies on Buffer Exchanges and sub-models, you are bound to run across a stray part or arrow in your renderings occasionally. This could be due to an improper use of Buffer Exchanges (see Chapter 12), or it could be because you forgot to ghost the parts added in a sub-model. Double-check these things if you encounter stray parts or arrows.

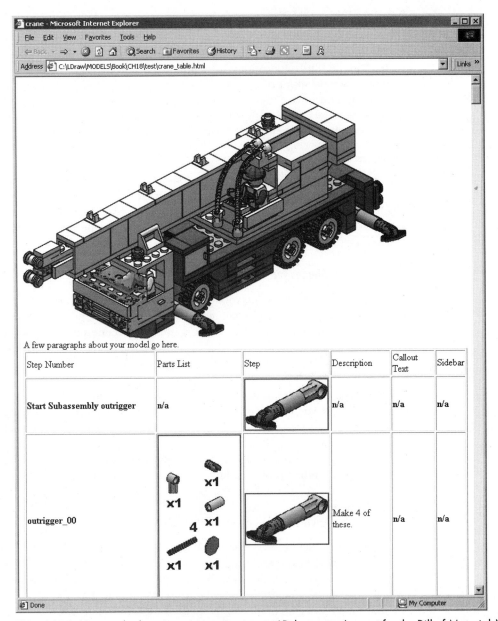

Step Number	Parts List	Step	Description	Callout Text	Sidebar
Start Subassembly outrigger	n/a		n/a	n/a	n/a
outrigger_00			Make 4 of these.	n/a	n/a

Figure 18-3: You can look at every instruction image LPub creates (except for the Bill of Materials) by creating a "Web Page Table" via LPub's Generate menu.

Parts List Images: Again, if your model contains Buffer Exchanges, you need to pay close attention to your parts list images. As we discussed in Chapter 15, you need to comment out duplicate instances of parts that are used to create exploded views or callouts via the LPub meta-commands **PLIST BEGIN IGN** and **PLIST END**. These tell LPub to ignore certain parts when creating parts list images. To learn how to add these, see the "Avoiding Duplicate Parts in Parts List Images with Models that Use Buffer Exchange" section in Chapter 15. Another thing to look out for in Parts List Images is parts that are normally whole that appear broken into components. You can use the **PLIST BEGIN SUB** meta-command we taught you in Chapter 15 to solve parts list image issues that arise with these parts.

Rotation Steps: Since rotation steps are rather tricky, double-check your uses of them. Depending on how you use rotation steps, they could cause undesired images to be rendered. Discard these images in your layout preparation; just because LPub rendered them, it doesn't mean you need to use them.

NOTE   *Depending on how you use rotation steps, MLCad will also generate these extra steps if you preview the model in View Mode.*

These are common issues we found when rendering our instructions. Sometimes when creating files you can make mistakes, which lead to bugs like these in the final output. As you gain experience creating your own instruction images and layouts, you will run into similar issues. We rendered a good handful of test renders for the crane before putting out the final print resolution images, in our attempt to solve these bugs. As you learn, you can develop a checklist for your instruction creation process. This will help you spot key details you may have missed as well as problems that exist when you produce instructions of your models.

Rendering Your Final Images

Once you've debugged your LDraw file, render your final set of building instruction images in LPub. If your goal is to print these instructions out, render in Print Resolution (**LPub > Mode > Style > Print**). If your goal is to put these on the Web, render in screen resolution. You probably want to stick to a resolution of 800×600 for the Web, or 1024×768 at the most, considering your audience probably runs those low to medium screen resolutions.

Choosing an Image Format

LPub can output images in BMP and JPG, and it can tell POV-Ray to output in PNG and TGA (Targa — compressed or uncompressed). How do you know which image format to use for the Web? JPG and PNG are both common web image formats. We recommend PNG because it is lossless (you can select these options via **POV > Output > Format**). For print resolution, LPub gives you the choice of BMP or JPG. We recommend sticking with BMP for LPub output. After you assemble the images, save them as TIF or EPS files for best results when printing.

Finding Your Rendered Images

Your rendered images are stored in the same directory along with the LDraw file you used to generate them. As we mentioned at the beginning of the chapter, we included raw print resolution bitmaps of the crane instruction images on the CD-ROM under **C:\LDraw\ VLEGO\CH18\crane-raw-printres**. You can look at these for examples of output, or generate the crane images yourself in LPub by using the ready-to-render LDraw file under **C:\LDraw\VLEGO\CH18\crane-inst.mpd**. Our config file is stored there as well: **crane-screenres.lpb** as well as **crane-printres.lpb** (the MegaPOV settings for line thickness differ between the screen resolution and print resolution configurations).

Sequencing Your Building Instructions

If your LDraw model contains sub-models, sequencing your instruction images properly is critical for the builder to understand the model. In the case of the crane, sub-models were referenced four deep in one area, as shown in Figure 18-4. The main sub-model, crane.ldr, references superstructure.ldr, which references boom.ldr, which references boom_extension.ldr. Why are there so many sub-models? Each of those components can move independently of each other in the physical crane. According to the lessons we taught you in Chapter 8, they each need to be different sub-models.

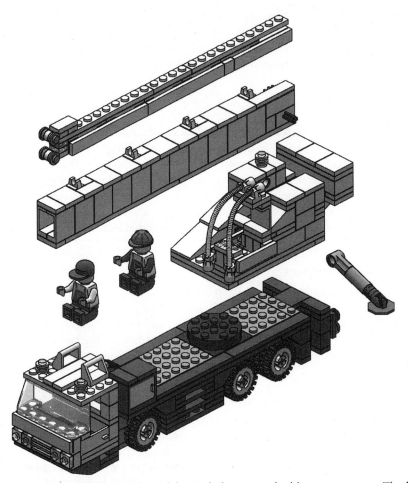

Figure 18-4: The crane sub-models needed to create building instructions. (The file contains some sub-models that should not appear in the instructions. They are not pictured here.)

Why is sequencing important? You need to ensure that you've had the builder construct a low-level sub-model before they construct the higher-level sub-model that references it. If you don't, and you go ahead and present a finished sub-model the builder doesn't know how to build, it will confuse them.

Linear vs. Hierarchical Sequencing

When sequencing your building instruction images, you can arrange the presentation of sub-models in different ways: *linear*, or one complete sub-model after another; or *hierarchical*, meaning sub-models are presented in the instructions right before you need to insert them into a main model. (This could be a loose interpretation of "hierarchical" in this case, but we think it's the word that fits best.)

In reality, what often happens is a combination of both of these. Many factors come into play: the size and complexity of a sub-model, the logical flow of building the model, etc. Some sub-models are small, such as the crane's outrigger. This sub-model only shows one exploded view to tell the builder how to assemble it (see Figure 18-5). We placed this sub-model in a callout immediately before the main model step that included it. In other cases, sub-models such as the superstructure warrant an entire block of pages dedicated to building that one particular model (as shown previously in Figure 18-4).

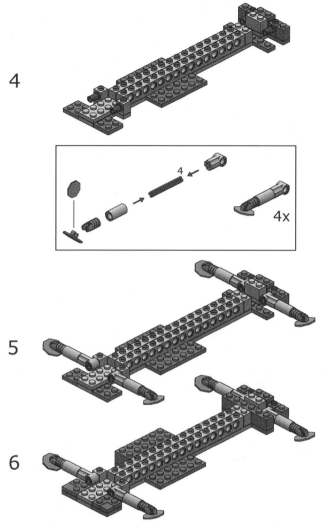

Figure 18-5: The outrigger is not presented as an entire separate sub-model like the superstructure is. Instead, it receives a callout right inside of the main model's instructions.

In the end, we arranged the instructions using a combination of linear and hierarchical sequencing. The instructions begin with an image of the finished crane and its minifigs. The builder is instructed to begin building the main model. The second page has the callout for the outrigger, and several times throughout the main model's instructions appear explosions or small callouts for sub-assemblies that are not separate sub-models. After the entire crane truck is constructed, the instructions begin for the superstructure. The superstructure contains two sub-models: the boom and the boom extension. These are both included as large offset panels (in the same style as a callout) *within* the instructions for the superstructure, *before* they are needed in the superstructure's construction. In that sense, the boom and boom extension are included in a linear fashion, whereas the superstructure is included hierarchically.

Arranging Your Images

Here are a few observations we have for arranging images into page layouts. We are not going to instruct you in detail on the exact techniques behind these procedures, since that goes away from the focus of the book and there are plenty of other materials specific to image editing programs that cover these topics. See the "Learn How to Use Needed Software" section later in the chapter for references to manuals and tutorials for the various programs we recommend.

Using Guides and Grids

We suggest using your image or page layout editor's guides feature. Guides are lines you can place on the canvas (the image or page you are editing), which are not seen when you view the file. You can use them to align the various components of the page you are trying to assemble very easily. To see an example of the guides we used for a page, see Figure 18-6.

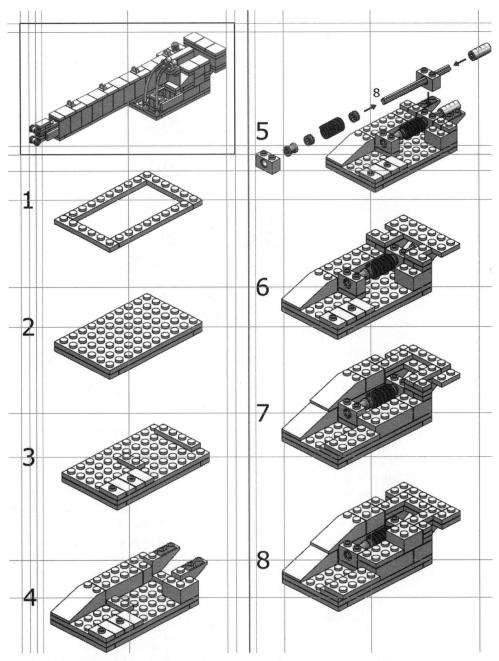

Figure 18-6: Guides help you align the various components you are assembling on one page.

Numbering and Adding Text to Your Images

When numbering or adding text to your images, you should be careful to make a note of the font and size you are using. This will ensure uniformity throughout your instructions. We chose a large font size for steps, a smaller one for sub-steps (steps inside of callouts), and an even smaller size to denote the length of Technic axles (see Step 5 in Figure 18-6). LEGO usually provides a 1:1 gauge in their instructions for measuring Technic axles and notes their length in studs in the parts list image.

Placing Callouts Effectively

Carefully consider where to place your callouts. If you think this through, the time spent in the planning will pay off with stunning building instructions. Consider marking off margins around the callout box's border so you ensure a consistent distance between the border and the step images inside and outside of the box.

Printing Your Building Instructions

If you plan to print your building instructions and arrange them into instruction books on your own, there are additional considerations above and beyond layout tasks. When printing, issues such as resolution, paper, and binding come into play. We will introduce these issues in brief to make you aware of them, but if you are serious about printing instructions, we encourage you to explore these issues fully on your own.

Resolution

Print resolution is measured in Dots Per Inch (equivalent to pixels per inch). The more pixels a printer prints per inch, the higher quality the image will be. Generally, you don't want to print an image smaller than 150 DPI, and 300 is considered ideal. At 300 DPI, the 2048×1536 image LPub generates at its Print Resolution setting would measure 6.8×5.12". This is not very large for building instructions. You can see how renderings need to be very large in order to print high-quality large format building instructions. Make sure you understand the resolution you want your printouts to be before you render your instruction images! If you do not, you may find you have wasted time and have to re-render larger images. Remember that you can always scale down an image (removing pixels) and it will look fine. You cannot scale the same image up (adding pixels that were never there in the first place). If you do, it will lose its crispness and even look blurry depending on how much you scale it.

Paper

Consider the type of paper you use for printing building instructions. Different weights and finishes can effect how your instructions look, how bright the colors are, and so forth. Go to a copy shop or office supply store and ask to see samples of different kinds of printer paper.

Binding

If you decide to print instructions, how will you bind them? One of the examples to follow in Steve Barile's Freight Train Instruction Book uses a spiral-binding system. The pages of Larry Pieniazek's Drop Center Flatcar are inside clear plastic page protectors that fit in three-hole binders. Then he binds them with a "report cover" like this one: http://www.staples.com/Catalog/Browse/Sku.asp?PageType=1&Sku=ess53306. This style of binding generally allows pages to lie flat when opened. You should look at various binding options at your local office supply store. (We've provided the preceding link as an example only; we are not intending to endorse a particular vendor or brand name.)

Examples of Finished Instructions

Here are some examples of finished building instructions for you to see. These three examples use the same elements, but with different styles. The two fan-created examples use different methods to create the layouts.

Official LEGO Building Instructions

The best source to look at for finished building instructions is the LEGO Company itself. They have been creating instructions and assembling pages for them for far longer than individual LEGO fans have. Building instructions for the relatively recent Sopwith Camel (#3451) are especially well done. In the instructions, you can see many callouts and exploded views. Callouts are essentially sub-steps in building instructions, showing a small assembly (even smaller than most functional sub-assemblies) being assembled and then inserted into the model as a whole. We introduced you to the concept of explosions in Chapter 12, showing parts in an "exploded view," very similar to auto mechanics' manuals that show how the parts of a car fit together.

Example: Sopwith Camel

Let's take a look at three pages from the official LEGO set #3451 Sopwith Camel. Because of the complexity of this model, it requires intricate building instructions to communicate to the builder how to assemble it. The first two pages are sequential, while the third one was taken from a different part of the instructions.

Take a look at the first example (Figure 18-7), and you will see several key elements. First, the upper left-hand corner displays graphics of what you are building. In recent years, LEGO has broken up models into components in their instructions. You build different components and then assemble the components together to achieve the final model. The first component you build in the Sopwith model is the engine/landing gear assembly. Another element is the Parts List images, very much like the ones created by LPub. LEGO does not add these on all models, but instructions for more complex models like the Sopwith usually have them. Within a step you see the main step diagram as well as the callouts, which are the boxes showing sub-steps or exploded views outside of the main step diagram.

Looking at the second example (Figure 18-8), you can see even more complex elements on the page. The callout for Step 6 is actually an arrangement of sub-steps, one even with its own callout. Step 6:1 (Step 6, sub-step 1) includes a guide for the two pulleys that are attached together by the half-pins. The lines on each of the pulleys act as an axis, telling you to align the axle hole in the middle of the pulley before inserting a pin. For Step 6:5, you see a guide to measure the location of a bushing on an axle. The side view of a 1×4 plate indicates that the bushing should be placed 4 studs in from the left side of the axle before putting the plane's radial engine block in place.

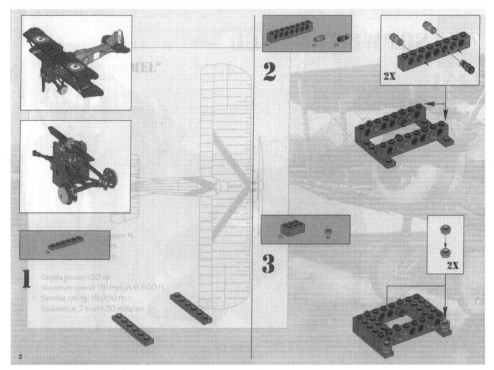

Figure 18-7: Page 2 (inside cover) of the #3451 Sopwith Camel's building instructions.

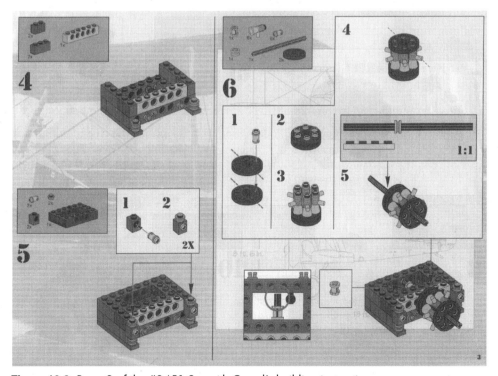

Figure 18-8: Page 3 of the #3451 Sopwith Camel's building instructions.

This third example (Figure 18-9) shows additional basic steps. Step 1 of the tail component (note the component graphic up in the top left corner) shows an explosion in the main step diagram. Step 2 shows an explosion inside of a callout; the sub-assembly is placed on the other parts in the main step dialog. What you should note about this callout (and others in the previous two examples) is there is a "2X" in the corner, telling the builder to build two copies. While you've probably come across these countless times before, you need to be conscious of them when you are creating your own instructions.

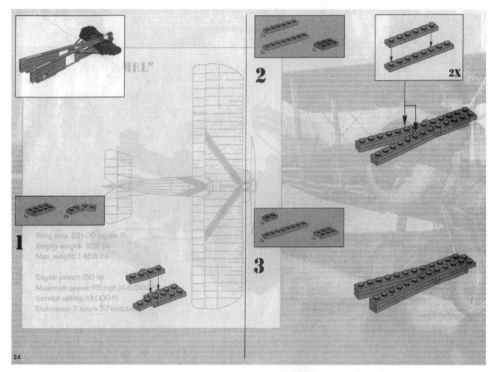

Figure 18-9: Page 24 of the #3451 Sopwith Camel's building instructions.

The Sopwith Camel instructions represent some of the best building instructions made. Their complexity is mind boggling from the perspective of LDraw users because of the implications it has for LDraw files to arrange, images to render, and pages to lay out. LEGO can do this because 1) they create building instructions commercially and 2) they have custom professional building instruction software to help them. We are using community-developed freeware tools; for the amount of money spent on them (none), they are capable of doing an incredible job of creating instructions.

Fan-Created Building Instructions

Now let's take a look at some totally fan-created building instructions. These examples are more realistically achievable than the official instructions, though slightly simpler.

Example: BricWorx™ 1001 Freight Train Instruction Book

Steve Barile, whom we discussed in Chapter 12, created his Freight Train Instruction Book to be sold through the custom kit company BricWorx (www.bricworx.com), which he operates with fellow builder Dwayne Towell. This particular example represents some top-quality fan-created building instructions, created using LDraw tools and MegaPOV for rendering the images. The Freight Train Instruction Book was created *before* LPub was written, meaning Steve had to use L3PAO's Stepclock and create several different LDraw files to render all the images needed for this book. With LPub, it's possible to consolidate everything you need to render down to one LDraw file, and LPub does the rest!

Take a look at the first example of the Freight Train Instruction Book (Figure 18-10). Steve put a lot of work into making this example of similar quality to official LEGO building instructions. One thing it lacks is Parts List images. There are probably two reasons for this. First, before LPub, creating Parts List images would have been incredibly time-consuming and laborious, no doubt deterring building instruction creators from taking the time to do so. Second, the models in the Freight Train are more simply constructed than the Sopwith Camel is. LEGO often includes Parts List images on Technic sets, or sets with Technic parts like the Sopwith, but LEGO doesn't usually include them on System sets that are designed to minifig scale. Why is this so? Technic models are far more complex than System models, and contain many small pieces. Steps in Technic sets often add many small pieces in each step, and you may find it difficult to read the parts that have been added to the Technic construction recently. Even with the absence of these parts list images, you still see the component graphic for the log car in the upper left corner, and a callout with sub-steps for Step 4.

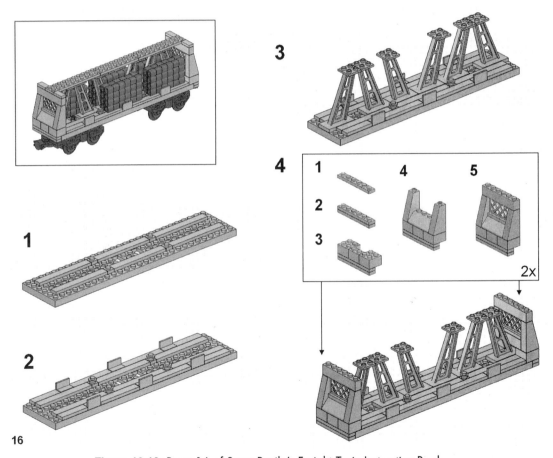

Figure 18-10: Page 16 of Steve Barile's Freight Train Instruction Book.

Our second example of Steve's instructions, the Hopper Car (Figure 18-11), features the same elements, but includes many more steps per page. Note how the model that is assembled is long, and therefore allows for placing subsequent steps in close quarters. By placing more steps on a page, printing the instruction book becomes more economical because you don't have to use as much paper. When you're doing a high-quality print job like Steve did, any way to keep costs down helps. By doing this, you are also conserving natural resources.

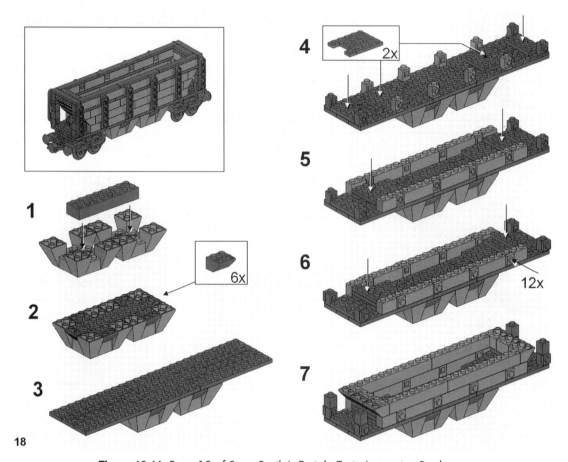

Figure 18-11: Page 18 of Steve Barile's Freight Train Instruction Book.

For the last example from the Freight Train Instruction Book (Figure 18-12), you can see an explosion in a callout. Note how these parts are placed. There doesn't have to be an arrow tying each one together; their connection is implied. Use your judgment when assembling exploded views and determine when and where arrows are needed to clarify what a callout points to, or where a callout's sub-assembly should be attached to the main model.

Steve created his building instruction layout using Microsoft PowerPoint, a part of the Microsoft Office suite. He uses it because he is familiar with it from work, and it handles all the requirements he needs. This shows you that it doesn't necessarily take a professional image editor or page layout program to create amazing instruction layouts; you can make them with commonly available software.

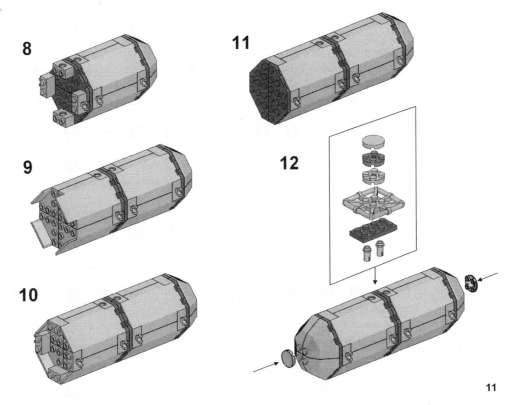

Figure 18-12: Page 11 of Steve Barile's Freight Train Instruction Book.

NOTE   *Check out Steve's production notes for the Freight Train Instruction Book on his site at http://www.bricworx.com/1001-how.php.*

Example: Milton Train Works™ 1004 Drop Center Flatcar

Larry Pieniazek also sells his own custom LEGO kits, through Milton Train Works (www.miltontrainworks.com). His instructions are another example of excellent layout, with a slightly different style, created using Microsoft Word. You can use the tables in Word to create callout boxes, instead of having to work in an expensive and complex image editor or page layout program. Larry doesn't use LPub or MegaPOV to render his instruction images; instead he uses Paul Gyugyi's LDLite. We mention LDLite briefly in Chapter 19, "LDraw and the Web: Viewing and Publishing Models Online."

NOTE   *Since LDLite doesn't support MLCad's Buffer Exchanges and Rotation Steps, we are not covering creating instructions with it in depth in this book. If you want to explore LDLite on your own, you are welcome to at http://www.gyugyi.com/l3g0/ldlite/ (that is a zero in l3g0) or participate in its open source project at http://sourceforge.net/projects/ldlite.*

Let's take a look at two pages from Larry's Drop Center Flatcar kit instructions. You can see the first page in Figure 18-13 on the following page. Here you see all the necessary callout boxes that have been created in Word.

The second page, seen in Figure 18-14 on page 326, shows a more complex page layout. Here he is instructing the builder to create the Bolster/Load Bar sub-assembly, and provides some guidelines in text beyond the images alone. Whether or not you want to include text in your instructions is a question of your audience. Milton Train Works sells kits online, from an English-only website. You can assume that most if not all of the MTW customers understand English. LEGO does not use text in their instructions, save for important guidelines for electrical elements. When they do, they publish in many different languages, so people all around the world can understand it. If your audience is English-speaking, use text at your own discretion. If you are unsure about which language your audience speaks, make sure your instruction images include everything that is needed for the builder to create the model successfully.

Another thing you will notice is the photographs at the bottom of this page of instructions. While some might argue that it dilutes the message of this book, we think it enhances it. Here Larry went as far as was practical using the available freeware tools, but when it proved difficult to instruct the builder how to attach the string, he turned to a tried-and-true method of visualization: photography. We think you can communicate 99 percent of what you need by using LDraw tools. In fact, we think 99 percent of the building instructions you create will never need photography as an aid, if you do them right. However, it is not taboo or out of line at all to resort to your trusty camera to solve a problem such as the one that is shown in Figure 18-14.

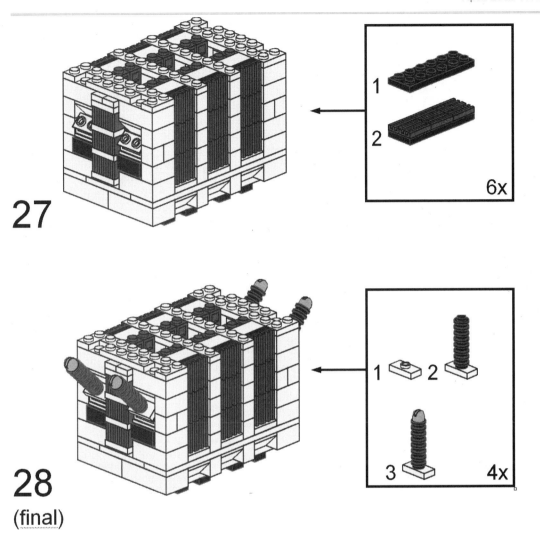

27

28
(final)

Figure 18-13: Page 8 of Larry Pieniazek's Drop Center Flatcar.

Subassembly: Bolster/Load Bar

*(2 required)*This is the truck bolster (where the truck kingpin on the bogie plate plugs in) part of the frame of the flatcar. It also has an adjustable rigging winch built in. Some steps use photos for additional clarity to get the string threaded right, as this is the trickiest part of the entire assembly process.

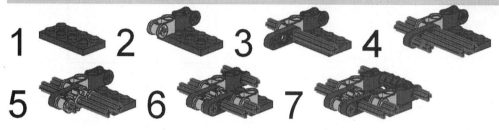

1   2   3   4

5   6   7

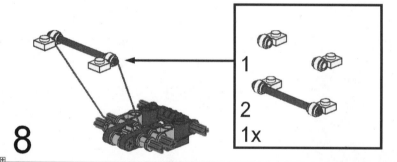

8

1

2

1x

Step 8 notes

Thread the string through one bushing as shown at left, with the string entering the slit and exiting the axle hole. Then thread it through the load bar's blue tubing. Then thread it through the other bushing, again exiting out the axle hole.

Now carefully fit one bushing onto the axle with the gear. Have the string end face out. It will take some extra effort but it wedges securely. You may want to melt the string end with a small flame before you start so you don't get the fraying I did in this picture.

Finally, fit the other bushing onto the other end of the axle, again with the string end facing out. I like to line all the slits up the same way but that's not required. If you did it right you should now have an assembly that looks like step 8, above.

Figure 18-14: Page 8 of Larry Pieniazek's Drop Center Flatcar.

NOTE   *For more information on the Drop Center Flatcar, see its page on the Milton Train Works website at http://www.miltontrainworks.com/item_info_1004.html#MTW-1004-by.*

Summary

In this chapter, you learned about the techniques of laying out building instruction images. We discussed the various elements of these instructions in order to familiarize you with them for when you create your own instructions. Next, we discussed the various processes for laying out instructions and provided examples in order to inform and inspire you to create your own. Finally, we showed you examples of complete instructions, both from the LEGO Company and from individual LEGO fans.

19

LDRAW AND THE WEB: VIEWING AND PUBLISHING MODELS ONLINE

The online LDraw community has been a central theme throughout this book. As we said early on, LDraw grew out of the Internet. Whether or not you are already familiar with the online LDraw community, this chapter introduces you to some ways to view other models on the Web, and how you can publish your own models for others to see.

Viewing LDraw Models Online

Some LDraw-based programs are devoted to viewing models, either directly from the Internet or from files saved on your hard drive. These viewers do not allow you to edit LDraw files, but specialize in displaying them, and give various options for how they do so. People place their LDraw files on their websites for others to download and see. It is convenient to view these models in a viewer, such as LDView, because it renders quickly and allows you to rotate the model with your mouse. The viewer LDLite does not let you rotate a model with your mouse, but allows you to draw the model step-by-step, pausing until you click a mouse button to continue. This way you can see how a model is constructed.

NOTE   *LDLite is similar to MLCad's View Mode; however, LDLite renders LDraw models more nicely than MLCad does. You may choose to view models step-by-step in MLCad out of convenience.*

Models Posted on LUGNET

LUGNET™, the LEGO Users Group Network, hosts a collection of discussion boards that allow users to post LDraw files directly to the LUGNET server. The newsgroup category lugnet.cad.dat.* (http://news.lugnet.com/cad/dat/) has several subgroups that allow you to upload LDraw files with your posts. You can use an LDraw viewer to see the models posted to these groups.

NOTE   *Anyone can post to LUGNET for free. You can learn about LUGNET newsgroups at http://news. lugnet.com/news/ and set yourself up to post at http://news.lugnet.com/news/post/setup/.*

Figure 19-1 shows a post to the newsgroup lugnet.cad.dat.models with an attached LDraw file. The top field in the message header, "Special," contains the link to the LDraw file. To date, the groups still reference the old LDraw .DAT extension, although LUGNET also allows you to post .LDR and .MPD files. If your Windows file associations are properly configured, you can view the file by clicking the [DAT] link (we discuss techniques for configuring viewers in the next section).

Figure 19-1: You can browse LDraw files embedded in LUGNET posts made to the lugnet.cad.dat.* newsgroups.

When you select this, perhaps your browser will prompt you and ask if you want to open the file from its current location or save it to disk. To use the viewer, select **Open file from current location**. Once you have done this, the default LDraw viewer should open the file, as shown in Figure 19-2.

NOTE *File Associations*
*Due to horribly inconsistent browsers, different browsers or even different versions of the same browser will handle this in different ways. To ensure it is successful, you should be familiar with Windows file types and how to configure them (**Tools > Folder Options > File Types** in Windows Explorer). The MIME-Type for LDraw files is* application/x-ldraw.

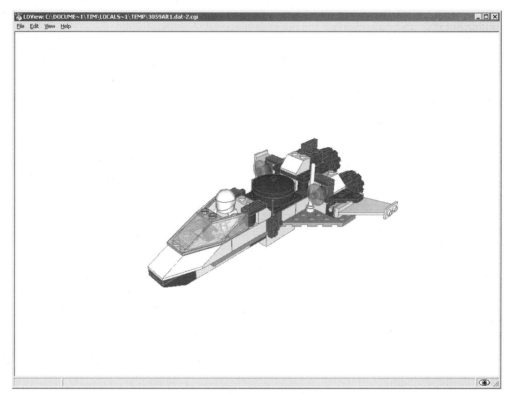

Figure 19-2: LDView is one of the LDraw viewers available. This image shows Steve Bliss' alternate model for LEGO set #3059 Mars Mission.

LDraw Viewers

Here we list three popular LDraw viewers and discuss each one. You can select one to be your default LDraw viewer by configuring your Windows file associations.

L3Lab

L3Lab, by Lars C. Hassing, is designed as a diagnostic program for parts authors. You an also use it to view LDraw files. It renders models very fast, has a simple interface, and can automatically register LDraw file types to itself. You can use your mouse to rotate a model. However, L3Lab does occasionally pop up error messages, and it does not allow you to save your settings (how you like a model rendered, and so forth).

LDLite

LDLite, by Paul Gyugyi, offers the best instruction-style rendering of the three viewers we are discussing. However, we do not recommend it as an everyday viewer because it does not have mouse interaction (you cannot spin your model around with the mouse). It has a command line interface, and you can read the files that are distributed with the program to learn how to use this.

LDView

LDView, by Travis Cobbs, can render models very nicely. It includes special shading and lighting features that give models a photo-realistic look. LDView does offer mouse rotation. However, LDView loads models slowly and may overload your system if you are rendering a large model.

Publishing Your Models Online

It is common to see LDraw files available for download alongside photos of peoples' LEGO models. These powerful tools enable tens of thousands of people, if not more, to share incredible amounts of LEGO building knowledge with others all over the globe. The small size and text-based nature of the LDraw file format makes all of this possible.

NOTE *The LDraw tools originally caught on partially because ,at the time, digital cameras were not readily available. People could use LDraw tools to document and share models with others across the Internet when taking and scanning photos was too expensive and/or time consuming. Now, using and publishing LDraw files allows the publisher to present detailed instructions, beautiful renderings, and 3D model files. When you download an LDraw file, you have direct access to all the details of the model.*

Finding a Host for Your Models

In order to display your models online, you need to find a place to host them. *Web hosting* is the term commonly used for placing files on a computer, called a *server,* which is accessible to web surfers via Hypertext Transfer Protocol (HTTP). This is the protocol used to transfer files over the Internet to your web browser. With little or no knowledge of how to make a web page, you can place your files on the Internet using some of the LEGO community's free services. If you already know (or are willing to learn) how to create your own web pages using HTML, you can choose a free web host (which pays for its expenses by serving your visitors advertisements), or you can pay for access to a web host if you do not want your site to feature ads. In addition, most Internet service providers (ISPs) offer a small amount of free web space to account holders; check with your provider to see if they offer this service.

We will now spend some time introducing you to some of the various available options.

LEGO Community Services

We like to highlight the amazing services and sites that exist within the online LEGO community, so we will discuss these resources first. Currently, there is only one actual host for files in the LEGO community. There are several websites that serve as directories of models and files (but don't actually allow you to place files on the server), that even allow you to create pages online. What these directories and page creation services do let you do is reference images and LDraw files hosted elsewhere. Here's a quick guide to the various LEGO community resources available to you.

BrickShelf™ — www.brickshelf.com

The BrickShelf™ Gallery allows you to host files in an online gallery, free of charge. You use a simple web interface to create an account (directly off the main page), and once your account has been activated you may upload files by logging in and using the links on the screen. Visitors to BrickShelf can browse the Gallery's recently added items, where all of the most recently uploaded folders appear, or they can browse the gallery by theme. A simple search can be used to find folders or images. The advantage of BrickShelf is that it allows for off-site linking to files. This means you can host your images on BrickShelf's gallery, while creating a page or entry on other community services like LUGNET's member pages, the Building Instructions Portal, or MOCpages.

Building Instructions Portal — www.bricksonthebrain.com/instructions/

The Building Instructions Portal is an online directory for building instructions of all formats. You can sign up for a free account and then add links to your building instructions. Instruction links are displayed by theme, and you have the option of specifying a preview image (which must be hosted elsewhere, along with your instruction files). The BI Portal accepts links to instructions in any format, either as images or as an LDraw file download. Upon clicking your link, site visitors will be directed to the place you host your building instructions so they can view them online or download the files.

MOCpages™ — www.mocpages.com

MOCpages™ is a free service that allows you to create web pages for your LEGO creations. MOC is an acronym that stands for "My Own Creation" (a model you built, as opposed to an official LEGO set). MOCpages account holders can create individual pages for their creations and link the pages to images stored elsewhere online (such as on BrickShelf). Visitors can rate and review creations posted. This is a basic page creation service. The pages it creates are all a certain style. They offer some limited customization, such as changing the theme (or *skin*) and allowing you to insert HTML formatting into your model description. It is targeted at people who either want a quick and easy way to display models, or those who are not skilled in website creation and design. As of this printing, MOCpages is a relatively new service. Many features are in the idea stage for the future, and this site promises to be a wonderful community resource as it grows!

LUGNET™ — www.lugnet.com

Even though LUGNET™, the LEGO Users Group Network, is the quintessential LEGO fan site, we are listing it last in terms of web hosts for a reason. The only actual hosting of your files it can provide you is storing your LDraw files in its CAD newsgroups, as we talked about earlier in our discussion on viewers.

Create Web Pages on LUGNET

If you are a paid LUGNET member, you can create your own pages on LUGNET using FTX, the site's own easy-to-use markup language. This provides you more customization options than MOCpages, but LUGNET does not list pages in any sort of searchable database or even a static index page. This means someone visiting your site needs to know of the URL or web address beforehand by clicking a link from one of your newsgroup posts or from others' websites. Still, for $10 (which goes to support LUGNET and keep it online), it's an excellent solution. You can find out about becoming a LUGNET member at http://members. lugnet.com. You can also learn how to create your own pages on LUGNET there.

Free Web Hosts

Outside of the LEGO community, you can find many free web hosting services. Before choosing a free host, we recommend you check with your ISP to see if they offer a hosting package with your account. This is preferable to a free host, which is likely to serve advertisements along with page views.

Free Online Services

The chief disadvantage of these services is the advertisements they serve over your pages. Often these ads are invasive and annoying pop-ups or pop-unders that launch a new window above or behind the current browser window. These ads are extremely annoying, and a surefire way to irritate your website guests — not to mention that most (if not all) of these ads also collect data on surfers to target advertising at them, an ethically questionable practice. If you do not mind subjecting potential visitors to this, or if you do not have the means to pay for your own space, choose a free web host. However, for a few bucks a month you can get a nice web space package that does not serve advertisements. Among the many free hosts out there, we recommend GeoCities (http://geocities.yahoo.com/). This is not to say we like them, but they just seem to be the least annoying of the free hosts on the Internet.

Paid Web Hosts

You can find a web host for any budget. The good thing about paid web hosting is most (not all) do not serve advertisements to your visitors. We've picked two hosts that LEGO fans have used successfully, that fit different price ranges. If you're looking for a budget host, you can get an excellent deal for $30 per year at http://www.atlnetworks.com. If you want to create a more robust site, we suggest Pair Networks (www.pair.com). This is a favorite among the large LEGO sites like LUGNET and Peeron (www.peeron.com) and is extremely reliable. They also serve as a relatively cheap domain registrar (www.pairnic.com). You can also check CNET (www.cnet.com) for good reviews of web hosts under their "Internet" section.

If you are planning to build a website on an existing service such as the ones we listed in the preceding sections, you will need to know a bit about HTML, FTP, and other web development topics. A good resource for tips and tricks is WebMonkey (www.webmonkey.com).

Respecting Intellectual Property Online

When you are posting LEGO models on the Internet (or any other content for that matter), it is important to know and respect the boundaries of intellectual property law. What is intellectual property? Quite simply, it refers to the ability of individuals or corporations to own ideas, and control how those ideas are represented. If you are using someone's intellectual property, you should always at least credit the source, and ideally seek out the owner's published guidelines for representing it. Generally, a legal disclaimer on the bottom of web pages does the trick, but sometimes content you might want to post goes outside the boundaries of what IP owners want you posting.

The LEGO Company has developed a set of guidelines called "Fair Play," which it uses to inform LEGO fans online of how they can handle the company's trademarked and copyrighted properties.

Fair Play

The LEGO Company's Fair Play document is available on the Web at http://www.lego.com/eng/info/fairplay.asp. This does an excellent job of explaining the issues around the use of their trademarks online as a LEGO fan. The document has a lengthy introduction and is followed by a section under the header "How LEGO® Enthusiasts May Refer to LEGO Products on the Internet." When you are posting LEGO items on the Internet, please follow these guidelines.

Many people in the LEGO Company have begun to form relationships with the LEGO fan community. These relationships are built on trust and a common interest in seeing the LEGO hobby grow. The LEGO Company has begun to open itself up and support fan community-driven efforts. This would not happen if first the LEGO Company did not trust the LEGO fans online with their brand. The continuation of such support depends on how the fans treat the company's intellectual property online. Please respect the relationship that exists and follow the Fair Play guidelines.

Finding an Audience for Your Models

After finding a web hosting solution for your models, you need to present them to an audience. If you do not publicize your efforts, chances are few will find you. Fortunately, there are plenty of outlets online for announcing your LEGO website, for discussing LEGO building techniques, and for connecting with other LEGO fans.

Websites such as MOCpages, BrickShelf, and the Building Instructions Portal provide an audience for your models by design. Thousands of LEGO fans browse these sites looking for what's new. Having a presence on these sites in addition to a personal website on a server can increase your exposure.

Beyond that, get to know the various LEGO sites that feature discussion forums. Sign up to post on these groups and become involved with the communities there. People visit these sites to see others' announcements, but they also visit to socialize and discuss LEGO. Here is a list of the most popular sites for news and discussion.

LUGNET News — http://news.lugnet.com

Already mentioned earlier in this chapter, LUGNET is probably the largest component of the online LEGO community. The news.lugnet.com sub-domain is home to discussion groups for everything from themed building to regional and local LEGO clubs. For the largest (and most important, according to its users) audience, go to LUGNET. LUGNET's discussion system is very powerful. It combines a web-posting interface, NNTP newsgroups, and mailing lists. You can choose how you want to read and post; users on all three protocols can read and post to the same groups and see the same messages. Many Internet users have repeatedly praised LUGNET's discussion system as among the best online. This community is mostly English-speaking, with the exception of local area sub-groups that are set up for every major country, geographic region, and metropolitan area in the world.

From Bricks to Bothans™ — www.fbtb.net

From Bricks to Bothans (abbreviated "FBTB" by the online community) is hands down *the* source for LEGO Star Wars™ information and discussion. FBTB offers discussion forums for talking about LEGO Star Wars, Star Wars custom creations (or MOCs), communication between fans and LEGO Company representatives, non-Star Wars LEGO discussion, non-LEGO Star Wars discussion, and topics of general community interest. If you are building Star Wars themed models, make sure you introduce them on these forums to be recognized by the bulk of the LEGO Star Wars fan base.

BZPower™ — www.bzpower.com

Just as FBTB is the source for LEGO Star Wars, BZPower (a merger between two former websites, BIONICLE Zone and Kanohi-Power) is the number one site for LEGO BIONICLE® news, reference, and discussion. Their massive forums have more than 5,000 members! In particular, the BZPower BIONICLE-Based Creations Forum offers BIONICLE MOC builders the most widely read venue to display and critique creations. If you're looking for a place to show off a BIONICLE creation or want to learn how to incorporate the unique elements of the BIONICLE world in your designs, BZP is the place to be.

NOTE   *Unfortunately for LDraw users, there is not a large selection of BIONICLE parts available yet. As time goes by, parts authors will document them and they will become more plentiful.*

1000steine.de — www.1000steine.de

1000steine.de (1000 Bricks) is a large German language LEGO fan site. If you can read German, this site is a must-see. It also has a very popular forum system where you can announce your creations to German-speaking LEGO fans.

Summary

In this chapter, we introduced you to LDraw viewers, viewing LDraw models via LUGNET's cad.dat newsgroups, and finding places to host your models online. We also described several LEGO community websites that you can use to publicize the models you have created. We encourage you to visit these sites and view others' models, as well as share your own with other LEGO builders.

20

BUILDING INSTRUCTIONS: MOBILE CRANE

These building instructions are provided for your building enjoyment, as well as a reference for Chapter 18. The crane model was built in mid-2002 by author Tim Courtney.

Builder's Notes

(Please read these notes if you're interested in some background information about the crane model. This background information is not necessary to build the model.)

I was inspired to build this model because I needed a mobile crane for a project of mine. At the time, Jennifer Clark's Demag AC-50-1 Technic crane (see Figure 1-3 in Chapter 1) was all the rage online, so I decided to give homage to it in minifig scale. My crane captures the major features of Jennifer's crane, except mine is about two studs longer than it should be to maintain the same proportions.

I built the physical LEGO crane long before I documented it with LDraw tools. I constructed it over the course of a couple days, after playing with a failed 7-stud wide crane design. The project I needed the crane for required a virtual model, and I also thought it would be fun to LDraw the crane for inclusion in the book. You can see the physical model being lifted by Jennifer Clark's Technic crane in my Brick-Shelf Gallery folder at http://www.brickshelf.com/cgi-bin/gallery.cgi?f=22060.

The model here in the instructions is very close to finished, though I would like to revise it someday by adding a few features. Specifically, the outriggers are not functional (they cannot be held in place, do not prevent the model from tipping, and cannot lift the model's wheels off of the ground), and there is no winch system for a cable to run the length of the boom. Currently, I attach a piece of nylon cord to the crankshaft used to raise the boom, but this cannot change in length and is intended for display purposes only.

Parts of the boom were designed by Ondrew Hartigan.

My physical model features custom designed stickers with logo markings (ZACKTRON™ Rental) and danger stripes on the front corners of the cab and the rear corners of the counterweight. The stickers themselves were designed to parody the markings on Jennifer's crane, as well as make humorous reference to Brandon Grifford's and my LEGO Space sci-fi universe, the Zacktron™ Alliance (http://www. zacktron.com). There's not much there right now, but we do have big plans for it, someday.

I hope you enjoy building the model (see Figure 20-1) as much as I enjoyed designing it and creating the instructions.

—Tim Courtney

Mobile Crane
Version 1.0
April 2002
Tim Courtney

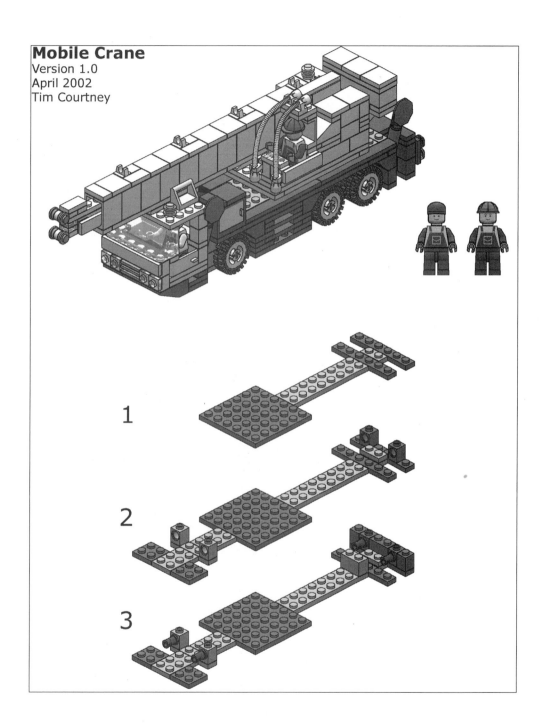

1

2

3

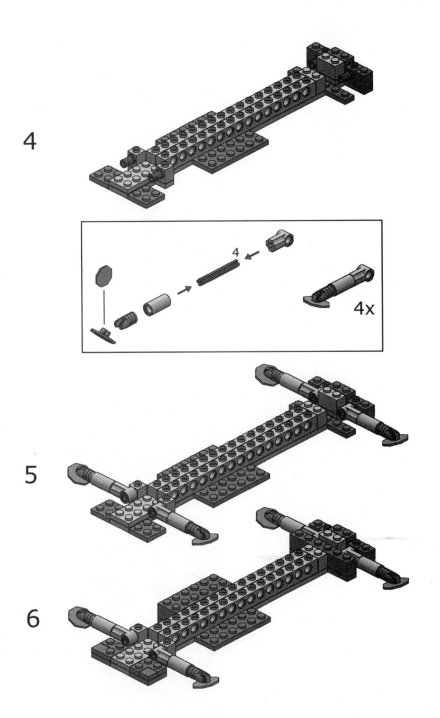

4

5

6

4x

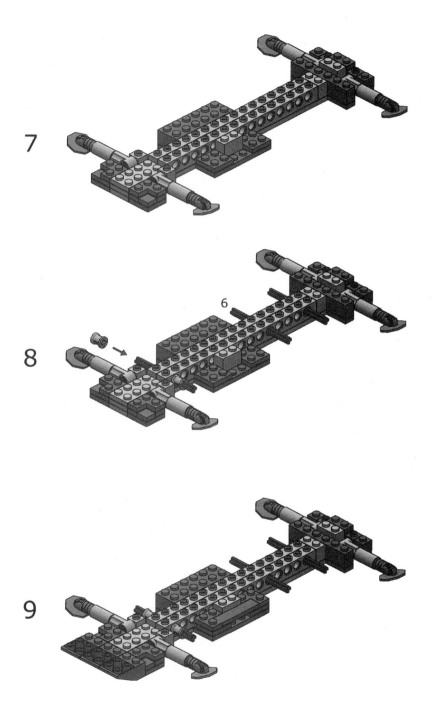

7

8

6

9

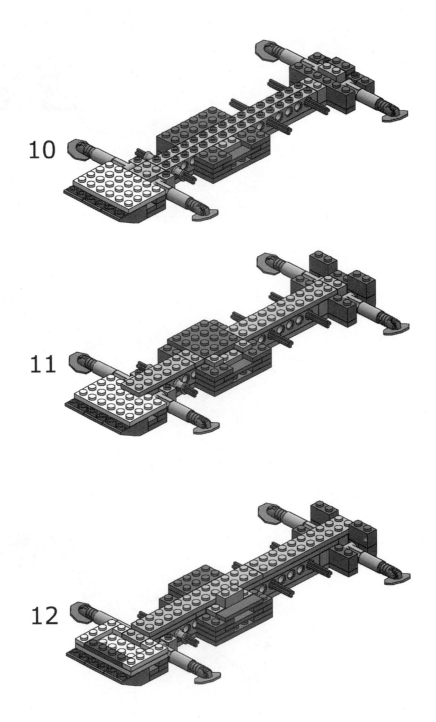

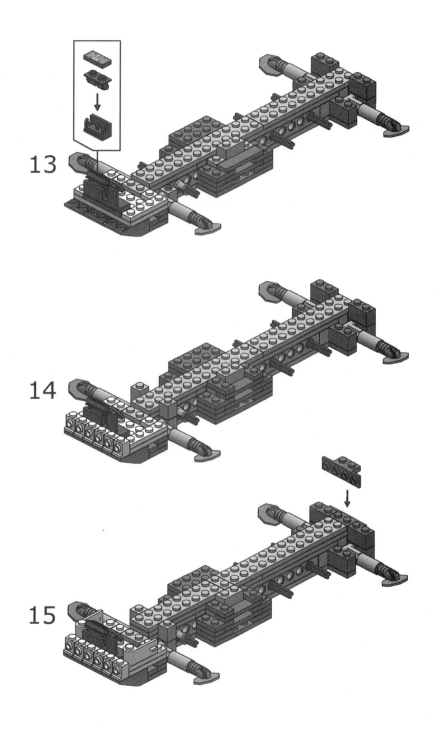

13

14

15

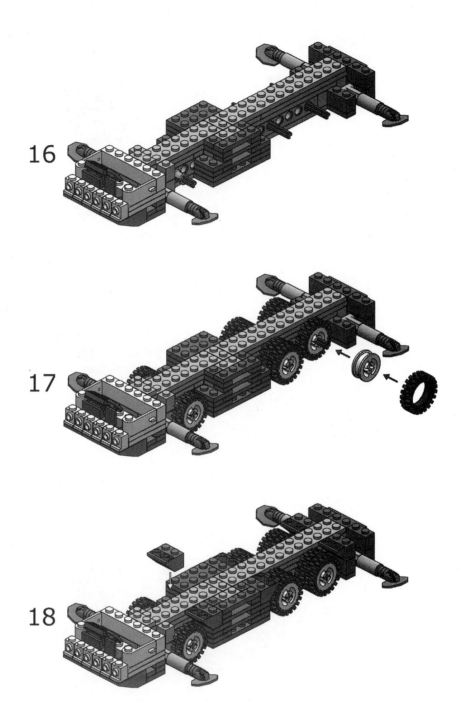

16

17

18

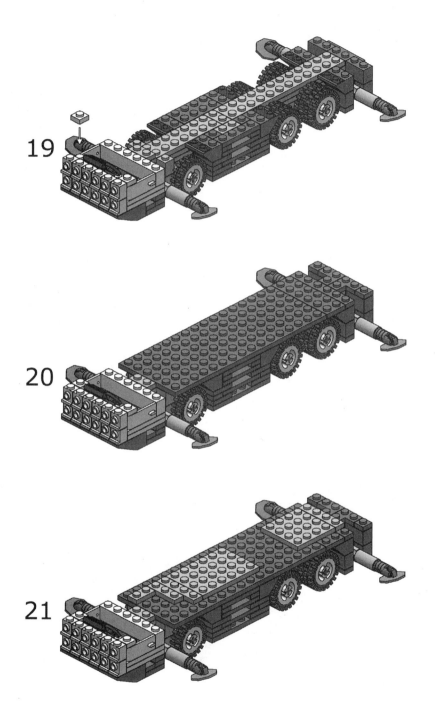

19

20

21

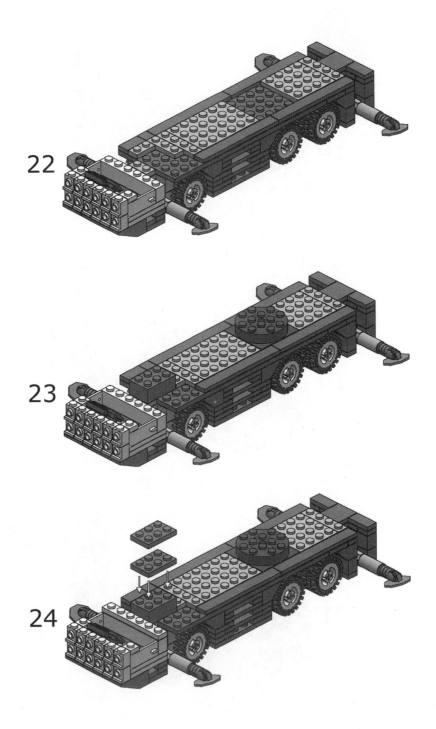

22

23

24

25

26

27

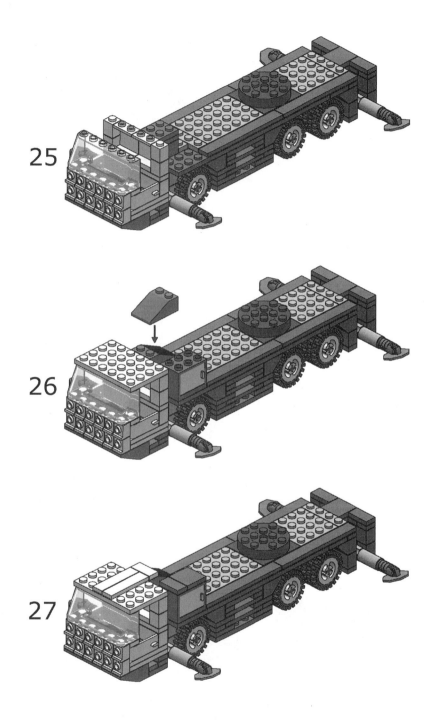

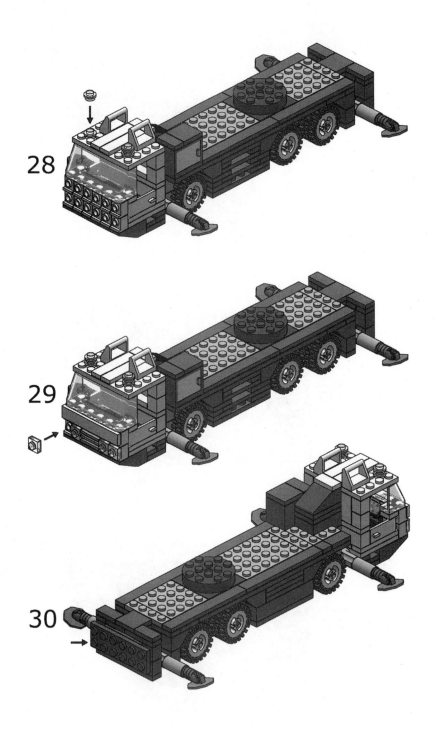

28

29

30

31

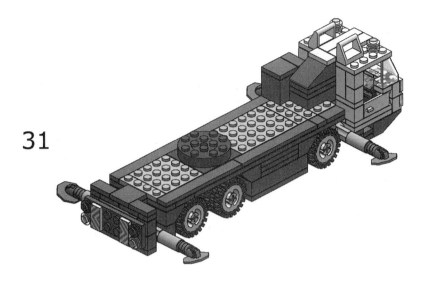

32

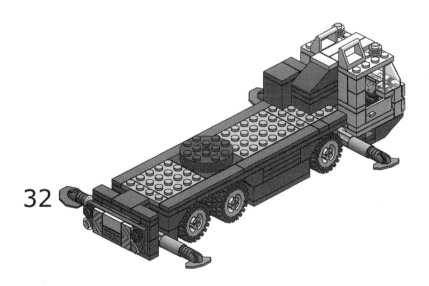

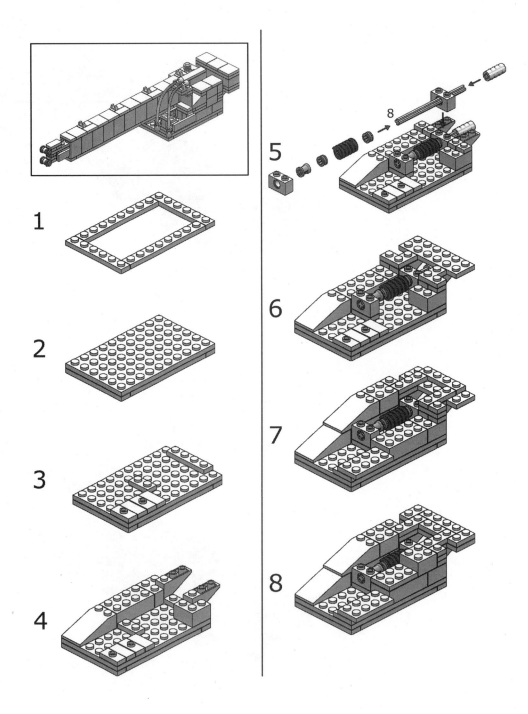

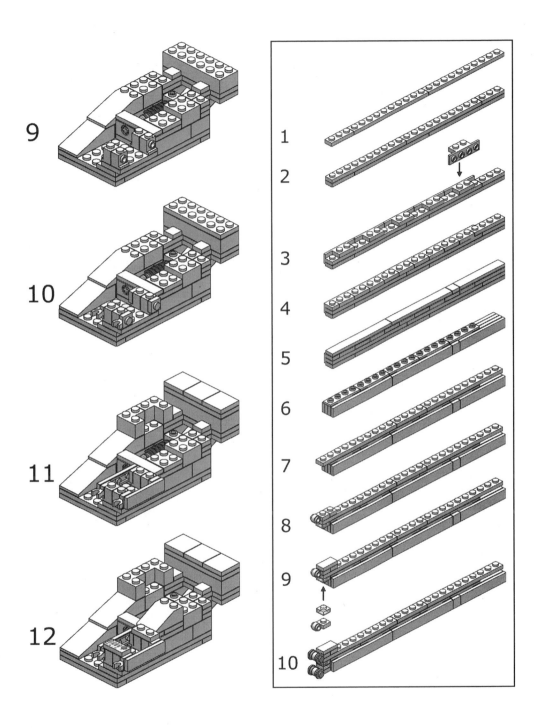

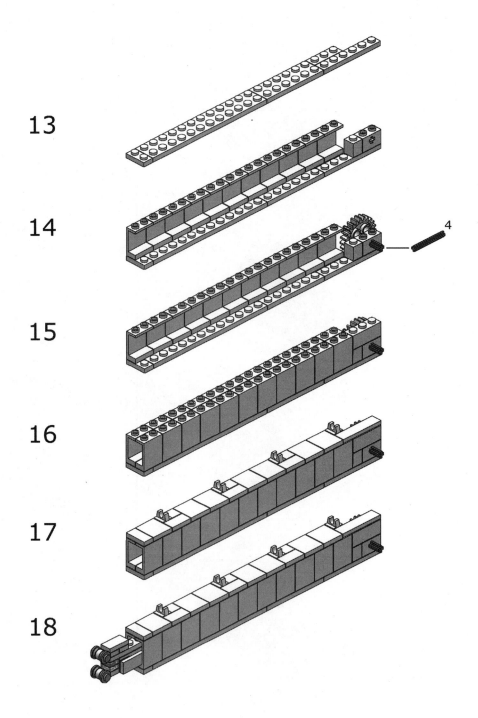

13

14

15

16

17

18

4

19

20

21

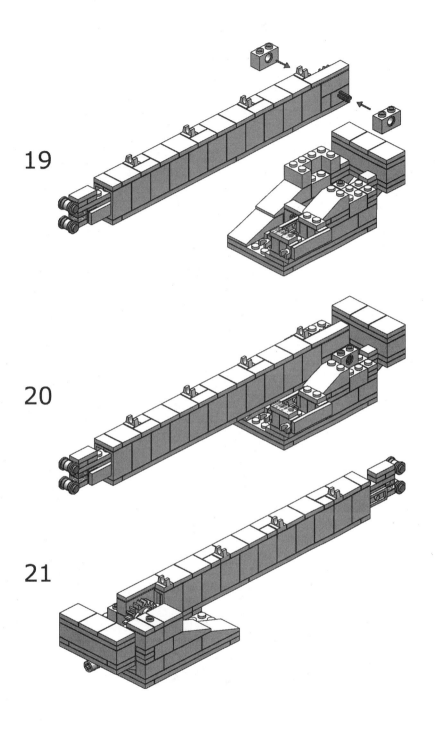

22

23

24

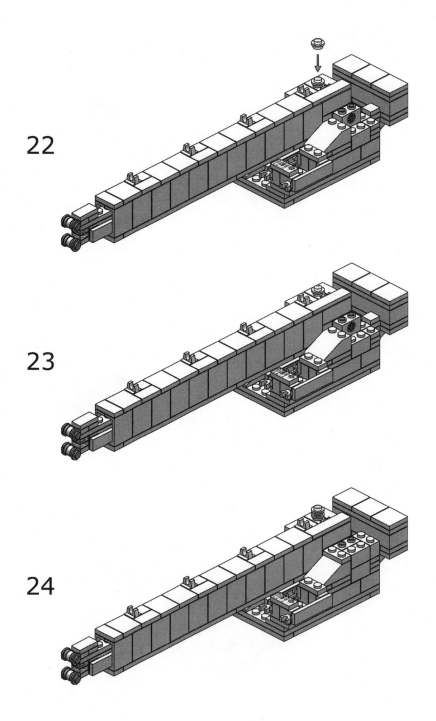

25

26

27

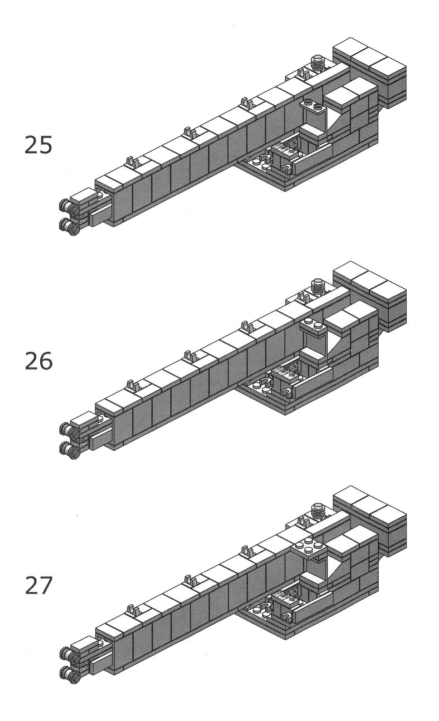

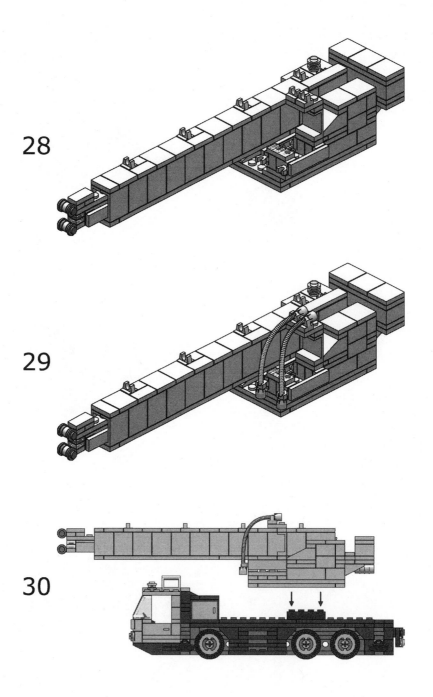

28

29

30

Mobile Crane

Version 1.0
April 2002
Tim Courtney

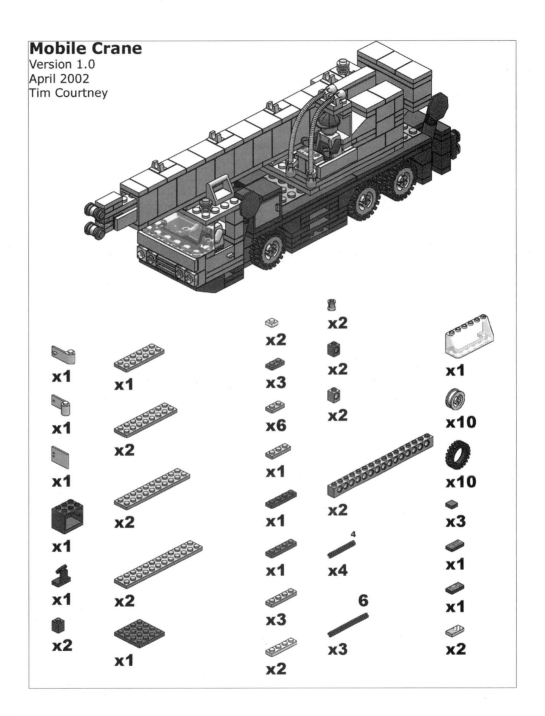

x1

x1

x1

x1

x1

x2

x1

x2

x2

x1

x2

x1

x1

x2

x2

x1

x1

x3

x6

x1

x1

x1

x3

x2

x2

x2

x2

x2

x2

4
x4

6

x3

x1

x10

x10

x3

x1

x1

x2

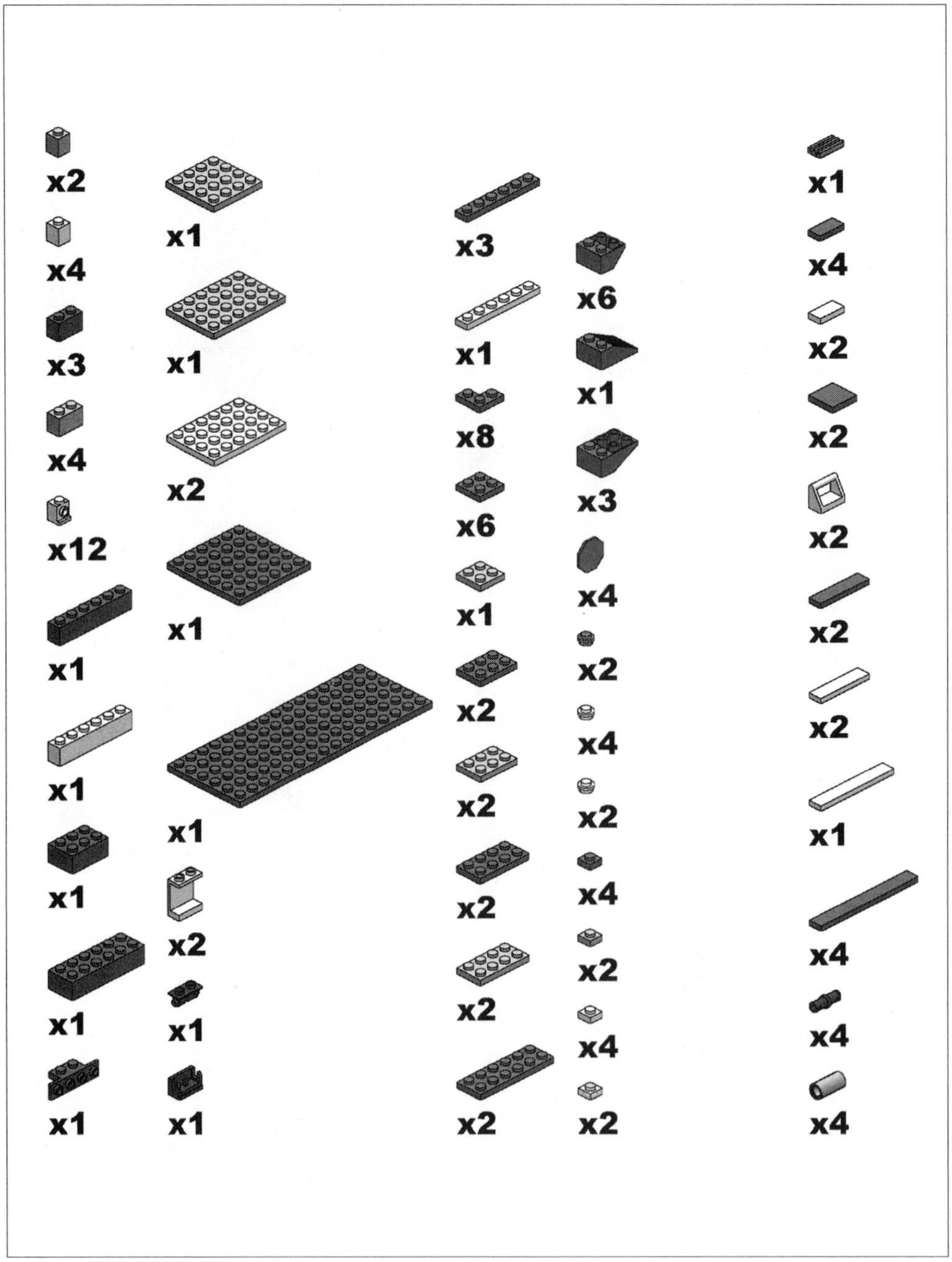

x2

x4

x3

x4

x12

x1

x1

x1

x1

x1

x1

x1

x2

x1

x1

x2

x1

x1

x3

x1

x8

x6

x1

x2

x2

x2

x2

x2

x6

x1

x3

x4

x2

x4

x2

x4

x2

x4

x2

x1

x4

x2

x2

x2

x1

x4

x4

x4

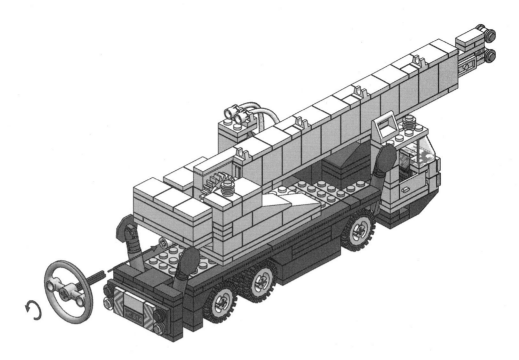

Operating the Boom

To raise the boom, insert a Technic axle in the axle extender at the rear of the crane as shown in the above figure. It helps to add a piece such as a gear or a pulley to the axle to help you turn the crank shaft. To raise the boom, turn the axle counter-clockwise as shown. To lower the boom, turn the axle clockwise.

21

CREATING YOUR OWN LDRAW PARTS

Throughout this book, we've dealt with LDraw files primarily as models and as collections of parts. In this chapter, we'll explore the power of the LDraw *graphics description language* (GDL) in depth by teaching you how to create a part file from scratch. We'll also go over techniques and tools involved in the process of creating part files.

Reasons to Make a Part File

People make their own LDraw parts because the LDraw parts library doesn't have the part they are looking for. Because the LDraw system is a fan effort, individuals must create new parts from scratch if they want to be able to use them when building 3D LEGO models using LDraw tools. Most of the time, fans are looking to use an official LEGO part that isn't in LDraw, but sometimes people want to create non-LEGO parts, such as other brands of building bricks, or other construction systems entirely. We mention some of those (unofficial) collections in Chapter 22 in the "Non-LEGO LDraw Parts" section.

By exploring the lower levels of the LDraw GDL, you'll also be able to enhance your models by adding non-part elements, such as polygons for backdrops, or frames around sub-assemblies. You could also use LDraw to model and design things other than LEGO models. People have used LDraw to model bunk beds, for basement remodeling projects, and craft corner work areas. Perhaps you're interested in 3D coordinate systems and want to use LDraw as an easy system to learn from.

Gearing Up

Before diving into the parts authoring tutorial we have prepared, let's look at a few pieces of background information that you will need to use throughout the chapter.

Tools to Use

You can use a number of different sets of tools to create part files. Every tool has strengths and weaknesses. For most of this chapter, we are going to edit the LDraw files using a program called LDraw Add-On, and view them with L3Lab. Both of these programs are included on the CD-ROM that came with this book; they should already be installed on your computer if you installed the LDraw software from the CD-ROM.

LDraw Add-On (LDAO)

Steve Bliss, one of the co-authors of this book, wrote LDraw Add-On (LDAO). It was one of the earliest "third party" tools for LDraw, written back when the only LDraw program was LDraw itself. Its original purpose was to make it easier to run LDraw from Windows, by providing a GUI interface for all of LDraw's command-line options. Over time, Steve added more features to LDAO so people could use it to work on LDraw files. The important feature for this chapter is LDAO's text editor. It's a basic editor, but it has a number of special functions for working with LDraw files, especially part files.

L3Lab

Lars Hassing, the author of L3P, also wrote L3Lab. L3Lab is useful for parts authoring in a number of ways, as you'll discover. One key feature is that LDAO and L3Lab will work together. When you open LDAO's Editor, an L3Lab window will open as well, with a rendering of your file.

Software Check

Before we dive into creating parts, there are a couple of housekeeping chores. First, you need to make sure that LDAO knows where to find L3Lab. To do this, launch LDAO. Once LDAO starts up, click the **Edit File** button, which should be on the far right side of the LDAO window. This will open an Editor window. If an L3Lab window opens as well, then you are all set, and can skip the rest of this section.

If an L3Lab window doesn't appear automatically, you will need to tell LDAO to use L3Lab. In the Editor window, select **Options > Viewer Settings** to open the Modeler Viewing Options dialog. Select the **View with L3Lab** option and click **OK**. Most likely, this will bring up a warning message telling you that LDAO doesn't know where L3Lab is located. Click **Yes** to go into the Settings dialog. Click the **...** button by the L3Lab location; this brings you into the Locate File dialog. L3Lab *should* be in the **C:\LDraw\Programs\L3Lab** directory. Navigate to the directory; you should see a smiley face in the status box, as shown in Figure 21-1. Click **OK**, and you should be all set.

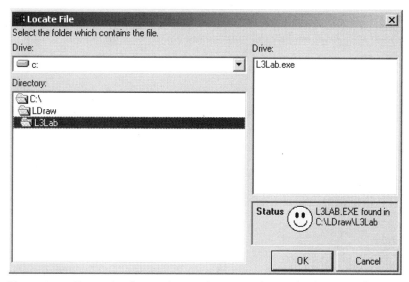

Figure 21-1: The smiley face indicates that LDAO knows the location of L3Lab.

LDraw Dimensions, a Review

When you work with models by themselves, you can often get by with only a rough understanding of the LDraw "world," the 3-dimensional virtual space in which an LDraw creation exists. When you create part files, you need a comprehensive understanding of this world.

The basic unit of measurement in LDraw is called an *LDraw Unit*, or LDU for short. In the original LDraw programs, 1 LDU was roughly equal to one pixel on a computer screen, and parts were rendered on screen as roughly life-sized. In real-world terms, 1 LDU is almost exactly 0.4 millimeters. A 1×1×1 LEGO brick is both 20 LDU wide and deep, and 24 LDU tall, as shown in Figure 21-2. The figure contains a number of other useful basic LDraw dimensions.

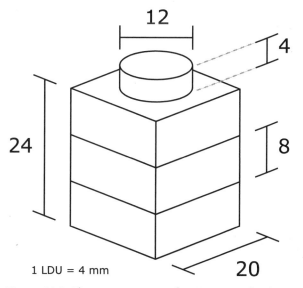

Figure 21-2: The measurements of a 1x1 LEGO brick, in LDraw Units (LDU).

Graphics Terms

There are a few 3D modeling terms you need to be familiar with before moving forward:

Polygon: A 2D shape with straight edges. LDraw allows you to draw triangles and quadrilaterals.

Quadrilateral: Any four-sided polygon.

Axis: One direction, one dimension. 3D space has 3 dimensions, so there are 3 axes; these are identified as X, Y, and Z. Figure 21-3 shows a diagram of the axes in LDraw.

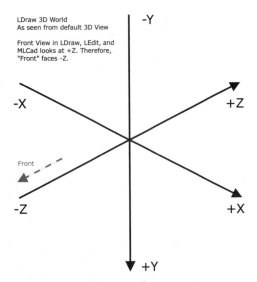

Figure 21-3: A diagram of LDraw "space."

Point: Single location in space. A point is identified by its position along each of the axes, so a position is given as three numbers. For example: (100, -50, 75) is 100 units (in the positive direction) on the X-axis (to the right in LDraw), -50 units on the Y-axis (up in LDraw), and 75 along Z (into the screen or towards the back of a model in LDraw).

Origin: The position (0, 0, 0).

Vertex: An endpoint on a polygon or line.

Matrix: A 2D array of numbers. These are very useful in 3D modeling, and we will discuss them later.

Transformation: The act of changing an object by performing operations through a transformation matrix. Typical changes are translation, rotation, or scaling.

Translation: Moving an object from one position to another.

Rotation: Turning an object.

Scaling: Changing the size of an object. This may be done so that the object keeps the same shape, or the object may be stretched (or compressed) along one or more axes.

Creating a Simple Part

In this section, we will walk you through writing a simple part. We will pick one of the simplest parts, the 1×1 brick (part number 3005). An LDraw file for this part already exists, so you can even check your work when you are finished.

Analyzing the Part

Before you actually write a part, you need to analyze the physical part you are modeling and take measurements. For this tutorial, we have sketched a diagram with the part's measurements (see Figure 21-4).

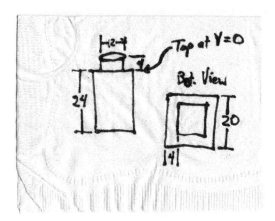

Figure 21-4: The design for your first part.

As you can see from the figure, the design of the part doesn't need to be complicated. You just need to gather the basic information so you can create your part. Some parts require more design work than others do. Since the design is just for your use, you can do as much or as little with it as you like.

Measuring the Part

In order to create your part file, you will need to measure the actual part. It's usually a good idea to use a ruler, or better yet, calipers, to make measurements. Your ruler or calipers should be marked in metric units. However, you can also measure a part by comparing it to other LEGO parts. If you know the size of one part, you can use that as a gauge to compare to the new part. If a part is very tall, you might want to stack up several bricks and plates to check the height of the part (use Figure 21-2 as a reference for brick and plate heights). In many cases, measuring your part against other parts is a very good way to determine its size. For complex or irregularly shaped parts, you're better off using calipers.

In this case, we already have the necessary measurements for the part, so let's start modeling.

How Much Detail is Enough?

When creating files for detailed or complex parts, you'll find yourself frequently having to decide whether a certain small feature needs to be included in the file. The general rule of thumb is to model any feature that is 1 LDU or larger.

Starting a New File

To start a new file, open LDAO's Editor by starting LDAO, then clicking the **Edit File** button on the right side of the LDAO window.

Naming the File

Give your part a name by saving the file (**File > Save**). Type in the name for your file and click **OK.** Remember that the LDraw system names parts by their numbers. A 1×1 brick has the number 3005. We suggest you give your part a name like **3005tut.dat** in order to distinguish the part you are creating for this tutorial from the official LDraw part.

Generally, you can use any name you want for your part files. Just make sure you don't use spaces in the file name; LDraw programs can't reference files with spaces in their name, because spaces are used to delimit the parameters in the LDraw commands.

Give It a Title

The first line in every LDraw file should be a title. The file will render fine without a title, but every LDraw program expects to find a title line. Programs like MKLIST.EXE read the first line to compile the list of parts that programs like MLCad use.

LDAO probably supplied some code for your new file, like this:

```
0 Untitled Model
1 16 0 0 0 1 0 0 0 1 0 0 0 1 3001.dat
0
```

You can delete this code, since it's not useful for what we are doing.

The title line is written in the form of a comment. Comment lines start with a 0, then a space, then the comment text. Type a title like this:

```
0 Brick  1 x  1 - Tutorial
```

An official file would use this same format for the part title.

Write the Part

Now we will begin writing the part. We follow a logical order when placing the lines and polygons that we use to construct the part.

Frame It with Edge Lines

As we start writing our part, the first command we're going to use is the Line command. The code for a Line command is 2. Besides the linetype code, the Line command also includes a color code, and two vertices. Because each vertex is identified by three coordinates, there are six coordinates total. A Line command looks like this:

```
2  0  0 0 0  100 0 100
```

This Line command (linetype 2) draws a black (color 0) line between the points (0, 0, 0) and (100, 0, 100). To understand this more clearly, refer to the LDraw Language Reference that is located inside the back cover of this book. The reference contains a diagram of all of the LDraw line types that indicates what each number of the line means.

So far, we have a file with just a single comment, and nothing displayed in the L3Lab viewing window. To get the part started, we need to add some edge lines. First, however, you should read this important note.

NOTE *Positioning Parts in LDraw Space*
To create a model, you have to set it in its own "space," that is, its own set of the 3D LDraw dimensions. Normally, parts are positioned so that the part is centered on the X- and Z-axes, and the part is placed vertically so that the top is at Y=0.

It's usually a good idea to draw parts from the bottom up, and from the inside out. That reduces the chances that the section you're creating now will be hidden by a section you created five minutes ago. Based on the napkin sketch, we know the bottom of our brick is 24 units below the top. The Y-axis in LDraw is upside down, so vertical measures *increase* as they go down. The top of the part is at Y=0, so the bottom is at Y=24. The part is centered on the X- and Z-axes, and each side is 20 LDU wide. So each of the four vertices at the base of the brick must be located 10 LDU away from the center. This means the four base vertices are located at (10, 24, 10), (10, 24, -10), (-10, 24, -10), and (-10, 24, 10). Four lines connect the vertices, forming a square.

Type four Line commands into your file. Each command should include a pair of the vertices. The first four Line commands should read:

```
2 24   10 24 10   10 24 -10
2 24   10 24 -10   -10 24 -10
2 24   -10 24 -10   -10 24 -10
2 24   -10 24 10   10 24 10
```

When you finish, your screen should look similar to Figure 21-5. Notice that L3Lab shows a nice square. If L3Lab has given you an error message, click **OK** to clear it away.

NOTE   *As you are working on your model, L3Lab will occasionally pop up error messages. This happens because LDAO is continually updating L3Lab with the changes you make, even if they sometimes aren't complete. Whenever an L3Lab error pops up, you can clear it away. If you're doing some typing, you'll probably want to finish your typing before clearing the L3Lab message, because if you don't, it will just display the same message again.*

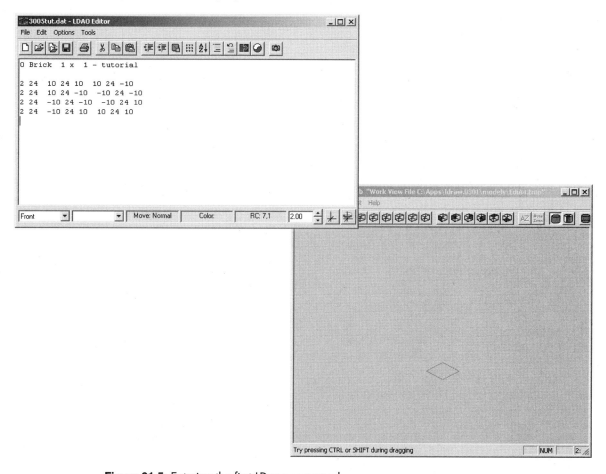

Figure 21-5: Entering the first LDraw commands.

Colors in Part Files

When writing parts, you don't use defined colors like you do when you're creating a model. This is because parts need to be written in such a way that allows the user to define the color each instance of the part will be. The LDraw language provides two special color codes: 16 and 24. 16 is a generic color for polygons, and 24 is a generic color for edge lines.

Notice that we used color code 24 to type our first four lines. Color 24 is a special code that tells the rendering program to use the edge-color that corresponds to the main color that you are using to draw the part. This allows the edge line color to change on the basis of what the color of the part is, which is necessary to guarantee the edge line will contrast with the surfaces of the part.

When creating parts, you will use color codes 16 and 24 almost exclusively. The only exceptions are for drawing printed decorations (which have defined colors by their nature), and for occasional bits in complex parts, like the metal rails on train track, or control buttons on an electronic brick. Line and Conditional Line commands are the only commands that use Color 24. Polygon (line types 3 and 4) and Subfile (line type 1) commands use color 16.

Customizing Your Editing Environment

You can make a few customizations to your editing environment.

Zooming in L3Lab

You can zoom in or out on the part you are creating. At the bottom of the LDAO window, there is a text field with up/down buttons next to it. Figure 21-5 shows the number 2.0 in the field, but yours likely reads 1.0. This is the Zoom control. If you click the **up** button, the scene in the L3Lab window will increase in size. You can also control the zoom from the L3Lab window, but it is easier from the LDAO window. Go ahead and set the Zoom to whatever works for you.

Show Axes in L3Lab

L3Lab can display the axes for you as well. You can only control this setting directly through the L3Lab interface. Right-click on the L3Lab scene and choose **Show Axes**. This will draw blue, green, and red arrows along the X, Y, and Z axes. The position and length of the arrows show you the size of your part. Because you have only a flat square so far, the green arrow for the Y-axis is just a dot in the center of the square.

Drawing the Inside Edge

The first four lines you wrote laid the outside edges of the brick. However, the bottom of the brick also has inside edges. In LDraw, brick walls are always modeled as being 4 LDU thick (that is, the plastic wall of a brick is interpreted to be 4 LDU thick). Each outside edge is 10 LDU away from the center, and each inside edge is going to be 4 LDU closer to origin, so the inside edges will all be 6 LDU away from the center. Their Y value will still be 24, because they're on the bottom of the brick. So the inside corner vertices are (6, 24, 6), (6, 24, -6), (-6, 24 -6), and (-6, 24, 6). Add four Line commands for the edges between these points.

```
2 24   6 24 6   6 24 -6
2 24   6 24 -6   -6 24 -6
2 24   -6 24 -6   -6 24 6
2 24   -6 24 6   6 24 6
```

Boxing in the Brick

Now, let's draw the outline of the top of the brick. You need to find the position of the four top corners. These are directly above the four corners that we've drawn already (that is, the X and Z values stay the same; only the Y value needs to change). We already know the top of the brick is at Y=0, so the four top vertices must be (10, 0, 10), (10, 0, -10), (-10, 0, -10), and (-10, 0, 10). Type in the Line commands to connect these four vertices in a square.

```
2 24   10 0 10   10 0 -10
2 24   10 0 -10   -10 0 -10
2 24   -10 0 -10   -10 0 10
2 24   -10 0 10   10 0 10
```

To finish the outside frame of the brick, draw the four vertical edges that connect the top and bottom. When you are finished, your screen should look like Figure 21-6.

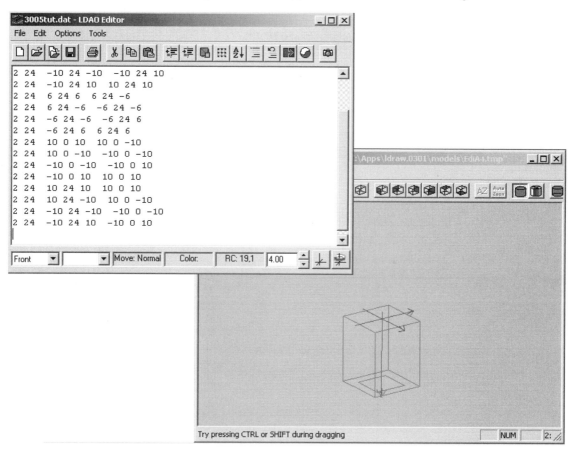

Figure 21-6: The completed frame of your brick.

Normally, you would want to draw the inside of the part as well. In this case, we will come back to it. However, we'll show you an easier method soon.

Filling in the Surfaces

Your brick has a frame, but it has no solidity. Let's change that by drawing some polygons.

In this tutorial, we're only going to use Quad commands, linetype 4. A Quad command draws a quadrilateral, a 2D shape with 4 vertices. The Triangle command (linetype 3) works the same as the Quad, but it has only has 3 vertices. Both commands are structured like the

Line command: The linetype is first, followed by a color code, then ends with enough sets of coordinates to specify either three or four vertices. Here is an example of a Triangle and a Quad:

```
3  16   10 0 10   10 0 -10   10 0 50
4  16   10 0 10   10 0 -10   -10 0 -10   -10 0 10
```

These two commands would draw two adjacent polygons, using the special color code 16.

For our first polygon, let's fill in one of the backsides of the part. We will use the side where all the vertices have X=-10. Type in the command:

```
4   16   -10 0 -10   -10 0 10   -10 24 10   -10 24 -10
```

Now your file should resemble Figure 21-7. Notice that the vertices are specified in order around the polygon. This is called *winding*, and is a good way to write polygon commands. For more information, see the "Vertex Winding" section later in this chapter.

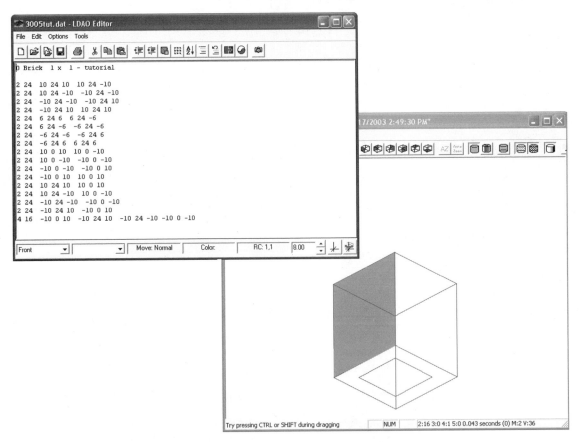

Figure 21-7: The first polygon.

Next, let's fill in the narrow frame on the bottom side of the brick. Because the brick is open on the bottom, we'll need to draw polygons all the way around. It works out that we'll need one Quad for each side of the brick. You can lay the Quads out several ways; the best way is to have adjacent polygons share common vertices. In other words, the vertices in one Quad should also be used by the next Quad. Figure 21-8 on the next page illustrates an improper and a proper layout.

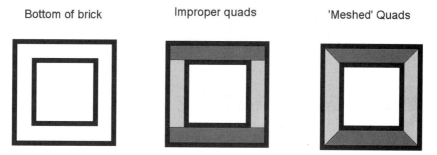

Figure 21-8: The bottom of the brick.

The "Meshed" Quads in Figure 21-9 illustrate the proper way to align the vertices of the Quads. Type the four Quad commands to fill in the bottom of the brick. To make it easier to see any mistakes, temporarily use contrasting colors for each Quad. You can change the colors to Color 16 when you are finished. The code should look like this:

```
4  15   10 24 10   -10 24 10   -6 24 6   6 24 6
4  14   -10 24 10   -10 24 -10   -6 24 -6   -6 24 6
4  13   -10 24 -10   10 24 -10   6 24 -6   -6 24 -6
4  12   10 24 -10   10 24 10   6 24 6   6 24 -6
```

Notice the vertex winding in each Quad.

Double-check your work against Figure 21-9, then change all the color codes to 16.

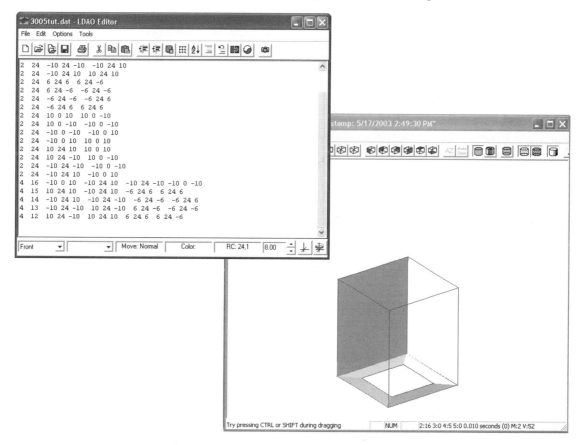

Figure 21-9: Well-meshed polygons form the base of the brick.

To finish drawing the surfaces, add Quads for the three remaining sides, and another Quad for the top. Your part should now resemble Figure 21-10. You can rotate the part in L3Lab by dragging it with your mouse. Use the illustration to check your part from other angles, and to ensure everything aligns properly. If you hold down the **CTRL** key while you drag, L3Lab will continue to do full-detail, real-time renderings while you drag.

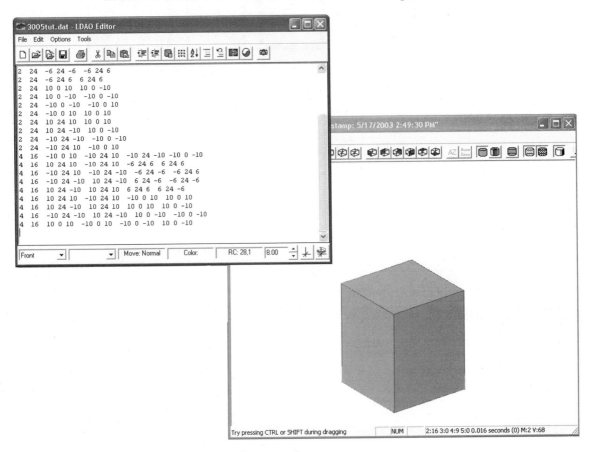

Figure 21-10: All outer surfaces complete.

Using the Subfile Command to Add the Stud

One of the most powerful features of the LDraw GDL is the Subfile command, linetype 1. This is the primary command in model files, where it's used to place parts in the model. However, linetype 1 can reference any LDraw file, and any LDraw file can include linetype 1 commands. This recursive chain of nested references can continue as deeply as you like. Besides positioning a subfile within a model or part, the Subfile command can also rotate the subfile object, and change its size and shape.

The Subfile command is the most complex LDraw linetype. In addition to the linetype code and color code, there's also a set of coordinate values to set the position of the object, a set of nine parameters that make up the transformation matrix, and finally the name of the subfile. An example Subfile command looks like this:

```
1  16  0 0 0  1 0 0  0 1 0  0 0 1  4-4edge.dat
```

This command will draw the file 4-4edge.dat at position (0, 0, 0), in a normal orientation (that is, right side up, default size, and not rotated).

Part files use Subfile commands to pull in ready-made sections of parts. The LDraw library includes a special set of files, including a wide variety of 3-dimensional shapes and standard segments found in many LEGO parts. You can use these files when writing parts. These files are called *primitive files*, because many of them model basic, or primitive, geometric shapes. There are several groupings of primitives, including circles, discs, cylinders, cones, cross-axle shapes, hinge fingers, and several styles of LEGO studs. There is not enough room here to discuss them all; however, we've included the Primitive Reference found on LDraw.org in Appendix F.

"Primitives?"

Primitive is a common term in 3D modeling. However, the LDraw system uses primitive in a sense opposite of the normal definition. In most GDL's, a primitive is a core command of the system. In LDraw, this corresponds to the drawing linetypes, such as Lines, Triangles, Quads, and Conditional Edges. LDraw's primitives would probably be called "macros" in other languages.

One of the most frequently used primitives is the stud.dat file. This file models a single standard LEGO stud. We need to place a stud on the top center of our brick, so the position of the stud will be (0, 0, 0). Likewise, we'll need the stud straight up and of normal size. This is the command to use:

```
1  16  0 0 0  1 0 0  0 1 0  0 0 1  stud.dat
```

Add this line to your file, and you should match up with Figure 21-11.

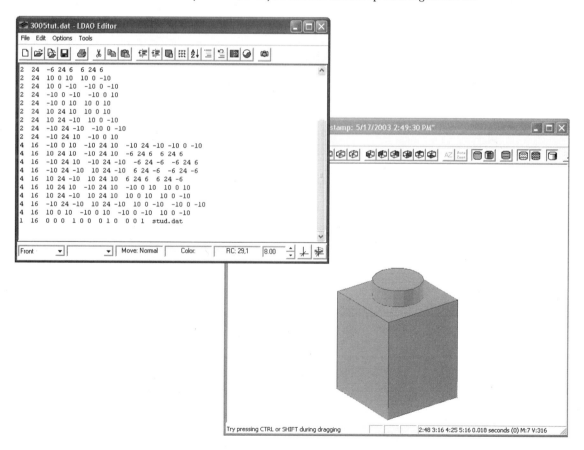

Figure 21-11: Adding the stud primitive to the 1x1 Brick.

Getting Primitive on the Inside

Remember that earlier in this chapter, we skipped modeling the inside of the brick. Let's go back and fill it in now. Instead of drawing individual lines and Quads, we can use a pre-made primitive file.

Using the Comment Block

In order to see what we're doing on the inside, we'll need to temporarily remove the polygons we've drawn for the outside surface. LDAO makes this operation easy with the Comment Block toolbar button. This feature toggles selected lines into comments, by placing a 0 in front of each line. To use Comment Block, highlight all of the polygon command lines in your file, and click the button on the toolbar. This will remove all the polygons from your model (except for polygons in the stud.dat) and leave just the frame. Don't worry. Your work is still there, but it is just "commented out" for the time being.

Inserting the Primitive

We're going to use a primitive called **box5.dat**. This primitive makes a simple box that is missing one face, the top. This is very similar to what we need: The inside of the brick is a basic rectangular box, with one face missing. However, box5.dat is a different size and in a different position and orientation than we need; because of this, we'll need to make some changes in the Subfile line in order to make the primitive fit properly. Go ahead and enter a basic Subfile line that references box5.dat, similar to the command we used for stud.dat (you can even copy and paste it, if you want):

```
1 16  0 0 0  1 0 0  0 1 0  0 0 1  box5.dat
```

At this point, you won't be able to see the box because it's very small and hidden by the stud.dat.

The exact dimensions of box5.dat are 2 LDU × 1 LDU × 2 LDU. The top of the object (the missing top face) is located at the origin. Whenever you manipulate a subfile by scaling or stretching it, the origin of the subfile will always stay at the position specified in the Subfile command. So let's get the origin of box5.dat into the proper location in the brick. The missing face is centered at the bottom of the brick; position (0, 24, 0). Change your file to put box5.dat at (0, 24, 0):

```
1 16  0 24 0  1 0 0  0 1 0  0 0 1  box5.dat
```

Now we need to make our box5 wider. The box5 object is 2 LDU wide; we need a box that is 12 LDU wide (one side is at -6, the other is at +6). So we need to scale up the object by a factor of 6 (12 / 2 = 6) in both the X and Z directions. Because we want to change the sizes along the axes, and we haven't rotated our subfile, resizing is an easy operation. You can apply your scale factor by multiplying specific matrix parameters. Looking at our line above, the matrix parameters are the block of numbers that read "1 0 0 0 1 0 0 0 1". The first three numbers ("1 0 0") control the scaling of X. The second three numbers control Y, and the third three numbers control Z. We're scaling X and Z, so multiply the first and third sets of numbers by 6, giving:

```
1 16  0 24 0  6 0 0  0 1 0  0 0 6  box5.dat
```

NOTE   *Take a look at the LDraw Language Reference card. The matrix parameters are clearly subscripted to show how they are grouped.*

Now we can scale the Y dimension. The final height of the interior box is going to be 20 LDU, leaving the top of the brick as a 4 LDU wall. The current height of box5.dat is 1 LDU, so we need to scale up by 20:

```
1 16  0 24 0  6 0 0  0 20 0  0 0 6  box5.dat
```

Your box is now scaled properly, but it is still hanging upside down. A quick fix would be to change the 20 parameter to -20, but a better solution is to rotate the box by 180 degrees. There's a simple way to rotate 180 degrees: Change the signs of two groups of parameters. That is the exact effect given by rotating around the other axis by 180 degrees. Rotating our box5 around the X-axis will give us what we want; go ahead and change the signs of all the parameters in the Y and Z groups:

```
1 16  0 24 0  6 0 0  0 -20 0  0 0 -6  box5.dat
```

Refer to Figure 21-12 to double-check your progress.

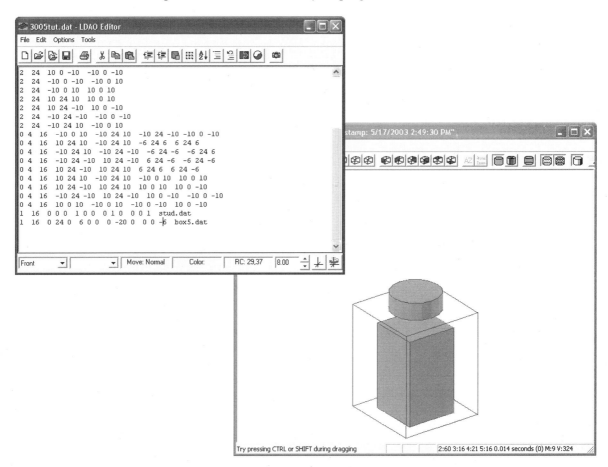

Figure 21-12: The interior box5 is finalized.

Finally, let's restore the polygons we hid previously. The Uncomment Block button is right next to the Comment Block button. Select all the polygon lines, then click the button.

Your part is now complete! Take a moment to bask in your achievement. Save your file, rotate the view in L3Lab, and zoom the view in and out to check out your new part.

If you want some more practice, you can replace the edges and polygons that make the outside of the brick with another reference to the box5.dat primitive.

Two More LDraw Commands and Other Important Details

There are two commands that we didn't discuss in the preceding tutorial: Comment and Conditional Line.

Comment

The comment line is very simple: The linetype is 0, then you type your comment. It's considered good style to use a marker at the start of your comment. A *comment marker* is one or more punctuation marks, such as "//". Comment markers prevent rendering programs from confusing your comments with meta-statements. A comment line looks like this:

```
0 // This is my comment.
```

LDraw files tend to be hard to read because they are made primarily of numbers. Adding comments to your code can make it much easier to find your way through your part when you go back to modify it, especially if it has been a long time since you wrote it.

In the tutorial earlier in the chapter, we told you how to hide commands in a part file by temporarily turning them into comment lines.

Conditional Line

If the Subfile command is the most complex linetype, the Conditional Line is the hardest to understand. The format of the Conditional line is 5 for the linetype, then the color code (usually 24), then four sets of coordinates for four points. The first two points are the endpoints of the line, and the second two points are control points. The command looks like this:

```
5   24   x₁ y₁ z₁   x₂ y₂ z₂   x₃ y₃ z₃   x₄ y₄ z₄
```

The Conditional Line exists to provide a method for drawing outlines around curved surfaces. LDraw was created originally for drawing LEGO parts, models, and instructions in a schematic style, with lines marking all edges and outlines of objects. It's easy to draw edges for flat surfaces. All you have to do is draw a line wherever two surfaces meet. However, curved surfaces typically only show edges along their outlines. So LDraw needed a way to describe edges that are only drawn when actually needed. To make the problem even harder, LDraw doesn't actually model curved surfaces; it can only draw flat polygons. To model curved surfaces in LDraw, you have to combine a number of small polygons that meet at small angles. The "cylindrical" stud we used in the tutorial is actually a 16-sided prism.

The way LDraw handles the problem of drawing outlines on curved surfaces is simple, but confusing. Each "curved" surface also has a Conditional Line drawn along each edge where two polygons meet. However, these lines are only drawn if a specific condition is met. That is what is meant by the name "Conditional Line." The condition, or rule, for Conditional Lines can be written as:

The line between points 1 and 2 is only drawn if points 3 and 4 would both appear on the same side of that line, as the line and points (would be) drawn onscreen.

To see how this works, let's take a closer look at stud.dat. Figure 21-13 shows a close-up of stud.dat. This image has all of the Conditional Lines drawn around the sides of the stud. The ones you'd normally see, on the left and right edges of the stud walls, are drawn in black. The lines that normally would not be drawn are shown in white. Additionally, the points for two Conditional edges are marked. The two endpoints are connected by the Conditional Line, and the control points lie on either side of the upper endpoint. Additionally, gray lines are drawn between the control points, to help illustrate their placement. Notice that for the left-hand line, the two control points both lie on the same side of the conditional line. However, for the center-front line, the control points lie on different sides of the conditional line. By applying the rule for Conditional Lines, we know that the left-hand line should be drawn, and the center-front line should not be drawn.

You might imagine this magnified stud slowly rotating clockwise. At the left side, the figure would reach a point where the next Conditional Line would line up with the current left-most Conditional Line. They would appear to be a single vertical line. At that point, they would both satisfy the rule for Conditional Lines, and both of them could be drawn. In the next instant, as the stud continued its rotation, the current left-most line would pass to the back of the figure, and would no longer satisfy the rule. The second line would now be the left-most line, and would continue to be drawn. At least, it would continue to be drawn until the third line caught up with it, and the entire process would repeat.

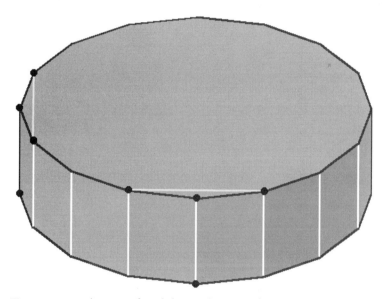

Figure 21-13: Close up of stud.dat. At this magnification, you can see how curves are simulated by straight lines and flat surfaces.

The Transformation Matrix

The nine parameters in the Subfile linetype make up a transformation matrix. In the tutorial, we manipulated these parameters to control the scaling and rotation of primitive files. If you are familiar with matrices and their use in 3D modeling, you can modify the parameters yourself. Here are how the Subfile parameters make up a matrix. The parameters from this command:

```
1  16  x y z  x_x y_x z_x  x_y y_y z_y  x_z y_z z_z  subfile.dat
```

map into a 3x3 matrix like this:

$$\begin{bmatrix} x_x & x_y & x_z \\ y_x & y_y & y_z \\ z_x & z_y & z_z \end{bmatrix}$$

There isn't room in this chapter to explain all there is to know about transformation matrices, but you can find out about them through any of a number of mathematical or computer graphics books or websites.

Vertex Winding

When setting up polygon commands, it is generally useful (and considered good style!) to list the vertices in either clockwise or counter-clockwise (anti-clockwise) order. This is called *winding* the polygon. There are three ways you can wind a polygon, all dependent on which order you choose to plot the points. Figure 21-14 illustrates this order.

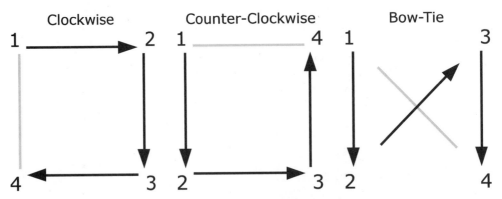

Figure 21-14: Proper and improper ways to wrap polygons. The Bow-Tie is incorrect.

The first two examples are the right way to wind a polygon. However, the third example, the "bowtie," is a poor way to write a Quad; it is "unwound."

NOTE *Triangles are always "wound," which means there is no way to get the vertices in a bowtie.*

The direction of winding is determined by looking at the front side of the polygon. The "front" of a polygon is the side you can see when the part is completed. If a polygon is on the bottom of a part, the front of the polygon is the bottom, which is the side that is visible on the completed brick.

Why is winding important? LDraw does not require winding polygons, but it is a good idea. Some LDraw programs can make more use of polygons with clean winding, and it is easier to use the information in some graphics operations. The LDraw winding convention is to use counter-clockwise winding, but either direction is acceptable. All polygons in a file should use the same direction of winding. This uniformity is more important than the direction of the winding.

Software for Part Files

Different programs have different capabilities to work with part files. Here is a short rundown of programs you might ordinarily use, and why you might use them.

Using LDAO's Edit Functions

In the tutorial, we used LDAO's Editor to work on our part file. However, we didn't make much use of its special capabilities, except to tie into L3Lab to view our file while we edited, and to turn lines of LDraw code into comments, and back.

The editor in LDAO is not as powerful as MLCad, in terms of manipulating your file. However, LDAO does have standard movement and rotation functions, as well as some special tools. By selecting a line, or a range of lines, you can use the functions on the Edit and Tool menus to relocate, rotate, or modify blocks of code. This can be a tremendous time-saver.

Using LDDesignPad

LDDesignPad is another text editor for LDraw files, especially part files. In a number of ways, it exceeds the LDAO Editor in functionality. It is also open source, and features a plug-in architecture that allows members of the LDraw community to contribute new functions to the program.

Creating Parts in MLCad

It is certainly possible to create and edit part files in MLCad. You do this using the functions on the Expert Bar toolbar (see Figure 21-15). The Expert Bar allows you to create Lines, Triangles, Quads, and Conditional Lines, as well as to access the parameters of the subfile commands directly, all through MLCad's user interface.

Figure 21-15: MLCad's Expert Bar.

Another huge advantage of editing parts in MLCad is that you are able to use your mouse in the editing process, just as you do when you build a model. You can click objects in the viewing windows to select them, drag objects around, or lasso a number of objects at one time. When you're trying to track down one specific Quad in a large part file, the mouse can save you a lot of time and frustration.

Using Spreadsheets

In some cases, you may need to move around blocks of points. Maybe you need to turn some Quads into Lines or calculate new points that are based on existing points. It's pretty easy to open a part file in a spreadsheet, and use the spreadsheet's grid to manage the columns of numbers in the file. When you are done, you can either copy or paste your code back into a text editor, or simply save the file in text format. You might need to do some minor cleanup afterwards, but it could be worth the time that you've saved.

Hints for Pasting from Excel to the LDAO Editor

Use the mouse to paste, not the regular keyboard shortcuts. If you paste from Excel to LDAO using the keyboard, you'll get a table in the middle of your LDAO file. By using mouse or menu, you'll get text, with tabs instead of spaces. Second, to change the tabs to spaces, use the *Replace* command. In the *Find What* field, type ^t to match tabs. Put a space in the *Replace With* field. Then click *Replace All.*

Summary

In this chapter, you learned the basics of creating part files. We provided some background for designing parts, including how to measure parts, basic graphics terms, and the coordinate system LDraw uses. While building a simple part file from scratch, we explored the LDraw graphics description language (GDL). We discussed and used the Comment, Subfile, Line, Triangle, Quad, and Conditional Line linetypes.

As you probably suspect, we have only scratched the surface of part creation. Topics such as creating patterned parts, making complex parts, and submitting your files to the LDraw.org Parts Library will be left for you to explore. Here are a few tips and web resources to help you on your way:

http://www.ldraw.org/library/tracker/ref/: This web page offers a great deal of material on authoring "official" parts, that is, files that are added to the LDraw.org Parts Library.

C:\LDraw\Parts\: The entire LDraw.org parts library is already installed on your hard drive. If a part file exists that is similar to one you need, you can explore the file to see how it has been built. You can also copy a part file (or portions of it) to use in your own files.

http://news.lugnet.com/cad/dev/: The CAD Development newsgroup on LUGNET is often a good place to read about part files and to research questions you have about creating parts.

22

LDRAW AND LEGO BRICKS: TAKING THE HOBBY FURTHER

This chapter explains how you can take your LEGO hobby further than you have before. Whether you become involved with the LDraw development community, create parts, write software, or decide to connect yourself with the LEGO builders in your local area, there is something for you. The LEGO hobby is more than dumping a pile of bricks on the floor in your room, yet we all at one time enjoy that simplicity. The LEGO community is about learning, building, sharing, connecting, and growing. The LEGO community is fun.

The LDraw Community

Let's start with our own little corner of the LEGO world, the LDraw community. This group of users and developers interacts mostly via the Internet. Although the central resource site is LDraw.org, most development and user discussion takes place on the LUGNET CAD newsgroups, as it has since LUGNET started. As of the time of this book's printing, LDraw.org does not provide its own discussion groups, simply because up until now the LUGNET newsgroups have provided a more than adequate outlet for LDraw discussions.

LDraw.org

LDraw.org is the central site for the LDraw file format and parts library. It was preceded by Terry Keller's James Jessiman Memorial, which was erected shortly after James' death in 1997 for the same purpose. The founders of LDraw.org saw a need for a website that collected LDraw resources in one location and provided a unique online presence for the LDraw system of tools. Since its founding in 1999, LDraw.org has served that purpose, hosting resources and introducing tens of thousands of LEGO fans to LDraw tools.

Today, LDraw.org continues to strive towards the goal of improving the available LDraw resources and serving as a foundation of the LDraw system of tools. LDraw.org holds the benchmark: James' original software combined with the LDraw file format and language description. Individual developers build off of that foundation, and what emerges are programs such as MLCad, LPub, LDDesignPad, and more.

Getting Involved

As we mentioned earlier, there are a few key areas where you can get involved with the LDraw community. Here we'll talk about each in a little more detail.

Discussion

Most LDraw discussion takes place on the lugnet.cad.* newsgroups on the LUGNET server. You can access these groups via the Web at http://news.lugnet.com, or via NNTP news at nntp://news.lugnet.com. There are sub-groups to discuss development, raytracing, posting LDraw files, and more. If you have any questions or need help with a problem, the LUGNET CAD groups are the place to go, since chances are the developer of the tool you have a question about posts there too. Beyond that, the CAD groups are populated with many LDraw users of varying experience levels. No matter what question you have, someone should have the answer there. If you want to get involved and "break in" to the LDraw community, introduce yourself on LUGNET CAD and start participating. Perhaps you don't need help yourself, but have enough expertise to offer it to someone else? A good way to earn the respect of LDraw contributors is to participate productively in discussion.

Parts, Authoring, and Reviewing

In the previous chapter, we introduced you to parts authoring and submitting parts to the LDraw.org Parts Tracker. Parts authoring isn't for everyone, but parts authors are a very necessary group of people in the LDraw community. Without them, the library wouldn't grow and new parts wouldn't be added.

Programming

Perhaps you picked up this book as a computer programmer and a LEGO fan. Members of the LDraw community are eager to try out new programs designed to enhance the LDraw experience. There is tremendous potential for new utilities — especially a multi-platform program that combines editing, high-quality rendering, and batch processing of instruction images. We encourage you to get involved and use your programming skills to develop software for the LDraw community.

Taking LDraw Further

Several services in the LEGO fan community use the LDraw parts library, part numbers, and part naming conventions to build reference databases for inventorying LEGO sets, personal collections, and parts for commerce. Without LDraw, these services would not be possible, at least to the level they exist at today.

Partsref — http://guide.lugnet.com/partsref/

Steve Bliss compiled the Partsref database, with information, images, and links for many LEGO parts. It is hosted for anyone to use the data as needed. However, this site only provides information on LEGO elements that are represented by official LDraw parts, leaving it far from complete.

Peeron™ LEGO® Inventories — http://www.peeron.com

Peeron LEGO Inventories, created by Dan and Jennifer Boger, contains a partial listing of the parts that are included in official LEGO sets. With these parts lists, or "inventories," you can more easily keep track of your own personal collection, or use the data to help you when buying and selling parts online. The site provides illustrations for the individual parts when they are available, with either LDraw renderings or photographs if a part is not available in the LDraw library. You can search the database for parts or sets. Peeron provides extensive cross-references for sets and parts to other LEGO databases online, and also displays a record of each (currently inventoried) set a part occurs in, and in what quantity.

BrickLink™ — http://www.bricklink.com

BrickLink is an online global bazaar that enables LEGO fans worldwide to buy and sell parts and sets. Because BrickLink specializes in individual parts sales, the LDraw parts library enables it to display an image of each part that is for sale, that is, each part that is currently in the LDraw library. BrickLink buyers can browse online "shops" maintained by other individuals for parts. They place orders, make their payment using one of the methods the seller accepts, and the seller ships the parts to the buyer. Both the buyer and the seller can leave "feedback" ratings (Positive, Negative, or Neutral) for each other, along with comments.

BrikTrak™ — http://www.briktrak.com

BrikTrak is free advanced inventory software for Windows. Created by Richard Morton and currently in beta version, it provides a very powerful system for inventorying (cataloging the individual parts) your collection and BrickLink store. Among other things, you can import and export your BrickLink store contents, view the average BrickLink selling price for each part, create inventories of LDraw files, and more. Look at the BrikTrak website for more information on this incredible piece of software, which is compatible with the LDraw file format.

Non-LEGO LDraw Parts

Several people have created LDraw libraries of non-LEGO construction systems. These libraries and the building systems they represent are not officially endorsed by LDraw.org. The LDraw Parts Library only includes official LEGO elements. Nevertheless, we thought to include them in this book for your reference and enjoyment.

LDraw Apocrypha — http://members.bellatlantic.net/~drteeth1/clones/clone5.htm

Dave Schuler maintains this collection of "non-canonical," if you will, LDraw parts. The site begins with an FAQ for those familiar with LDraw. You can proceed to the rest of the site via an arrow at the bottom.

Attack of the Clones

What does "clone bricks" mean? This is another term arising from the online LEGO fan community, referring to imitation LEGO parts made by other companies. For example, MegaBloks, TYCO, and Tente all fall in this category. Also, direct rip-offs like the Chinese Shifty bricks are clones (albeit very poorly made ones).

Clone Bricks — http://home.swipnet.se/~w-20413/mybricks/warning.htm

Tore Eriksson hosts collections of various types of "clone bricks," which he places in the public domain. The site begins with a rather humorous warning before entering, playing off of most LEGO fans' hatred for imitation bricks.

Virtual Construction — http://users.ifriendly.com/fourfarrs1/construction.htm

Ryan Farrington has produced LDraw parts for K'NEX elements as well as Lincoln Logs. He is also planning a 3D library based on the Metropolitan Museum of Art's book *Fun With Architecture.*

Idea for the Future: The Artemis Project — http://www.karimnassar.com/design/artemis/

Karim Nassar puts forth the Artemis Project, a design study of an ideal LEGO CAD program. This study was published in 2000, and though lauded by the community, to our knowledge no effort has been made to implement these ideas. Mr. Nassar is a designer, not a programmer. He published his study for the community's benefit. This document is a must-read for anyone looking at the future of LDraw.

The LEGO Hobby

The LEGO hobby is as vast and diverse as the people who participate in it. LEGO fans can be found all over the world. Though factors like age, race, culture, religion, and political borders may divide LEGO fans, the love for the brick unites them. This hobby is unique simply because of the strong online community that supports it and encourages it to grow. The online resources, some of which we've already introduced, provide a vast support network for LEGO hobbyists of all interests.

Meeting People

You can meet others who share your interests and form friendships via online LEGO discussion sites, local and regional LEGO clubs, and LEGO fan events. The three authors of this book met each other first through the online community on LUGNET, and later in person at LEGO events and through visits to each others' houses. You too can find people who share your interests and build friendships as well as LEGO models with them.

Online Discussion

In Chapter 19, we talked about a few LEGO discussion sites. Here we'll briefly recap, but for a more thorough introduction, see the "Finding an Audience For Your Models" section in Chapter 19. We'll also add a few more that weren't previously mentioned.

> **LUGNET: The LEGO Users Group Network** (http://www.lugnet.com): Discussion of all LEGO topics through extensive newsgroups, available via web, email, and NNTP news.
>
> **FBTB: From Bricks to Bothans** (http://www.fbtb.net): A community for LEGO Star Wars fans.
>
> **BZPower** (http://www.bzpower.com): The largest BIONICLE website on the Internet. BZPower provides the latest BIONICLE news, reference, and discussion.
>
> **1000steine.de** (http://www.1000steine.de): A German-language online LEGO community.
>
> **LEGO Message Boards** (http://boards.lego.com): Moderated discussion for members of LEGO's Online Club.
>
> **BrickFilms** (http://www.brickfilms.com): A community for fans of LEGO stop-motion animation.

Local LEGO Clubs

The online communities above, most notably LUGNET, paved the way for the face-to-face LEGO communities to flourish. In fact, the words that make up LUGNET's partial acronym, the **LEGO U**sers **G**roup **Net**work, illustrates its focus on "user groups," or local clubs that center around playing with or "using" LEGO. LUGNET made a big contribution to the start of local LEGO clubs through the local discussion groups it provides (http://news. lugnet.com/loc/). Today, local LEGO clubs are common across the United States, Europe, and Australia. There are also significantly sized LEGO organizations in Japan and Brazil.

Finding a Local LEGO Club

If you live near a major metropolitan area, there's probably a LEGO club nearby. You might want to try searching the Internet for information, including terms such as your city name and "lego club" or "lego users group." There are also a couple of sites that you can use to find out more about a LEGO club in your area. We list them below.

LUGMap — http://www.lugnet.com/map/

To find a LEGO club near you, visit the LUGMap at http://www.lugnet.com/map/. Browse the site by continent, country, and region, until you've found a link to your local club. Where applicable, the LUGMap will link to local club websites.

International LEGO Train Club Organization — http://www.iltco.org

LEGO Train Clubs are very popular and very active worldwide. You can find out about a local LEGO Train Club via the ILTCO website — http://www.iltco.org. The International LEGO Train Club Organization serves as a representative body for these LEGO train clubs worldwide. The organization coordinates LEGO train club support with the LEGO Company. It also serves as a medium for LEGO train clubs (also known as "LTCs" to insiders) to communicate with each other, comparing notes and coordinating efforts.

Starting Your Own Local LEGO Club

If there isn't a LEGO club in your area, why not start one? First, find out whether other LEGO fans live in your area.

If you're a kid looking to find others your age for a club, look no further than your street, school, scout group, or church. Chances are, you'll find other kids who enjoy playing with LEGO just as much as you do. There are some benefits to forming a club over just getting together and playing — you can create your own identity, put on displays for others, participate in activities like FIRST LEGO League (http://www.firstlegoleague.org) and more! For more organized activities, we recommend you find an interested adult sponsor to help you, such as a parent or teacher.

If you're looking for adult LEGO fans, we recommend posting in your own local LUGNET newsgroup. The main page for LUGNET's local groups is http://news.lugnet.com/loc/. This section is subdivided by country, and divides the individual countries by state, province, region, or city. Browse this section of the LUGNET site to find the group that corresponds to your local area. When you get there, post to the group introducing yourself, state that you are interested in forming a LEGO club, and ask for others who live in your area to speak up.

NOTE   *Tim Courtney lives in the Chicago suburbs, so his local LUGNET newsgroup is lugnet.loc.us.il.chi. Jaco van der Molen, one of the people who helped us while we were writing this book, lives near Amsterdam, the Netherlands, so his local LUGNET newsgroup is lugnet.loc.nl.ams.*

Holding Your First Club Meeting

Once you have a group and know where the members live, pick a spot to hold your first in-person meeting. If you are nervous about meeting people from the Internet, pick a neutral, public meeting spot such as a mall or a library. While you should always be careful about who you meet, LEGO fans, although they can be a bit nutty, are pretty much harmless. In fact, thousands and thousands of LEGO fans worldwide have met in person after communicating online, only to form stronger friendships around their favorite toy!

LEGO Fan Events

Local clubs and the Internet communities have fostered yet another level of in-person LEGO fan interaction. Regional, national, and even international LEGO fan events have been organized and hosted by LEGO fans like you in many parts of the world. These events, along with the activities of local LEGO clubs, have caught the eye of the LEGO Company, which is now aiming to support such events in the future. In fact, LEGO Direct, the LEGO Company's direct-to-consumer and community division, makes a point to send representatives to most of the events we list. We've included the most notable LEGO fan events below, categorized by region.

United States

BrickFest™ — http://www.brickfest.com

BrickFest was the first of the large LEGO fan events in the United States. Founded by the Washington Metropolitan Area LEGO Users Group (WAMALUG), this event is held annually for one weekend during the summer on the George Mason University campus in Arlington, VA (just outside of Washington, D.C.). BrickFest is oriented towards adult LEGO fans. It provides a venue for fans to come together, meet in person, display their creations, and learn from each other in unique conference sessions and presentations. BrickFest has seen steady growth, from the first event in 2000 with around 50 people, to 2002's 250 in attendance. The people who travel from all over the U.S. to attend this event are among the most serious LEGO fans in the world.

BricksWest™ — http://www.brickswest.com

BricksWest held its first annual event in February of 2002, as the West Coast response to BrickFest. This event coupled a public display inside of the LEGOLAND California theme park with conference sessions for adult fans at a nearby hotel. BricksWest aims to be both an adult fan event and a family event. The public display component inside of LEGOLAND is very important to BricksWest organizers as well as the park, and is responsible for putting many a smile on kids' faces. Kids go to LEGOLAND all the time and see enormous models built by LEGO's Master Builders with their seemingly endless resources. BricksWest shows kids what amazing models can be built with individuals' personal collections and a little imagination. For the die-hard fan, BricksWest holds conference sessions and lectures just like BrickFest in the private event at a nearby hotel. BricksWest occurs annually in February on Presidents' Day weekend.

Northwest BrickCon — http://www.sealug.org/nwbrickcon.html

This is another flavor of fan event that is located in the United States. Seattle, Washington hosted the first annual Northwest BrickCon in early October 2002. This event was held free to the public in the Seattle Center. Adult fans from around the region set up their LEGO creations to display for the public, while they socialized with one another. This event integrated the public display component much more than BricksWest did, with its entire venue open to the public.

UK and Europe

LEGOWORLD — http://www.legoworld.nl

LEGOWORLD is the largest LEGO event in Europe, and quite possibly the world. Held in Zwolle, Netherlands, this family event combines models and activities brought in by the LEGO Company, adult fans from the Dutch LEGO club De Bouwsteen (www.debouwsteen.com), and concerts by Fox Kids Network. The adult fans display large train layouts, models from virtually every LEGO theme, historical set displays, and more. If you are at all able to attend, LEGOWORLD is an event not to miss. It is held annually in mid-to-late October.

1000steine-Land — http://www.1000steine.de

1000steine-Land is a new event in 2003, organized by the German LEGO fan website 1000steine.de. The first annual 1000steine-Land was held in Berlin in early July 2003. This was the first German LEGO event designed for families who are not involved with the online LEGO community. The event included building workshops for kids, model displays, and presentations from the LEGO Company.

Annual LEGO Festival — http://www.brickish.org

The UK's LEGO club, the Brickish Association, hosted the first Annual LEGO Festival at the end of April 2003, at the Dilhorne Village Hall in Stoke-on-Trent. The event featured model displays, building competitions, and more.

Australia

LugOz BricksMeet — http://news.lugnet.com/loc/au/

Australian fans host a national LEGO gathering annually around June. Admittedly, they aren't as organized as some of the other events in the world. Their regional and state LEGO clubs rotate who hosts the meeting each year. There's no central website for the BricksMeet (or whatever they decide to call it each year), but if you live in Australia and are interested, get involved via LUGNET's Australia local group, shown above.

LEGOLAND Theme Parks

The LEGO Company owns four LEGOLAND theme parks worldwide. These are family attractions, with a focus on exploration, creativity, and imagination. All four parks feature enormous models and sculptures made entirely from LEGO bricks, as well as breathtaking "Miniland" displays, featuring scale models of various cities, real-world landmarks, and rural life. The models at LEGOLAND parks are incredibly detailed. The parks also feature rides, hands-on play activities, and more. LEGOLAND parks are a must-see for all LEGO fans, and are great fun for kids of all ages. For more information on the four parks, visit http://www.legoland.com.

LEGOLAND Billund

LEGOLAND Billund is the original LEGOLAND Park, constructed in the LEGO Company's hometown of Billund, Denmark. The park opened its doors in 1968, and since then has more than doubled in size. Today the park attracts visitors from all over the world.

LEGOLAND Windsor

LEGOLAND Windsor was the second LEGOLAND Park to be built. It opened in 1996 and is located in Windsor, England.

LEGOLAND California

LEGOLAND California opened in 1999 as the first LEGOLAND park in the United States. It's located in beautiful Carlsbad, California, only a couple of miles from the Pacific Ocean. Carlsbad is between two large population centers, San Diego and Los Angeles, and its location makes for a perfect family vacation spot.

LEGOLAND Deutschland

LEGOLAND Deutschland opened in 2002 in Gunzburg, Germany. This is Europe's third LEGOLAND park.

Reference

This chapter has already highlighted a few LEGO reference sites. Here are a few more, which aren't LDraw-based, but are useful nonetheless. For more reference sites, see the "Taking LDraw Further" section earlier in the chapter.

BrickSet™ — http://www.brickset.com

BrickSet, created by Huw Millington, is an excellent online LEGO set reference. It cross-references the LUGNET set database, eBay auctions, BrickLink sales, AuctionBrick auctions (www.auctionbrick.com), and the MISBI (Mint in Sealed Box) price guide (www.misbi.com). BrickSet also allows you to post reviews of individual sets and rate them.

Technica — http://w3.one.net/~hughesj/technica/technica.html

Technica is a reference site that focuses on the history of LEGO Gear Wheel, Technical, Expert Builder, Technic and MINDSTORMS sets and elements. It includes parts references as well as historical pages, with images of patents, box art, and sets throughout the ages. The historical section of the site also provides an in-depth commentary on the history of this genre of LEGO elements. Technica's Registry classifies elements and provides images of them. This site is not only fun to browse, but useful for the LEGO Technic fan.

LEGO Lexicon — http://members.chello.nl/~f.buiting/lego/lexicon.html

Frank Buiting maintains an encyclopedia of LEGO-related words, called the "LEGO Lexicon." This site is useful for all of you who want to familiarize yourself with the jargon online LEGO fans use when discussing the hobby with each other. It catalogs words related to the LEGO Company, the online community, and non-LEGO words that are used frequently in the community.

Summary

In this chapter, we took you through some aspects of the LDraw community today, and introduced you to the available resources and activities for LEGO fans interested in the LEGO hobby. We discussed some of the nature of LDraw.org, and how you can get involved with the LDraw community in discussion, as a developer, or as a parts author. We listed various online reference guides derived or partially derived from the LDraw parts library, and we shared ideas for the future of the LDraw system. We discussed the online LEGO hobby, and also introduced you to various outlets for meeting other fans, via local LEGO clubs, events, and LEGOLAND's theme parks. Overall, this chapter was a quick primer about the diverse LEGO hobby.

A

GLOSSARY OF TERMS

Animation Tools   Category of LDraw-based tools that specializes in creating animations based on LDraw model files.

Axis   A line used as a reference to measure distances or angles in a coordinate system. The LDraw coordinate system has 3 dimensions, and therefore 3 axes (X, Y, and Z). (Figure 21-3 displays a diagram of LDraw's 3D coordinate system.)

Backface Culling   Detecting backwards-facing elements in a 3D object for the purpose of eliminating them from the rendering task. These backwards-facing elements are unnecessary because the camera does not see them, and eliminating them speeds up render time.

Bezier Curve   A mathematical curve defined by control points (Chapter 10).

BFC   See *Backface Culling*.

Bill of Materials   For our purposes, a Bill of Materials is an LPub-produced image that contains graphics of all the parts needed to build a specific model, along with the required quantity of each (Chapter 15).

Buffer Exchange   MLCad extension that allows you to save a "buffer" or record of a model at one step and recall the model as saved in a later step. Useful for creating exploded views in building instructions (Chapter 12). Buffer Exchanges are also recognized by LPub.

CAD   Computer Aided Drafting/Design.

Callout   Box in a building instruction page that contains steps for a sub-assembly (Chapter 18).

Center of Rotation   The location in 3D space that a part or sub-model rotates around. By default, a part's center of rotation is its own origin. In MLCad, you can set custom centers of rotation to suit your needs (see Chapter 5 for details).

Color Code   A value representing a color in the LDraw file format. Each color code represents a color used by LEGO parts. Color codes do not match RGB values. You can read more about LDraw colors in Chapter 6 and see the colors they represent in Appendix E, "Extended and Dithered Color Information."

Color, Dithered   A color that is rendered by mixing pixels of two other colors. In LDraw, color codes 256 through 511 represent dithered colors. You can read more about LDraw colors in Chapter 6 and in Appendix E, "Extended and Dithered Color Information."

Color, Extended   A color code that has been defined by LDraw.org, but is not recognized by the original LDraw program. You can read more about LDraw colors in Chapter 6 and in Appendix E, "Extended and Dithered Color Information."

Color, Standard   A color code that the original LDraw package recognizes.

Comment   A line type in the LDraw file format that allows the author to insert a comment. Comment commands are denoted by a 0 in the line type column. Additionally, members of the LDraw community have begun using two slashes (//) to further separate comments from meta-commands, since they use the same line type (Chapter 6, Appendix D). Example: `0 // Hi, I'm a comment!`

COMPLETE.EXE   This file is distributed via LDraw.org, and contains all of the official LDraw parts in the library, minus James Jessiman's original parts. These parts are packaged with LDRAW027.EXE. If you are simply updating your parts library with the latest update, you don't need COMPLETE.EXE. Instead, LDraw.org provides individual update files for you to download, with only the most recent parts added to the library.

Constraint Part   An LSynth-specific part used to guide the path a flexible element will take when created by LSynth (Chapter 10).

.DAT File   .DAT was the original extension for all LDraw files. Today, .DAT is used to define parts. The extension, although ambiguous, is still used, because James's LDraw and LEdit look for parts with the .DAT extension. Today, LDraw models use .LDR as their extension, and Multi-Part LDraw files use .MPD.

Edge Color   The color an LDraw viewer or editor produces on the edge where two planes of a part intersect. This color is determined by the part's color, and is usually a complement of the part's color.

Editors   Class of LDraw-based tools that allow you to interact with an LDraw file in a graphical or text-based environment. The purpose of an editor is to create or modify an LDraw model or part.

Element   Another word for *Part*.

Exploded View   A series of unconnected parts that are aligned to display them being put into the proper place. Exploded views are often accompanied by arrows, pointing to the location a part will connect to (Chapter 12).

Explosion   See *Exploded View.*

Fair Play   The LEGO Company's document that explains their intellectual property claims, and sets guidelines for fans' use of their intellectual property (Chapter 19).

File Format   See *LDraw File Format*.

File Format Converters   Class of LDraw-based tools that convert an LDraw file into another 3D file format.

Functional Building   Construction and documentation approach that separates movable components of models into separate LDraw sub-models, so that all of the parts in a component can be moved easily as one; using this technique, you can pose models easily (Chapters 8 and 12).

Ghost   MLCad extension that allows you to hide a part from view when viewing a sub-model via a higher-level model. Used in association with *Buffer Exchange* (Chapter 12). Ghost is also recognized by LPub.

Grid   Set of evenly spaced, invisible parallel lines running parallel to all three axes in a 3D world. In LDraw-based editors, the grid helps you align parts accurately. Coarse Grid generally allows you to move parts based in half-stud increments on the horizontal, and plate increments in the vertical. It also allows for 90-degree rotations. Medium Grid allows for quarter-stud and half-plate movements, respectively, along with 45-degree rotations. Fine Grid or Grid Off allows for 1 LDU movement in each direction, and 1- to 5-degree rotations.

Group   An MLCad-specific feature that associates a group of selected parts so that they can be moved and rotated as a single part.

Include Files   Files that contain code that can be referenced within another file and "included" as though their contents were a part of the file referencing it.

Initialization File   A file that is used to store settings that pertain to rendering in POV-Ray or MegaPOV (Chapter 16).

James Jessiman   Creator of the original LDraw package (LDRAW.EXE and LEDIT.EXE) along with the file format. James died on July 25, 1997 of flu complications, at the age of 26.

James Jessiman Memorial Page   The original James Jessiman Memorial Page was founded by Terry Keller soon after James's death. The site was a place for people to write letters expressing their loss, as well as celebrate James's work during his life. The site was the central location for LDraw related tools until the launch of LDraw.org on July 7, 1999. LDraw.org currently hosts Terry Keller's memorial page at http://www.ldraw.org/community/memorial/archive/.

L2P   L2P (LDraw to POV-Ray) was the first LDraw → POV-Ray file format converter. The program converts LDraw files to POV-Ray files using the LGEO POV-Ray library. L2P and LGEO were written by Lutz Uhlmann.

L3Lab   Windows LDraw viewer with many advanced options, written by Lars C. Hassing. Very popular among parts authors for its advanced features (Chapter 21).

L3P   L3P is a program that converts LDraw models to POV-Ray scenes without the need for a separate POV-Ray LEGO parts library like LGEO. However, it can make substitutions using higher-quality LGEO parts where they are available. It also has a file-check mode, useful for finding problems in part files (Chapter 14).

L3PAO   L3PAO is a Windows interface shell written for L3P. Because L3P is run on the command line, it can be cumbersome if you are declaring a lot of options. Using L3PAO allows you to set your L3P options in a Windows-friendly environment (Chapter 14).

LDLite   Windows LDraw renderer/viewer written by Paul Gyugyi. The first LDraw renderer for Windows. LDLite has features for making excellent instruction renderings.

LDLite Language Extensions   A set of graphic meta-commands that provide extra functionality in LDraw part files.

LDList   Utility written by Anders Isaaksson that searches the LDraw Parts Library by part name and keyword, allowing you to search for and preview parts in the library easily.

.LDR File   Default file extension for LDraw model files.

LDraw   1. LDraw: System of tools that is based on the LDraw Parts Library and File Format. 2. Original LDraw Package: Original software package written by James Jessiman in 1996. Included LDRAW.EXE, the original renderer, and LEDIT.EXE, the original graphical editor. 3. LDraw (verb): to document one's LEGO models using the LDraw file format. E.g. "Tim is going to LDraw his crane model for the book *Virtual LEGO*"; "Ahui LDrew his spa for the MegaPOV chapter of *Virtual LEGO*." ☺

LDraw-Mode   LDraw editor for Emacs.

LDraw Language   Extensible Graphics Description Language (GDL). The LDraw GDL uses a text-based file format and is common to both models and parts.

LDraw Unit (LDU)   Basic unit of measurement in the LDraw language. Approximately equal to 0.4 millimeters. See Figure 5-4 and the LDU chart on the inside back cover.

LDraw.org   The LDraw Organization, or LDraw.org, is the online community of people who support LDraw. LDraw.org is also the group's website.

LDraw.org Parts Library   Collection of virtual representations of LEGO parts made in the LDraw File Format that have been approved for official distribution from the LDraw.org website. The parts library is the common link between all of the various editors and renderers (Chapter 4).

LDView   Windows LDraw renderer/viewer written by Travis Cobbs.

LeoCAD   LeoCAD is a Windows CAD program that can read and write to the LDraw file format, but uses a separate proprietary file format as a default (http://www.leocad.org).

LGEO   Lutz Uhlmann's POV-Ray library of LEGO elements, which is used by L2P and L3P.

Line   A command in a raw LDraw file, or a type of command, that draws a line in an LDraw file.

Line Type   First number on each line of an LDraw file. This number denotes what type of line follows. Line types are one of the following: comment, meta-command, part-file reference, line, triangle, or quadrilateral (Appendix D).

Linetype   A code identifying the type of an LDraw command (Appendix D).

LPub   Applications for batch processing renderings of individual steps in an LDraw model file using L3P and POV-Ray/MegaPOV (Chapter 15).

LUGNET   LEGO Users Group Network. This is the website where LEGO fans come to communicate with each other concerning all facets of the LEGO hobby. It is a private news (NNTP) server along with a collection of LEGO-related databases. LDraw is discussed extensively in the LUGNET CAD newsgroups (http://www.lugnet.com).

Matrix   A 2D array of numbers (Chapter 21).

Minifig, Minifigure   A LEGO figure commonly found in System sets (see Figure 9-1). *Minifig* is the shortened version of "minifigure" used by the online LEGO fan community.

MKLIST.EXE (and variants)   DOS utility included in the original LDraw package (and since updated by LDraw.org) that generated the PARTS.LST file, a list of parts.

MLCad   A second-generation LDraw editor for Windows, written by Michael Lachmann.

MODELS (Folder)   Subdirectory of the LDRAW folder created when the ldraw027.exe archive is extracted. C:\LDraw\MODELS\ is the default location for all LDraw models.

Model of the Month Competition   Monthly competition hosted on LDraw.org that allows website visitors to select their favorite models from a list of entries.

.MPD File   Short for Multi-Part DAT. MPD is a file format with extensions for containing multiple LDraw models in one file (Chapter 8).

Origin   The position (0, 0, 0) on the 3D coordinate system.

Parts Authoring   Process of creating LDraw parts using lines, triangles, quads, conditional lines, and primitives (Chapter 21).

P (Folder)   Subdirectory of the LDRAW folder created when the ldraw027.exe archive is extracted. C:\LDraw\P\ is the default location for all LDraw part primitives.

Part   In the LDraw system, a part is a file that represents a single LEGO-style part. The collection of official parts make up the LDraw.org Parts Library. Parts reside in the PARTS subdirectory of the LDRAW folder upon extraction of ldraw027.exe.

Part Number   LDraw parts are identified by alphanumeric codes, which are the *part numbers*. Most often, the official LEGO part number is used if it is known (they are usually printed on the underside of a brick). When an official LEGO part number is not known, an arbitrary number is chosen. Because the identification codes are primarily numeric, they are called part numbers.

PARTS (Folder)   Subdirectory of the LDRAW folder created when the ldraw027.exe archive is extracted. C:\LDraw\PARTS\ is the default location for all LDraw parts.

Parts Library   See *LDraw.org Parts Library*.

PARTS.LST   File containing a complete list of the parts on a user's computer. This file is used by all LDraw editors to generate internal parts lists for part selection during the editing process.

Parts Tracker   Online utility that manages submission of parts, approval into the LDraw.org Parts Library, and packaging into Parts Updates. The Parts Tracker is located at http://www.ldraw.org/library/tracker/.

Parts Update   A release of new parts into the LDraw Parts Library. These updates are made available on a semi-regular (roughly bi-monthly) basis at http://www.ldraw.org/library/updates/.

Partsref   Graphical database of LEGO parts (currently limited to what is in the LDraw library) that contains cross-references to several other parts databases on the Internet (http://guide.lugnet.com/partsref/).

Point   A single location in space that has a coordinate on each of the three axes.

Polygon   A 2D shape with straight edges. Polygons can have as many edges as you want them to have; however, LDraw only supports triangles and quadrilaterals.

Post-Processing   Modifications applied to an image or a rendering after it has been produced (Chapters 15, 17 [raytracing], 18 [building instructions]).

.POV File   File format for the POV-Ray raytracing program (Chapters 13, 14, 16, and 17).

Primitive   Representation of an element that is frequently used in a LEGO part. An example is STUD.DAT, which models a single LEGO stud. Primitive files are used in place of creating the element each time it is used in a parts file. They reside in the P subdirectory of the LDraw folder.

Quadrilateral   A four-sided polygon.

Raytracers   1. Programs that use raytracing as their method of rendering. We taught you to use POV-Ray and MegaPOV in this book. 2. Programs that are not LDraw-based, but are used to render 3D images. LDraw files must be converted to be compatible with programs classified as "renderers" in the LDraw system of tools.

Rendering   1. (verb) The act of using a software program to generate an image or a view of a model. 2. (noun) A computer-generated image.

Rotation   Changing the angle of an object around an axis without moving the object in 3D space.

Rotation Step   Feature that lets you change the view angle of your model when you browse building instruction steps. Recognized by MLCad and LPub (Chapter 12).

Scaling   Changing the size of an object in one axis or multiple axes (Chapter 21).

Scene   Rendered composition featuring an LDraw model or models along with some form of background or environment. The environment can be created using LDraw parts or non-LDraw 3D elements.

Scene of the Month Competition   Monthly competition hosted on LDraw.org that allows website visitors to select their favorite 3D-rendered scene from a list of entries.

Second Generation Editor   Label for any editor of the LDraw file format that was written after LEdit. Most significantly, non-DOS editors. Examples: MLCad, LeoCAD, BrickDraw3D, and LDGLite.

SNOT   LEGO community acronym for "Studs Not On Top." It's a building technique that identifies the large portions of a model's studs that do not face in the up direction.

Step   Line that marks the break between one building-instruction step and the next. Steps are denoted by the meta-command 0 STEP (Chapter 6).

Stud   Basic connector in the LEGO system. A stud is the "bump" that protrudes from the top of bricks. Dimensions of LEGO parts are commonly measured in studs. In the LDraw system, 1 "stud" equals 20 LDU.

Sub-assembly   Component of a model that is distinct in the building process, but not necessarily separated from the main model by a unique sub-model (Chapter 12).

Sub-model   The use of one model file as a part in another model. This is a useful method for creating large models, or models with many moving parts, if the author wants to reposition the parts easily (Chapter 8).

Transformation   The act of changing an object by performing operations through a transformation matrix. Typical changes are movement (translation), rotation, or scaling (resizing) (Chapter 21).

Translation   Moving an object from one position to another in 3D space (Chapter 21).

Utilities   Class of LDraw-based tools that perform special functions on LDraw files, such as mapping textures and creating flexible elements. These programs are typically geared toward advanced users.

Vertex   An endpoint on a polygon or line.

Viewers   Programs that are able to view LDraw files, but not edit or write them.

B

WEB LINKS

This appendix provides a concise list of important LDraw- and LEGO-related websites, as well as additional links of interest.

LDraw Links

Resources

LDraw.org (http://www.ldraw.org): The official LDraw website. Go here to download software and parts updates, and to find the latest news.

LUGNET CAD Newsgroups (http://news.lugnet.com/cad/): Many LDraw users and developers use the lugnet.cad newsgroups to communicate with each other.

Datsville (http://www.geocities.com/Heartland/Fields/1864/datsville.html): Datsville was a community project that sprung out of the original L-CAD mailing list in 1997. It is a compilation of many individuals' buildings, arranged into a virtual LEGO town.

LDraw and LEdit Tutorial (http://library.thinkquest.org/20551/): Bram Lambrecht's original LDraw and LEdit tutorial, which was written for the ThinkQuest competition in 1998. This tutorial is a very useful tool for those looking to learn how to use the original software.

Software

MLCad — http://www.lm-software.com/mlcad/

L3P — http://www.hassings.dk/l3p/l3p.html

L3Lab — http://www.hassings.dk/l3lab/l3lab.html

L3PAO — http://l3pao.malagraphixia.com/L3PAO.htm

LPub — http://www.users.quest.net/~kclague/LPub/

POV-Ray — http://www.povray.org

MegaPOV — http://megapov.inetart.net

LDAO — http://home.earthlink.net/~steve.bliss/ldao/

LDList — http://user.tninet.se/~hbh828t/ldlist.htm

LDLite — http://sourceforge.net/projects/ldlite/

LDView — http://home.san.rr.com/tcobbs/LDView/

LDDP — http://www.m8laune.de/

Unofficial Libraries

These websites host unofficial LDraw parts libraries of non-LEGO parts.

LDraw Apocrypha — http://members.bellatlantic.net/~drteeth1/clones/clone5.htm

Clone Bricks — http://home.swipnet.se/~w-20413/mybricks/warning.htm

Virtual Construction — http://users.ifriendly.com/fourfarrs1/construction.htm

Non-LDraw CAD Systems

Here are links to a couple of non-LDraw LEGO CAD systems. These programs can import and export LDraw files.

LeoCAD — http://www.leocad.org

BlockCAD — http://user.tninet.se/~hbh828t/proglego.htm

LEGO Community Links

Here are links to some of the most popular fan-created LEGO sites.

LUGNET — http://www.lugnet.com

From Bricks to Bothans — http://www.fbtb.net

BZPower — http://www.bzpower.com

International LEGO Train Club Organization — http://www.iltco.org

Building Instructions Portal — http://www.bricksonthebrain.com/instructions/

Peeron LEGO Inventories — http://www.peeron.com/inv/

BrickShelf — http://www.brickshelf.com

BrickSet — http://www.brickset.com

BrickLink — http://www.bricklink.com

BrikTrak — http://www.briktrak.com

BrickFilms — http://www.brickfilms.com

Cool Brick Movies — http://www.coolbrickmovies.com

OzBricks — http://www.ozbricks.com

LEGO Company and Affiliates

These are the addresses for the LEGO Company's website and affiliates' websites:

Official LEGO Website — http://www.lego.com

First LEGO League — http://www.firstlegoleague.org

PITSCO LEGO Educational Division — http://www.pitsco-legodacta.com

Educational

Robotics Learning (http://www.roboticslearning.com): Provides workshops for hands-on learning with LEGO MINDSTORMS, and works with First LEGO League.

People

Here are links to several people in the LEGO community. Some of these people helped to make this book happen; others have done notable work and we feel they deserve a mention.

Tim Courtney — http://www.timcourtney.net

Steve Bliss — http://home.earthlink.net/~steve.bliss/

Ahui Herrera — http://www.ozbricks.net/jediagh/

Jaco van der Molen — http://home.zonnet.nl/jmolen/

Jake McKee — http://www.bricksonthebrain.com

Eric Olson — http://olsone.pair.com

Anton Raves — http://www.antonraves.com

Tamy Teed — http://www.tanarth.com

Larry Pieniazek — http://www.miltontrainworks.com

Steve Barile — http://www.bricworx.com

Jon Palmer — http://www.zemi.net/lego/

Dan Jassim — http://www.brickshelf.com/cgi-bin/gallery.cgi?m=DanJassim

Carsten Schmitz — http://m8laune.de

Orion Pobursky — http://www.pobursky.com

Kevin Clague — http://www.users.quest.net/~kclague/

Bruce Lowell — http://bruce.kus-numa.net/lego/

C

LDRAW COMMUNITY HISTORY

The LDraw community originated around 1996 when James Jessiman placed his LDraw software on his website for visitors to download freely. It is reported that the first release of LDraw only contained three parts. Early acceptance of LDraw was slow, probably because of James's low-key approach to presenting his amazing tool. LDraw also presented a steep learning curve for many users, even though it was easier to use than previously available virtual LEGO systems.

At about the same time, there were several attempts at virtual LEGO software, both free and commercial. Some notable free attempts were BriCAD, SimLego, and BlockCAD. Gryphon Bricks was a commercial application that allowed you to build using LEGO-like bricks (the parts didn't explicitly follow physical LEGO parts). All but BlockCAD eventually fell by the wayside. There were also several part libraries created for various raytracing tools. These libraries included L3G0, Anton Rave's parts library for POV-Ray, and LGEO. Of these, only Anton Rave's library is still under active development, although LGEO parts can be used in combination with L3P/POV-Ray renderings.

In the early days of the LDraw community, Leonardo Zide wrote the editor LeoCAD. Leo used the LDraw parts libraries but created his own native format to implement features that he desired. However, he designed LeoCAD to import and export LDraw files. LeoCAD was adopted by some, but not all found it intuitive; however, LeoCAD boasts a relatively small (compared to MLCad) but loyal user base.

The Minifig World Tour was an event that helped to increase acceptance of LDraw. In early 1997, a small group of online friends organized a tour to send two minifigs to visit willing participants from all over the world. This Minifig World Tour (as it was called) provided the catalyst for LDraw to become the de facto standard format for unofficial LEGO CAD tools.

As the Minifig World Tour gained popularity, so did the idea of LEGO CAD tools. The original LDraw tools James released were very accessible to the average online LEGO enthusiast of the day. Most people online at the time possessed an intermediate to advanced knowledge of computers, so LDraw's steep learning curve didn't intimidate many. Soon after Doug Finney set up the Minifig World Tour mailing list, there was also demand for a CAD mailing list. The original LDraw community grew out of the L-CAD listserv email list.

James Jessiman was active on the L-CAD list until his death on July 25, 1997. News of his passing reached the group quickly because a friend of James posted word on his website, which the community used to download LDraw. His death was a unique experience for list members, because for many — if not all — of them, it was the first time a friend they had never met died. Many did not realize the emotional connection they had with their online friends until this sudden event. There was an immediate outpouring of grief on the L-CAD list. Within days, Terry Keller erected the James Jessiman Memorial, a website devoted to James and to preserving his memory for LEGO fans around the world (the original site is archived at www.ldraw.org/community/memorial/archive/).

For almost two years following James's death, the James Jessiman Memorial served as the starting point for LDraw resources on the Web. The community developed a process of "official parts updates," which in simple terms meant that releases of parts met arbitrary quality standards. This set the precedent for the Official LDraw.org Parts Updates of today. Terry Keller coordinated these parts updates, serving as the first parts administrator. He manually compiled the various files sent to him, posting voting lists for members of the community to review.

In late 1998, the LEGO Users Group Network (LUGNET) was introduced. The members of the L-CAD mailing list agreed that it would be good for L-CAD discussion to happen on LUGNET. So the L-CAD mailing list was discontinued, and was replaced by several newsgroups on LUGNET, most notably lugnet.cad (for general discussion) and lugnet.cad.dev (for development discussion). To this day, the LUGNET CAD newsgroups are the center of LDraw community discussion.

The LDraw.org website was launched on July 7, 1999, almost two years after James Jessiman died. Tim Courtney spearheaded the website's creation; soon after its release, LDraw.org gained volunteers and quickly became the center for LDraw resources on the Web. Though the site started with lofty goals, it remains a work in progress, supported by a core group and numerous members of the LDraw community. About this time, Terry Keller passed the torch of maintaining the LDraw parts library to Steve Bliss, who is the library's current administrator.

MLCad's release in late 1999 revolutionized the LDraw system. The program quickly became a favorite among new users and proved helpful to those who had trouble grasping James Jessiman's DOS editor, LEdit. Fans could now create LDraw models in a Windows environment, using the mouse, instead of being limited to a keyboard. MLCad gave the LDraw system the boost it needed to become more widely accepted among online LEGO fans.

LDraw (and related tools) gained recognition at LEGO conventions as well as on the Internet. In the fall of 1999, John Van Zwieten presented a talk on LDraw at the Massachusetts Institute of Technology's Mindfest™, an event centered on robotics, education, and LEGO bricks. LDraw presentations were also held at BrickFest 2000 and the LEGO Maniac KidVention, a week-long family event held at LEGOLAND California in July of 2000.

Despite the gains in the LDraw community and the variety of tools that were developed during this time, there was still no significant Macintosh LDraw software. Neils Dueck founded the LDraw for the Macintosh Campaign in March of 2000. Their goal was (and is to this day) to raise awareness of the system in the Macintosh community, and to solicit Mac programmers to write LDraw tools. The LDMC has been successful in encouraging programmers to write Mac software, though LDraw tools for the Mac are still rather primitive compared to their Windows counterparts.

By the end of 2000, the LDraw system had so entrenched itself in the online LEGO community that the LEGO Company decided to recognize the system in a software project of its own. The company invited several individuals to meet with its representatives to discuss a 3D file format project in early 2001. Six individuals from the LDraw community — Erik Olson, Lars C. Hassing, Leonardo Zide, Michael Lachmann, Steve Bliss, and Tim Courtney — met with LEGO employees from various departments to give feedback on the company's developments towards a common internal format. The company has since come forward with plans to make public its software developer kit, which will enable LEGO fans to write software for the new format, as well as write programs to convert between the LEGO format and the LDraw file format.

In the summer of 2001, Don and Robyn Jessiman, James Jessiman's parents, traveled from their home in Australia to meet LDraw.org volunteers at BrickFest, held in Arlington, Virginia. This was an especially moving moment in LDraw history, and gave members of the community a very special opportunity to meet those closest to James. James's parents sponsored a Memorial Award in their son's name, and presented it to Steve Bliss, to honor his hard work keeping the memory of James alive in LEGO fans worldwide by maintaining the LDraw parts library. The award was designed to be given annually; however, no candidate was chosen in 2001, mostly due to the low LDraw representation at BrickFest that year.

In mid-2001, the Parts Tracker system was launched on the LDraw.org website. This web tool provides a system for managing the submission, review, and release of new part files into the parts library. Two previous websites known as the Parts Tracker allowed people to request the creation of specific parts and gave parts authors a place to record their progress. The new system has enabled more frequent and larger releases of parts updates, and has encouraged more people to become involved in the process of reviewing and certifying new part files. The system tracks several hundred unofficial files at any one time — far more than the old, manual system could handle.

Fall 2001 brought LEGOWORLD, a five-day exposition designed for families in Zwolle, Netherlands. The LEGO Company, Fox Kids, and De Bouwsteen (the Dutch LEGO club) co-produced the event. Jaco van der Molen of De Bouwsteen organized and staffed an LDraw area where people were able to come and try out LDraw tools and learn about them. LEGOWORLD drew about 25,000 visitors in 2001. Jaco described the event as a resounding success for the [LEGO] hobby and for LDraw.

2002 brought many advancements to the LDraw community. BricksWest 2002 featured the most diverse array of LDraw sessions at a LEGO fan event to date. Kevin Clague introduced LPub, his software for automatically generating building instruction images. Finally, LEGOWORLD 2002 was again a success; Tim Courtney traveled from the United States to help Jaco with the LDraw display, and Dan Crichton flew from the U.K. to do the same. This event saw 35,000–40,000 people, up from the previous year.

As we write this, in 2003, there are several notable happenings in the LDraw community.

The LEGO Company has announced its LXF (or LEGO Extensible Format), an extendable 3D file format that it will be using in future 3D software. They have invited key members of the LDraw community to participate in discussions, and plan to release the Software Developer Kit (SDK) for individuals to use. A common desire among members of the community and the company is to see programs to convert files between LDraw and the new LEGO format.

The LDraw community has moved toward creating a standards body to efficiently extend the LDraw file format. While this standards body has not yet been erected, the community has accepted a proposal by general consensus, and there are plans to implement it later in the year. Interest in creating a formal organization has also been revived in order to guide the LDraw community; key stakeholders are moving toward that goal.

The LDraw.org website recently moved servers from Jacob Sparre Andersen's machine in Denmark to Peeron.com's web space. A completely new, more automated website is in development. The new site will make posting and maintaining information more efficient for the administrators, and will allow more individuals to contribute directly to the site's success. For the surfer, this means the website will be updated more often, offering news items, tutorials, and more.

Aside from the pushes within the community, several opportunities are upcoming to continue to get the word out about LDraw to LEGO fans of all ages, and even to individuals in the computer graphics industry. This is an exciting time for the world of LDraw, and we're glad the system is as useful and as popular as it is.

D

LDRAW FILE FORMAT SPECIFICATIONS

The following specifications are reprinted from http://www.ldraw.org/reference/specs/fileformat.shtml. Make sure you check the LDraw.org website for an updated version of this document.

File Extensions

LDraw model files carry either the extension LDR (default) or DAT. LDraw part files carry the extension DAT.

Text-Based

LDraw files are formatted in plain text. Each line is a single command, and each line is independent. Only a few commands are used in the file format. Specific commands are identified by the first number on the line, which is called the line type. The contents and format of the rest of the line depend on the *line type*. It is not known whether command lines have length limitations.

Line Types

Each line in the file is a single command. The first number on each line is the line type. Valid values are 0 through 5, with the following meanings:

0 Comment or Meta-command
1 Part-file reference
2 Line
3 Triangle
4 Quadrilateral
5 Conditional Line (draws a line between the first two points, if the last two points are on the same side of that line)

If the line type for a line is not a valid value, the line is ignored.

Meta-commands require a keyword to follow the line type. These keywords must be in all caps and you should assume that if the meta-command doesn't require more information, you shouldn't give it any additional information.

Commands

Here is a complete list of LDraw drawing and meta-commands:

Model Title	Comment
Step	File Type
Write	Part
Clear	Line
Pause	Triangle
Save	Quadrilateral
Conditional Line	

A description of each command follows. Meta-commands are listed first, with the "real" commands after. The command names are made up, not "official".

Model Title

The descriptive name of the model- or part-file.

Line format:

```
0 title-text
```

where title-text is whatever you want to name your model.

This is a meta-command, sort of. If the first line in a file has line type 0, the remainder of that line will be considered the title for the file (at least for parts files). This overrides any meta-commands on the line.

Step Meta-Command

Marks the end of a building step.

Line format:

```
0 STEP
```

This command will cause LDraw (ldraw.exe) to stop until you press ENTER, and it will cause LDraw to save the current image into a bitmap file in the **ldraw\bitmap** directory. These two effects are controlled by the **-M** command line option (see **ldraw.doc** for info on **-M**).

Write Meta-Command

Displays a message at the top of the screen.

Line format:

```
0 WRITE message-text
```

or:

```
0 PRINT message-text
```

where message-text is whatever you want displayed on the screen.

The displayed message is not saved in the image-bitmap files.

Clear Meta-Command

Clears the screen.

Line format:

```
0 CLEAR
```

Useful for advanced model files. Don't forget to redraw all the parts you just erased by using this command.

Pause Meta-Command

Causes LDraw to stop until you press ENTER.

Line format:

```
0 PAUSE
```

This is like the Step command, but it does not save bitmaps and it is not affected by the **-M** command line option.

Save Meta-Command

Causes LDraw (ldraw.exe) to save the current image in a bitmap.

Line format:

```
0 SAVE
```

This is also like the Step command, but it does not cause LDraw to pause, and it is not affected by the **-M** command line option.

File Type Meta-Command

Indicates the type of file (model, part, and so forth) and whether it is part of the LDraw.org Parts Library.

Line format:

```
0 LDRAW_ORG type update-tag
```

or:

```
0 Official LCAD type update-tag
```

or:

```
0 Unofficial type
```

or:

```
0 Un-official type
```

This meta-statement should appear as the last line of the primary file header (that is, the first lines in the file, with the Model Title, Name, and Author lines). The first format (LDRAW_ORG) is the current standard for files in the official parts library. Older files still carry the second (Official LCAD) format. The valid values for the type field are:

- Part *
- Element -
- Subpart *
- Sub-part -
- Primitive *
- Shortcut *
- Alias *
- Cross-reference +
- Model -
- Submodel -
- File -
- Hi-Res Primitive –

> * = Values currently in use for official part files.
>
> + = Obsolete code, still appears on some official files.
>
> - = Only used in unofficial files

Some older official files do not include a type code on their File Type line. Unofficial files may have a variety of values after the initial tag. In particular, the type values of Element, Model, and Submodel may be used.

The exact format of the File Type line has changed over time. In general, this linetype should be considered case-insensitive and free-format.

Part Command

Inserts a part defined in another LDraw file.

Line format:

```
1 colour x y z a b c d e f g h i part.dat
```

where

- colour is a color code: 0-15, 16, 24, 32-47, 256-511.
- x, y, z is the position of the part.
- a through i are orientation and scaling parameters (see below for more on this).
- part.dat is the filename of the included file.

Parts may located in the **p**, **parts**, or **models** subdirectories (under the **ldraw** directory). They may also be located in the current directory. James arranged the directories so that **p** contains elements that are intended to be used as parts of elements (such as **stud.dat**). The **parts** directory contains finished parts, ready to be used in construction models. And the **models** directory contains construction models.

Part files can also include part commands. There doesn't seem to be a specific limit to how deep these references can go.

Line Command

Draws a line between two points.

Line format:

```
2 colour x1 y1 z1 x2 y2 z2
```

where

- colour is a color code: 0-15, 16, 24, 32-47, 256-511.
- x1, y1, z1 is the position of the first point.
- x2, y2, z2 is the position of the second point.

Triangle Command

Draws a filled triangle between three points.

Line format:

```
3 colour x1 y1 z1 x2 y2 z2 x3 y3 z3
```

where

- colour is a color code: 0-15, 16, 24, 32-47, 256-511.
- x1, y1, z1 is the position of the first point.
- x2, y2, z2 is the position of the second point.
- x3, y3, z3 is the position of the third point.

Quadrilateral Command

Draws a four-sided, filled shape between four points.

Line format:

```
4 colour x1 y1 z1 x2 y2 z2 x3 y3 z3 x4 y4 z4
```

where

- colour is a color code: 0-15, 16, 24, 32-47, 256-511.
- x1, y1, z1 is the position of the first point.
- x2, y2, z2 is the position of the second point.
- x3, y3, z3 is the position of the third point.
- x4, y4, z4 is the position of the fourth point.

Conditional Line Command

Draws a line between the first two points, if the projections of the last two points onto the screen are on the same side of an imaginary line through the projections of the first two points onto the screen.

Line format:

```
5 colour x1 y1 z1 x2 y2 z2 x3 y3 z3 x4 y4 z4
```

where

- colour is a color code: 0-15, 16, 24, 32-47, 256-511.
- x1, y1, z1 is the position of the first point.
- x2, y2, z2 is the position of the second point.
- x3, y3, z3 is the position of the third point.
- x4, y4, z4 is the position of the fourth point.

Colors

Colors are outlined in LEDIT.DOC. Two special values are 16 and 24. 16 is the "current color," that is, the color specified on the type 1 line that caused the current line to be executed. Color 24 is a complementary color to the current color, usually the bright/dark complementary shade; so if the current color is dark blue, color 24 would give bright blue.

Also, colors 256 through 511 are dithered, so if you want to combine colors J and K, figure your color value as

```
colour = (J * 16) + K + 256
```

The complementary color of J is used as the complementary color of the dithered value, so you can control the edge color (somewhat) by switching J and K.

Format of Linetype 1

The part command (linetype 1) has some unique features that require special attention. This section makes more sense if you're familiar with matrix math; it makes much more sense if you're familiar with using matrix math in graphic modeling.

Line format:

```
1 colour x y z a b c d e f g h i part.dat
```

Fields a through i are orientation and scaling parameters, which can be used in "standard" 3D transformation matrices. Fields x, y and z also fit into this matrix:

```
| a d g 0 |
| b e h 0 |
| c f i 0 |
| x y z 1 |
```

so that every point (x, y ,z) gets transformed to (x', y', z'):

```
x' = a*x + b*y + c*z + x
y' = d*x + e*y + f*z + y
z' = g*x + h*y + i*z + z
```

or, in matrix-math style:

$$| X' \ Y' \ Z' \ 1 | = | X \ Y \ Z \ 1 | \times \begin{vmatrix} a & d & g & 0 \\ b & e & h & 0 \\ c & f & i & 0 \\ x & y & z & 1 \end{vmatrix} \cdot$$

Explanation of Conditional Lines (Linetype 5)

Here is an example using the conditional line command. If we try to approximate a (cylindrical) stud with a hexagonal prism, we have something like this:

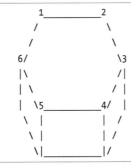

- The line below 1 should not be drawn because 6 and 2 are on different sides.
- The line below 2 should not be drawn because 1 and 3 are on different sides.
- The line below 3 should be drawn because 2 and 4 are on the same side.
- The line below 4 should not be drawn because 3 and 5 are on different sides.
- The line below 5 should not be drawn because 4 and 6 are on different sides.
- The line below 6 should be drawn because 5 and 1 are on the same side.

resulting in the following drawing:

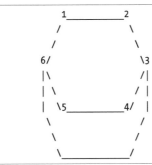

E

EXTENDED AND DITHERED COLOR INFORMATION

You can use the color explanations in this appendix, along with information on MLCad's Color Palette from Chapter 6 and the Color Chart on the front cover flap of this book.

Traditional LDraw Colors

The LDraw format, harkening back to the days of VGA graphics drivers, only allows for 16 colors! That's nothing compared to the multiple millions of colors you are used to enjoying on today's computer screens. Even when you're trying to create virtual LEGO models, so few colors are very limiting. We've seen the number of colors LEGO offers explode over the last few years with the introduction of Belville, Scala, licensed themes like Star Wars™ and Harry Potter™, and the newer non-licensed themes such as Life on Mars and Island Xtreme. With new colors such as tan, purple, and sand green, you can see how the LDraw color palette of 16 colors is inadequate for creating realistic 3D LEGO models.

Solving the Color Problem: Dithering

When we say VGA only supports 16 colors, we mean to say that it only supports 16 *solid* colors. Other colors can be achieved through a process called *dithering*, which is a way of mixing one color with another. LDraw (and MLCad, and the rest) create dithered colors by staggering pixels of two solid colors together like a checkerboard.

Transparent Colors

Transparent versions of the solid 16 colors are also available. The solid colors are *dithered* with undrawn pixels, allowing whatever is behind them to show through, giving a tinted transparent effect. Different programs use different approaches to provide transparent support.

Another Solution: Extended Colors

Because the original LDraw color set doesn't provide definitions for all possible LEGO colors, different programs offer various ways to define new colors. LDraw.org has also provided definitions for non-LDraw colors. These new color values can be thought of as an *extended* color set.

Often, these extended colors are defined in the range of dithered colors, allowing some backward compatibility with the original LDraw. An example of such a color is the extended definition for Sand Green: The LDraw.org definition for Sand Green is color 378, which is the dithering of Bright Green (color 10) with Gray (color 7). 378 gives a reasonably close color in LDraw, and newer programs can render 378 as a solid color with a good color match, instead of a dithered color.

Some other extended colors are defined from color values unused by LDraw. On LDraw.org's color specifications page, these colors are identified as "LDLite" colors. Using these color values allows programs to provide specific colors not supported by LDraw. Unfortunately, these colors are *completely* unsupported by LDraw. If you want to be fully compatible, you'll need to avoid these colors.

Why Stay with VGA?

The various LDraw tools are limited by the 16-color VGA palette because the file format will not allow for anything more. This loosely-organized community of developers has not created color standards that take them beyond the capabilities of the original LDraw software. This is both an organization issue and a standards issue, and until it is solved, LDraw colors will be based on this palette.

Most current LDraw tools recognize colors beyond LDraw's limited palette but these colors are not standardized across all packages. For example, MLCad defines 11 solid colors beyond LDraw's core 16 colors. Some of these extended colors are "semi-official." They are listed on LDraw.org's colors web page and are recommended for use. However, they are not recognized by LDraw or LEdit, so they are not fully official and standard.

Color by Number

The LDraw file format assigns each color a number. In the early days, with only LEdit available to create LDraw models, you would have to refer to each color's number and enter it by hand. Today, with MLCad's color palette feature, you can select a color visually and MLCad will enter the number into the file, so any LDraw-compatible program can understand it.

Dithered Colors

Dithered colors are assigned to begin at color code 256 (see Figure 6-24). Combining each of the 16 solid colors with every other solid color creates the range of dithered colors. Because the swatches are laid out in the fashion they are, each group of two rows represents one of the solid colors combined with every other solid color. We found the best explanation for this concept in a post by Steve Barile to LUGNET™, the LEGO Users Group

Network located in the site's CAD newsgroups: http://news.lugnet.com/cad/?n=9319. Here's an adaptation of Steve's explanation for your use. Scroll down MLCad's color palette until you get to 256, and read on.

Dithered Color Palette Layout

The formulas below explain the math behind the dithered color layout. This layout is based on Steve Bliss's formula: Color1 + (Color2 x 16) + 256 = Dithered Color Code. Here you can see the names of the colors being combined, and the formula used to get that color's number:

Row 1

The first two rows of colors are based on Black (color 0). The first row is color 0 dithered with colors 0 through 7:

- Black + Black → $0 + (0 \times 16) + 256 = 256$
- Blue + Black → $1 + (0 \times 16) + 256 = 257$
- Green + Black → $2 + (0 \times 16) + 256 = 258$
- Teal + Black → $3 + (0 \times 16) + 256 = 259$
- Red + Black → $4 + (0 \times 16) + 256 = 260$, and so on

Row 2

The second row is still based on Black (color 0), but dithered with colors 8 through 15:

- Dark Gray + Black → $8 + (0 \times 16) + 256 = 264$
- Light Blue + Black → $9 + (0 \times 16) + 256 = 265$
- Light Green + Black → $10 + (0 \times 16) + 256 = 266$
- Cyan + Black → $11 + (0 \times 16) + 256 = 267$
- Light Red + Black → $12 + (0 \times 16) + 256 = 268$, and so on

Row 3

The third row is based on Blue (color 1), and starts over, dithering Blue with colors 0 through 7:

- Black + Blue → $0 + (1 \times 16) + 256 = 272$
- Blue + Blue → $1 + (1 \times 16) + 256 = 273$
- Green + Blue → $2 + (1 \times 16) + 256 = 274$
- Teal + Blue → $3 + (1 \times 16) + 256 = 275$
- Red + Blue → $4 + (1 \times 16) + 256 = 276$, and so on

To utilize this formula, see the color chart and LDraw color codes on the first page of the color chart on the front cover flap. Figure E-1 shows the beginning of the dithered color section and the actual rows discussed above.

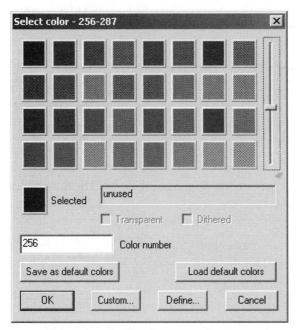

Figure E-1: Dithered colors begin at color 256 and continue through color 511.

Using Dithered Colors

Understanding dithered colors is one thing — now how do you use them? Realize that there is not always a physical LEGO color for a given dithered color. If you are aiming to represent actual LEGO colors and pieces, refer to the codes on the color chart. The colors are laid out logically based on the way LEGO has mixed colors to create new ones, and we include each LDraw color code for your reference.

Defining Custom Colors

Aside from the range of dithered colors, MLCad allows you to define custom colors. *Be warned that these custom colors are not compatible with any other LDraw tool, and are NOT recommended if you want to use your model outside of MLCad.* The reason for this is simple: Custom colors are defined via an unused swatch on the palette. This swatch has an assigned LDraw number, but no programs recognize it. You can define this color inside of MLCad, but once you save your file, MLCad cannot and does not encode that color in a way any other application will understand.

MLCad Defines Custom Colors Internally

When you define custom colors, MLCad stores the data inside its own configuration settings. Even if you open up your file in MLCad on a separate computer, the custom color definitions will be lost because they are not encoded in your LDraw file. If you want to play with strange colors in MLCad, custom colors are fine, but *be warned that you cannot use custom colors defined in MLCad outside of MLCad.*

Using the Custom Color Dialog

To define a custom color, find an unused swatch in the MLCad color palette and select it. You can access the custom color dialog (shown in Figure E-2) by clicking **Custom . . .** or **Define . . .** at the bottom of the color palette. From this window, you can set an HSV (Hue, Saturation, Volume) or RGB (Red, Green, Blue) color value manually, or you can pick your color using your mouse cursor.

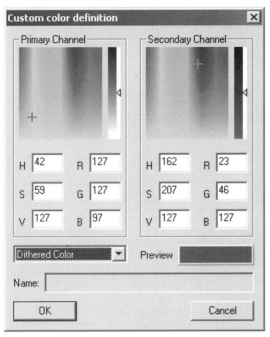

Figure E-2: You can define custom colors with the Custom Color dialog, but those colors can only be used within your local copy of MLCad and in not in any other LDraw-compatible software.

The drop-down menu to the left of the preview panel lets you select what type of color you want to create. You can choose a Solid Color, Transparent Color, Dithered Color, a 24-bit Solid or a 24-bit Transparent. If you select a dithered color, you can pick both the Primary and Secondary channel colors. The preview panel will show you the result. Any other selection disables the secondary channel, because you are not mixing colors at that point.

F

LDRAW PRIMITIVES REFERENCE

This appendix is copied from the Primitives Reference located at http://www.ldraw.org/ library/tracker/ref/primref/. This reference was compiled by Chris Dee, and is reprinted with permission.

This page is a source of reference for the LDraw primitives in the \LDraw\p directory. Primitives are defined as highly re-usable components of LEGO parts modelled for LDraw.

They serve several purposes:

- To speed up parts authoring by providing a library of components which can be incorporated into several parts.

- To allow rendering software to make substitutions for curved components.

Within this reference material the available primitives are categorized into :

Rectilinear primitives
Two-dimensional
Three-dimensional
Curved primitives
Two-dimensional
Three-dimensional
Special purpose primitives
Technic axle primitives
Technic bush primitives
Technic connector primitives
Stud primitives
Miscellaneous primitives

Each section contains an overview of the characteristics common to all primitives within that category. Primitives are grouped into classes within each category, one class of primitive serving a similar purpose at different sizes or resolutions. For each class of primitive, a brief description of the purpose of the primitive is provided, with notes on its coordinate origin, default size, and rules for scaling. A list of the available primitives is shown.

An understanding of the orientation of the coordinate axes is essential for authoring a part for LDraw. For reference, the axes and their directions are shown in this diagram.

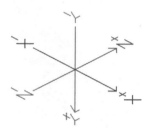

Rectilinear Primitives

These rectilinear elements may be scaled in the {x}, {y}, and {z} dimensions to make elements of any size. For example:

```
1 16  0 0 0  40 0 0  0 1 0  0 0 20  rect.dat
```

would generate a 80LDU × 40LDU rectangle in the {x,z} plane.

Although the default orientation of the rect.dat primitive is in the {x,z} plane, the LDraw language allows for this to be transformed:

```
1 16  0 0 0  0 1 0  40 0 0  0 0 20  rect.dat
```

would generate a 80LDU × 40LDU rectangle in the {y,z} plane;

```
1 16  0 0 0  40 0 0  0 0 20  0 1 0  rect.dat
```

would generate a 80LDU × 40LDU rectangle in the {x,y} plane.

Two-Dimensional

`rect.dat`

Rectangle with all edges

This primitive represents a rectangle in the {x,z} plane and the four edges that bound it. Its origin is at its center and by default has a size of 2 LDU in each of the {x} and {z} dimensions. To avoid matrix arithmetic problems in some renderers, the third dimension ({y} in the default orientation) should be given a scaling factor of 1:

`1 16  0 0 0  5 0 0  0 1 0  0 0 20  rect.dat`

Three-Dimensional

`box.dat`

Cuboid with all faces and edges

This primitive is used to define a cuboid. Its origin is the centre of the cuboid and by default has a size of 2 LDU in each of the three dimensions.

`box5.dat`

Cuboid with 5 faces and all edges

This primitive represents a cuboid missing the top {-y} face. Its origin is the center of the (missing) top face and by default has a size of 2 LDU in each of the {x} and {z} dimensions and 1 LDU in the {y} dimension.

`box4.dat`

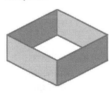

Cuboid with 4 faces and all edges

This primitive represents a cuboid missing the top {-y} and bottom {+y} faces. Its origin is the center of the (missing) top face and by default has a size of 2 LDU in each of the {x} and {z} dimensions and 1 LDU in the {y} dimension.

`box4t.dat`

Cuboid with 4 adjacent faces and all edges

This primitive represents a cuboid missing the top {-y} and front {-z} faces, but with all its edges. Its origin is the center of the (missing) top face and by default has a size of 2 LDU in each of the {x} and {z} dimensions and 1 LDU in the {y} dimension.

`box4-1.dat`

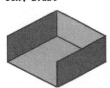

Cuboid with 4 adjacent faces missing 1 edge

This primitive represents a cuboid missing the top {-y} and front {-z} faces and the edge between those faces. Its origin is the center of the (missing) top face and by default has a size of 2 LDU in each of the {x} and {z} dimensions and 1 LDU in the {y} dimension.

`box4-4a.dat`

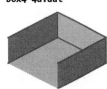

Cuboid with 4 adjacent faces missing 4 edges

This primitive represents a cuboid missing the top {-y} and front {-z} faces and all the edges of the missing front face. Its origin is the center of the (missing) top face and by default has a size of 2 LDU in each of the {x} and {z} dimensions and 1 LDU in the {y} dimension.

`box4-7a.dat`

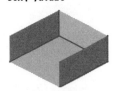

Cuboid with 4 adjacent faces missing 7 edges

This primitive represents a cuboid missing the top {-y} and front {-z} faces and all the edges of both missing faces. Its origin is the center of the (missing) top face and by default has a size of 2 LDU in each of the {x} and {z} dimensions and 1 LDU in the {y} dimension.

`box3#8p.dat`

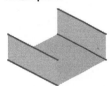

Cuboid with 3 adjacent faces missing 8 edges (in 2 parallel groups)

This primitive represents a cuboid missing the top {-y} and left {-x} and right {+x} faces and excludes all the edges of the left {-x} and right {+x} faces. Its origin is the center of the (missing) top face and by default has a size of 2 LDU in each of the {x} and {z} dimensions and 1 LDU in the {y} dimension.

`box3u2p.dat`

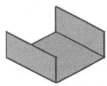

Cuboid with 3 adjacent faces missing 2 edges

This primitive represents a cuboid missing the top {-y} and left {-x} and right {+x} faces and excludes the top edges of the left {-x} and right {+x} faces. Its origin is the center of the (missing) top face and by default has a size of 2 LDU in each of the {x} and {z} dimensions and 1 LDU in the {y} dimension.

`tri3.dat`

Right-angled triangular prism with 3 faces

This primitive represents a triangular prism missing the top {-y} and bottom {+y} faces but including all edges. The left {-x} and front {-z} faces are perpendicular. Its origin is the right-angle corner of the (missing) top face and by default has a size of 1 LDU in each of the {x} and {z} dimensions and 1 LDU in the {y} dimension.

Curved Primitives

LDraw represents curved surfaces as polygons. For circular components, two series of primitives are provided.

All the circular primitives are orientated in the {x,z} plane with their origins at the center of the circle and a default radius of 1 LDU. Primitives are provided for complete circles and for commonly used fractions of a complete circle. Where the naming convention includes a prefix of the form *n-f*, this indicates the fraction (n/f) of the circle drawn by the primitive. Where this fraction is less than an entire circle, the primitive starts at {+x,0} and progresses in a counterclockwise direction when viewed from above {-y}.

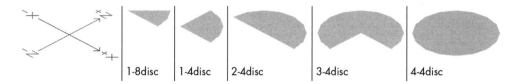

To avoid matrix arithmetic problems in some renderers, the third dimension ({y} in the default orientation) of two-dimensional primitives should be given a scaling factor of 1.

LDraw circles are normally formed of 16-sided polygons (hexadecagons) — the *regular resolution*. For larger elements, where scaling-up of hexadecagons would give too angular an appearance, a series of *high resolution* primitives based on a 48-sided polygon are available. These may also be used for parts not well suited to a 16-fold symmetry.

These circular elements may be scaled by the same factor in both the {x} and {z} dimensions to make circular elements of greater or less than 1 LDU radius. For example:

```
1 1  0 0 0  3 0 0  0 1 0  0 0 3  4-4edge.dat
```

would generate a circle in the {x,z} plane with a diameter of 3 LDU.

They may also be scaled asymmetrically in the x and z dimension to make ellipses.

Although the default orientation is in the {x,z} plane, the LDraw language allows for these to be transformed:

```
1 16  0 0 0  0 1 0  3 0 0  0 0 3  4-4edge.dat
```

would generate a circle in the {y,z} plane;

```
1 16  0 0 0  3 0 0  0 0 3  0 1 0  4-4edge.dat
```

would generate a circle in the {x,y} plane.

Two-Dimensional

n-fedge.dat

Circular line segment

This suite of primitives are used for edges that comprise an entire or part circle.

Currently available primitives:
Regular resolution (n-f) : 1-8, 1-4, 3-8, 2-4, 3-4, 4-4
High resolution (n-f) : 1-8, 1-4, 2-4, 4-4

---

n-fdisc.dat

Circular disc sector

This suite of primitives are used for surfaces which comprise an entire or part circle.

Currently available primitives:
Regular resolution (n-f) : 1-8, 1-4, 2-4, 3-4, 4-4
High resolution (n-f) : 1-8, 1-4

---

n-fndis.dat

Inverse of circular disc sector

This suite of primitives pad their matching n-fdisc.dat primitives out to the bounding square. They are used to integrate circular elements into rectilinear elements.

Currently available primitives:
Regular resolution (n-f) : 1-8, 1-4, 2-4, 3-4, 4-4
High resolution (n-f) : 1-8, 1-4

---

n-fringr.dat n-frinrr.dat ringr.dat ringrr.dat

Circular ring segment

This suite of primitives are used to generate circular rings or part rings. The numeric suffix r in the filename indicates the inner radius of the ring — the outer radius is 1 LDU greater.

For example, an n-fring4 primitive would create a ring with an inner radius of 4 LDU and an outer radius of 5 LDU.
These primitives are currently undergoing a change of naming convention — names of the form "ringrr.dat" are being migrated to "4-4rinrr.dat". Backward compatibility will be maintained.

Currently available primitives:
Regular resolution:
   r=1 : 1-4ring1, 2-4ring1, ring1
   r=2 : 1-4ring2, 2-4ring2, ring2
   r=3 : 1-8ring3, 1-4ring3, 2-4ring3, ring3
   r=4 : 1-4ring4, 2-4ring4, ring4
   r=5 : 1-4ring5, 4-4ring5
   r=6 : 1-4ring6, 4-4ring6
   r=7 : 1-4ring7, 2-4ring7, ring7
   r=8 : 1-4ring8
   r=9 : 1-4ring9
   r=10 : 1-4rin10, ring10
   r=11 : 1-4rin11
   r=12 : 1-4rin12
   r=15 : 1-4rin15
   r=16 : 1-4rin16
   r=18 : 1-4rin18
   r=19 : 1-4rin19
   r=38 : 1-4rin38
   r=39 : 1-4rin39
High resolution:
   r=3 : 1-4ring3
   r=4 : 1-4ring4
   r=9 : 1-4ring9, 4-4ring9

Three-Dimensional

n-fcyli.dat

Circular cylinder

This suite of primitives are used to generate cylinders or part cylinders.

Currently available primitives:

Regular resolution (n-f) : 1-8, 1-4, 3-8, 2-4, 3-4, 4-4

High resolution (n-f) : 1-8, 1-4, 4-4

---

n-fcyls.dat n-fcyls2.dat

1-4cyls.dat

Circular cylinder truncated by an angled plane

These primitives are used to generate cylinders or part cylinders that are truncated by a plane that is not perpendicular to the axis of the cylinder. The default angle of the plane is 45 degrees.

These primitives are notoriously difficult to describe; the user is encouraged to experiment in order to gain a full understanding of their geometry. One technique is to open the primitive in L3Lab, increase the zoom to 12800, deselect BFC and choose random colors. Selecting Show Axes also helps.

3-8cyls.dat

The 2-4cyls.dat primitive is orientated with the perpendicular bounding plane at the top {-y} and the angled bounding plane at the bottom {+y}. In common with the other 2-4xxxx.dat primitives, the {+z} semicircle is represented. By default the truncation plane is y=x-1. The 1-4cyls.dat and 3-8cyls.dat primitives are sub-sections of 2-4cyls.dat.

2-4cyls.dat

1-4cyls2.dat

The 1-4cyls2.dat primitive is orientated with the perpendicular bounding plane at the top {-y} and the angled bounding plane at the bottom {+y}. Unlike the other 1-4xxxx.dat primitives, the {-x,+z} quadrant is represented. The truncation plane is y=x.

The following observations may help:

1-4cyls and 1-4cyls2 are complements. If you put them together correctly, they make a 1-4cyli.

```
1 15  0 0 0  1 0 0  0 1 0  0 0 1  1-4cyls.dat
1 14  0 1 0  -1 0 0  0 -1 0  0 0 1  1-4cyls2.dat
```

1-4cyls2 is actually hidden in 2-4cyls, which could be constructed like this:

```
1 16  0 0 0  1 0 0  0 1 0  0 0 1  1-4cyls.dat
1 16  0 0 0  0 0 -1  0 1 0  1 0 0  1-4cyli.dat
1 14  0 1 0  1 0 0  0 1 0  0 0 1  1-4cyls2.dat
```

Currently available primitives:

Regular resolution: 1-4cyls, 3-8cyls, 2-4cyls, 1-4cyls2

n-fconr.dat
n-fconrr.dat

Circular cone

This suite of primitives are used to generate circular cones or part cones. The numeric suffix *r* in the filename indicates the inner radius of the cone — the outer radius is 1 LDU greater. By default, the cone is 1 LDU high in the {+y} dimension with the origin at the center of the outer diameter. For example, an n-fcone4 primitive would create a cone with an inner radius of 4 LDU and an outer radius of 5 LDU.

Currently available primitives:
Regular resolution:
  r=0 : 1-4con0, 4-4con0
  r=1 : 1-4con1, 4-4con1
  r=2 : 1-4con2
  r=3 : 1-4con3
  r=4 : 1-4con4
  r=5 : 1-4con5
  r=6 : 1-4con6
  r=7 : 1-4con7
  r=8 : 4-4con8
  r=10 : 1-4con10
  r=12 : 4-4con12
High resolution:
  r=5 : 1-4con5
  r=6 : 1-4con6
  r=9 : 1-4con9

---

tffirrrr.dat
tfforrrr.dat
tffqrrrr.dat

inner

outer

Circular torus

This suite of primitives are used to generate circular torus sections. By default, all these primitives produce a torus with a major radius of 1 LDU, so they typically need to be scaled up in the {x} and {z} dimensions. The second and third characters of the filename *ff* denotes the sweep of torus, as an inverse fraction (01=1/1, 02=1/2, 04=1/4, 08=1/8, 16=1/16, 48=1/48). The fourth character denotes the section of a torus (i=inner, o=outer, q=tube — the combination of two inner and two outer sections). The last four characters of the file name *rrrr* denote the tube radius in LDU (1333=0.1333, 3333=0.3333).

Currently available primitives:
Regular resolution:
  Inner: t04i0625, t04i1304, t04i1333, t04i3333
  Outer: t04o0625, t04o1304, t04o1333, t04o1429, t04o1538, t04o3333
  Tube: t04q0625

These diagrams (based on an idea of Paul Easter's) illustrate the relationship of the torus primitives to the major (orange) and minor (black) radii.

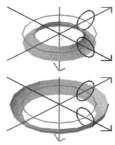

tube

inner (top); outer (bottom)

1-4ccyli.dat

Circular torus (obsolete)

This single primitive predates the creation of the tffirrrr and tfforrrr suite of primitives and will not be augmented. It produces a quarter torus with a major radius of 2.5 LDU and a tube radius of 1 LDU.
It should be superceded by t04q4000.dat.

---

n-fsphe.dat

Sphere section

Currently available primitives:
Regular resolution: 1-8, 2-8, 4-8, 8-8

1-8sphc.dat

Spherical corner

This primitive represents one octant (eighth) of a sphere, centered at the origin {0,0,0} of radius 1.414, and truncated by the sides of a cube with a vertex at {1,1,1}. The boundaries of the resulting surface are circular and fit with 1-4edge.dat.

Special Purpose Primitives

Technic Axle Primitives

These primitives represent various components of the Technic axle and its matching hole. They are orientated in the {x,z} plane. Except where noted below, and in the {y} dimension only, these primitives should not be scaled.

`axle.dat`

Technic axle section

This primitive comprises a 1 LDU long section of Technic axle, including its ends. It may be scaled in the {y} dimension to produce an axle of any length.

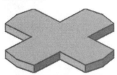

`axleend.dat`

Technic axle end

This primitive is used to produce the "plus-shaped" cross-section of a Technic axle.

`axlehole.dat`

Technic axle hole — closed

This primitive produces a Technic axle hole with the disc ends and all sides. It may be scaled in the {y} dimension as necessary.

`axleho11.dat`

Technic axle hole — reduced

This primitive produces a Technic axle hole with reduced teeth, including the disc ends and all edges. It may be scaled in the {y} dimension as necessary.

`axlehol4.dat`

Technic axle hole — open one side

This primitive produces a Technic axle hole with one side omitted. It may be scaled in the {y} dimension as necessary.

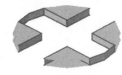

`axlehol5.dat`

Technic axle hole — open two opposite sides

This primitive produces a Technic axle hole with two opposite sides omitted. It may be scaled in the {y} dimension as necessary.

axlehol6.dat

Technic axle hole tooth

This primitive represents one tooth of a Technic axle hole.

---

axleho10.dat

Technic axle hole — tooth surface

This primitive produces the cross-section of a Technic axle hole.

---

axlehol7.dat

Technic axle hole — sides

This primitive comprises a 1 LDU-long section of the outer sides of a Technic axle. It may be scaled in the {y} dimension.

---

axlehol8.dat

Technic axle hole perimeter

This primitive is comprises a 1 LDU-long section of Technic axle without any ends. It may be scaled in the {y} dimension as necessary.

---

axlehol2.dat

Technic axle hole — side edges

This primitive produces the edges at the ends of the "plus-shaped" Technic axle.

---

axlehol3.dat

Technic axle hole — tooth outer edges

This primitive produces the outer edges of the teeth of a Technic axle hole.

---

axlehol9.dat

Technic axle hole — tooth inner edges

This primitive produces the inner edges of the "plus-shaped" Technic axle.

Technic Bush Primitives

These primitives are various representations of the castellated Technic bush. They are orientated with the Technic axle hole along the {y} axis and should not be scaled.

`bushlock.dat`

Technic 16-tooth castellation — long teeth

`bushloc2.dat`

Technic 16-tooth castellation — regular teeth

`bushloc3.dat`

Technic 16-tooth castellation — regular teeth indented

This primitive differs from bushloc2.dat in that the teeth are indented in four places to allow it to fit between four adjacent studs.

`bushloc4.dat`

Technic 16-tooth castellation — short teeth

`steerend.dat`

Curved end to Technic 1x*n* plate with 16-tooth castellation on underside

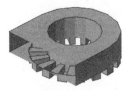

Technic Connector Primitives

These primitives are used to construct Technic connector pegs. They are orientated with the Technic axle hole along the {y} axis and should not be scaled.

connect.dat

Technic connector — long with collar

connect2.dat

Technic connector — long without collar

connect3.dat

Technic connector — short with collar

connect4.dat

Technic connector — short without collar

connect5.dat

Technic connector — long with collar and slot

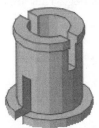

---

peghole.dat

Technic connector hole

---

peghole2.dat

Technic connector hole — 180 degrees

---

peghole3.dat

Technic connector hole — 90 degrees

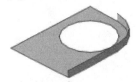

Stud Primitives

Each stud*xxx* primitive described below has a matching low resolution stu2*xxx* primitive, used by the fast-draw mode of renderers; these stu2*xxx* primitives should never be used in part files. An additional primitive (studline.dat) is used to substitute a single line for studs by the super fast-draw mode of LDraw.

All stud primitives are modelled with the solid disc uppermost, even if designed for the underside of parts (stud3, stud4, stud8). The origin is at the base of the stud and they extend for 4 LDU in the {-y} dimension. To use for the underside, the primitives need to be inverted in the {y} dimension, viz.

```
1 16   x y z   1 0 0   0 -1 0   0 0 1   stud3.dat
```

These primitives should not be scaled in the {x} or {z} dimensions. Ideally, they should not be scaled in the {y} dimension either, to allow the accurate substitution of chamfered studs by high-quality renderers (but this rule is flouted in the regular brick files, where stud4 is scaled by 5 to generate the underside tube).

stud.dat	**Regular stud**	
studp01.dat	**Regular stud with white dot**	
studel.dat	**Regular electric stud** The electric contact is on the {-x} {-z} corner.	
stud2.dat	**Hollow stud**	
stud2a.dat	**Hollow stud without edge around base**	
stud3.dat	**Small underside stud** This is modelled with the solid disc uppermost. To use for the underside of plates, reference like this: `1 16 x y z 1 0 0  0 -1 0  0 0 1 stud3.dat`	
stud3a.dat	**Small underside stud without edge around base** See note for stud3.	

stud4.dat

Ring underside stud

See note for stud3.

stud4a.dat

Ring underside stud without edge around base

See note for stud3.

stud5.dat

Scala stud

stud6.dat

Truncated hollow stud

For use on 2x2 round parts where the stud does not hang over the edge of the part.

stud6a.dat

Truncated hollow stud without edge around base

For use on 2x2 round parts where the stud does not hang over the edge of the part.

stud7.dat

Duplo hollow top stud

`stud8.dat`

Miscellaneous Primitives

This section comprises miscellaneous primitives that do not fit neatly into any of the other categories. All are highly specialized and represent components of parts that fit together with each other or other standard parts. As such these primitives are not intended to be re-sized.

`arm1.dat`

Cylindrical arm two-fingered hinge

This primitive produces a two-fingered hinge component of the cylindrical arm with a radius of 10 LDU.

`arm2.dat`

Cylindrical arm three-fingered hinge

This primitive produces a three-fingered hinge component of the cylindrical arm with a radius of 10 LDU.

`clip3.dat`

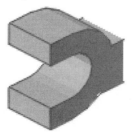

Horizontal clip

This primitive produces a horizontal clip designed to hold an 8 LDU diameter pole or rod.

clip4.dat

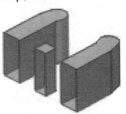

Vertical clip — three-fingered

This primitive produces a three-fingered vertical clip designed to hold an 8 LDU diameter pole or rod.

h1.dat

Two-fingered plate hinge

This primitive produces the standard two-fingered hinge with a depth of 8 LDU — the thickness of a plate.

h2.dat

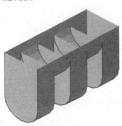

Three-fingered plate hinge

This primitive produces the standard three-fingered hinge with a depth of 8 LDU — the thickness of a plate.

INDEX

C

-c (color) switch, 218, 220
-ca (camera angle) switch, 215, 216, 279–80
CAD system, 30
Calista mosaic, 136
callouts, 309
camera angle box, 235
camera angle switch, 215, 216, 279–80
camera coordinates switch, 215–16
Camera Globe Position, 215–16
camera Look_At switch, 215, 216
camera option, 235
camera window switch, 214
cameras, 262–64, 279–80
 compensating for position of, 293
 focusing, 294
 lens angle, 279–80
 location for, 263
 orthographic, 293
 placement of, 223
Car model, 101–4
Cartesian coordinate system, 215–16
categories, 42–54
 bar, 42
 baseplate, 42
 bracket, 43
 brick, 43
 cone, 44
 confusion with, 39
 cylinder, 44
 Duplo, 44
 electric, 45
 glass, 45
 hinge, 46
 minifig, 46
 minifig accessory, 47
 panel, 47
 plate, 48
 round, 44
 slope, 48
 slope brick, 48
 sticker, 49
 supports, 49
 Technic, 50
 tile, 50
 train, 51
 turntable, 51
 tyre, 52
 wedge, 52
 wheel, 53
 window, 45, 53
 windscreen, 54
 wing, 54
-cc (camera coordinates) switch, 215, 216
-cc checkbox, 216
-cc option, 215
Central Perspective, 93
-cg (camera globe position) switch, 215
chains, 161, 163
City Creator, 271, 273, 282
-cla (camera Look_At) switch, 215, 216
Clague, Kevin, 121, 144, 209, *242*
Clark, Jennifer, 4
clip part, 40
clips, 70
clock feature, 259
clone bricks, 375
clubs, 377, 378
CNET reviewer of web hosts, 326
Coarse Grid setting, 63
Cobbs, Travis, 324
code, scaling, 278–79
Colefax, Chris, 271, 273, 282, 284
Color Bar, 77
color definitions, 292–93
color number, *152*
Color Number dialog, 87
Color option, 237
color palette, 22, 86–87
Color Palette window, *24*
color switch, 218, 220
colored lights, 224
colors
 calculating the difference between objects, 291
 changing, 133
 managing, 86–88
 in part files, 360
 scaling, 252
 setting the option, 139
 settings for, 139
 using multiple, 134
colors.inc file, 272
command line options, 260
commands, 152–53. *See also* names of specific commands
Comment Block, 366

dragging, 25
    from the Parts List window, 59
    from the Parts Preview window, 60
    from Select Part Dialog window, 60
    selecting multiple parts by, 83
Draw to selected part only box, 23
Draw to Selection option, 81, *82*
drawing, inside edge, 360
Drop Center Flatcar kit instructions,
    316–18
drop-down list, 32
drop-down menus. *See* specific types of
    drop-down menus
Duplo category, 44

E

e-cable.ldr file, 154–55
eBay auctions, 380
edge lines
    checking for in images, 303
    framing, 358–59
    settings, 243
edges
    drawing inside of, 360
    smoothing, 257
Edit Bar, 76
edit functions, 371
Edit Mode, 77, 78
editing
    groups, 35
    LSynth parts, 148
    manually, 169
    models, 81–86
editing environment, 360
editors tool, 7
electric cable, 146, 150, 154–56
electric category, 45
Electric Technic Micromotor Pulley,
    123
electric wires, 144
Element Bar, 26, *65*, 76
elements
    See also flexible elements
    in LDraw parts library, 29–31
Enable Minimum Distance, 243
endpoints, 166
-enp (exclude non-POV code) switch,
    218, 220
Enter Pos + Rot option, 62

Enter Position and Rotation dialog
    box, *61*
environment, editing, 360
Eriksson, Tore, 375
error message, L3Lab, *359*
errors, diagram, 129–30
European events, 379
events, 378–79
examples (finished building
    instructions), 310–19
    fan-created building instructions,
        312–19
        BricWorx 1001 Freight Train
            Instruction Book, 312–15
        Milton Train Works 1004 Drop
            Center Flatcar, 315–19
    official LEGO building
        instructions, 310–12
Excel to LDAO Editor, *371*
exclude non-POV code switch, 218, 220
.exe updates, downloading, 56
exercises
    L3P program, 222–26
    MegaPOV program, 295–97
    POV-Ray program, 279–88
        creating first scene, 281–83
        creating space backdrop,
            283–85
        creating space scene, 285–86
        using fog, 286–87
        using spotlights and
            pointlights, 280–381
        using the camera, 279–80
Expert Bar, 77
exploded views technique, 187
exporting sub-models, 110
extended colors, 86
Extras Bar, 77

F

-f (floor) switch, 218
fade_distance keyword, 269
fade_power keyword, 269
Fair Play document, 326–27
Fake color setting, 139
fan-created building instructions,
    312–19
    BricWorx 1001 Freight Train
        Instruction Book, 312–15

synthesizing flexible elements, 165
system of tools, 5. *See also* LDraw

T

T values, 130
tank treads, 144
Technic Axle Flexible, 146, 152
Technic Bush, 122
Technic category, 50
Technic Fiber Optic Cable, 146
Technic Link Chain, 147
Technic Link Tread, 147
Technic parts, 72
Technic Pin, 122
Technic Pulley Large, 123
Technic Shock Absorber, 127–28
Technic tubing, 144
Technic Wedge Belt Wheel, 123
Technica online reference, 380
techniques, building. *See* building
    techniques
terminology of parts, 40–41
Terragen program, 274–76, 278, 279
testing, building instructions, 192
text, adding to images, 309
text editors, 371
text file, saving to, 92
themed folders, 173
Thickness value, 124
tightness, 266
tiles category, 50
tires, aligning, 69, 72
titles, for parts, 31, 358
Toolbar Icons, 256
toolbars, 22
    on MLCad program, 76–77
    organizing, 76
    selecting multiple parts on, 82
tools
    for creating simple parts, 354
    development, 6, 8
    installing LDraw, 12
Total Height, 127, 128, 129–30
Towball Socket part, 41
trains category, 51
transformation matrix, 369–70
transformation term, 356
translate keyword, 282
translation, 356

transparent colors, 88
transparent parts, 220
triangles, *370*
Tschager, Willy, 188
tubing, 144–65
    adding LSynth constraint parts to
        MLCad program, 148–49
    parts LSynth creates, 145–47
    using LSynth, 156–58
        band-style parts, 159
        constraints, 151–52
        creating multiple flexible
          elements, 153–54
        electrical cable, 154–56
        file size warning, 158
        finding color name, 152–53
        hose-style parts, examples of,
          156–58
        placing constraints, 159–65
Turntable part, 41
turntables category, 51
Typical Installation Option, 14
tyres category, 52

U

United Kingdom events, 379
Ulhmann, Lutz, 208
undo feature, *64*
Ungroup icon, 85
ungrouping parts, 84, 85
United States events, 378–79
unused colors, 87
updating Parts List, 57
use colors checkbox, 142
Use Cylinder checkbox, 124
use LGEO library switch, 218, 220
use plates checkbox, 134, 139, 142
user groups, 377
utilities tool, 8

V

van der Molen, Jaco, 123–24, 378
version statement, 290
vertex term, 356
vertex winding, 370
vertical configuration, 60
View Bar, 76

GETTING STARTED WITH LEGO TRAINS

by JACOB H. MCKEE

Learn to build LEGO Trains, from setting up train tracks to building custom freight cars. Jacob McKee, an authority on LEGO Trains, teaches basic building techniques and shares some of his most fascinating and original train designs.

2003, 128 PP., 4-COLOR THROUGHOUT, $24.95 ($37.95 CDN)
ISBN 1-59327-006-2

JIN SATO'S LEGO MINDSTORMS

The Master's Technique

by JIN SATO

LEGO legend Jin Sato shares his way of thinking about designing MINDSTORMS robots. Includes illustrated, step-by-step instructions to build five fascinating robots, including the author's famous robotic dog, MIBO.

2002, 364 PP., $24.95 ($37.95 CDN)
ISBN 1-886411-56-5

STEAL THIS COMPUTER BOOK 3

What They Won't Tell You About the Internet

by WALLACE WANG

This offbeat, non-technical book looks at what hackers do, how they do it, and how you can protect yourself. The third edition of this bestseller adopts the same informative, irreverent, and entertaining style that made the first two editions a huge success.

"If this book had a soundtrack, it'd be Lou Reed's Walk on the Wild Side.*" —InfoWorld*

2003, 348 PP., $24.95 ($37.95 CDN)
ISBN 1-59327-000-3

THE OFFICIAL BLENDER GAMEKIT

Interactive 3D for Artists

by TON ROOSENDAAL *and* CARSTEN WARTMANN

Design interactive 3D animation and games with Blender, a fast, powerful, and free 3D creation suite. The book comes with a CD-ROM containing 10 playable and editable Blender game demos, all files needed for the tutorials, and copies of Blender, complete with game engine, for all platforms.

2003, 360 PP., $34.95 ($59.95 CDN)
ISBN 1-59327-004-6

HACKING THE XBOX

An Introduction to Reverse Engineering

by ANDREW "BUNNIE" HUANG

A hands-on guide to hardware hacking and reverse engineering using Microsoft's Xbox™ video game console. Covers basic hacking techniques such as reverse engineering and debugging, as well as Xbox security mechanisms and other advanced hacking topics. Includes a chapter written by the Electronic Frontier Foundation (EFF) about the rights and responsibilities of hackers.

2003, 288 PP, $24.99 ($37.99 CDN)
ISBN 1-59327-029-1

PHONE:

1 (800) 420-7240 OR
(415) 863-9900
MONDAY THROUGH FRIDAY,
9 A.M. TO 5 P.M. (PST)

FAX:

(415) 863-9950
24 HOURS A DAY,
7 DAYS A WEEK

EMAIL:

SALES@NOSTARCH.COM

WEB:

HTTP://WWW.NOSTARCH.COM

MAIL:

NO STARCH PRESS
555 DE HARO STREET, SUITE 250
SAN FRANCISCO, CA 94107
USA

UPDATES

Visit **http://www.nostarch.com/vlego.htm** for updates, errata, and other information.

CD-ROM LICENSE AGREEMENT FOR
VIRTUAL LEGO®

Read this Agreement before opening the CD package. By opening this package, you agree to be bound by the terms and conditions of this Agreement.

MLCad Keyboard Shortcuts and Hotkeys

Key/Combination	Function
A	Rotate selection clockwise around Y-Axis
C	Open Color Palette (edits color of selected part(s))
I	Insert new part (opens Insert Part dialog)
P	Modify selected part (opens Insert Part dialog)
Left Arrow	Move selected part(s) –X
Right Arrow	Move selected part(s) +X
Up Arrow	Move selected part(s) +Z
Down Arrow	Move selected part(s) –Z
Home	Move selected part(s) +Y
End	Move selected part(s) –Y
Page Up	Select previous line in file
Page Down	Select next line in file + Zoom In - Zoom Out
CTRL + Page Up	Select first line in file
CTRL + Page Down	Select last line in file
CTRL + C	Copy selected parts
CTRL + V	Paste selected parts
CTRL + X	Cut selected parts
CTRL + D	Duplicate selected parts
Del	Delete selected parts
CTRL+ M	Modify selected parts (opens Insert Part dialog)
F1	Help
F2	Toggle View Mode
F3	Toggle Edit Mode
F4	Toggle Move Mode
F5	Toggle Size Mode
CTRL + Left Arrow	Rotate selection counter-clockwise around Y-Axis
CTRL + Right Arrow	Rotate clockwise around Y-Axis
CTRL + Up Arrow	Rotate counter-clockwise around X-Axis
CTRL + Down Arrow	Rotate clockwise around X-Axis
CTRL + Home	Rotate counter-clockwise around Z-Axis
CTRL + End	Rotate clockwise around Z-Axis
CTRL + G	Group selected objects (prompts for group name)
CTRL + SHIFT + G	Snap selections to grid

Source: MLCad Help Documentation

LPub Meta-Commands

Here are some common LPub meta-commands you can use. Insert meta-commands into LDraw files as comments — use the **abl** icon in MLCad, or when hand-editing an LDraw file, use line-type 0.

Command/Syntax	Function
PLIST BEGIN SUB <filename.ldr>	Groups parts so they appear as one part in the LPub Parts List images — this is useful for parts that have sub-components which exist as separate LDraw part files (e.g. 3404.dat and 3403.dat). Use PLIST END to contain this group.
PLIST BEGIN IGN	Parts contained between PLIST BEGIN IGN and PLIST END will be ignored (not rendered) in the LPub parts list images.
PLIST END	Ends PLIST groups.
BI BEGIN GRAYED	Forces POV-Ray to gray-out parts contained within this command.
BI END	Ends BI groups.

LDraw Language Reference

Line Type 0 - Comment/Meta-command

Line Type

`0 // I am a comment!`

Comment Tag - used to explicitly denote a comment as opposed to a meta-command.

Line Type 1 - Part/Subfile Command

Line Type   Part Location   Transform Matrix Y   Part File

$1\ C\ x\ y\ z\ x_x\ y_x\ z_x\ x_y\ y_y\ z_y\ x_z\ y_z\ z_z\ $ `3001.DAT`

Color Number   X   Z

Line Type 2 - Line

Line Type   Endpoint 1

$2\ C\ x_1\ y_1\ z_1\ x_2\ y_2\ z_2$

Color Number   Endpoint 2

Line Type 3 - Triangle

Line Type   Vertex 1   Vertex 3

$3\ C\ x_1\ y_1\ z_1\ x_2\ y_2\ z_2\ x_3\ y_3\ z_3$

Color Number   Vertex 2

Line Type 4 - Quad

Line Type   Vertex 1   Vertex 3

$4\ C\ x_1\ y_1\ z_1\ x_2\ y_2\ z_2\ x_3\ y_3\ z_3\ x_4\ y_4\ z_4$

Color Number   Vertex 2   Vertex 4

Line Type 5 - Conditional Line

Line Type   Endpoint 1   Control Point a

$5\ C\ x_1\ y_1\ z_1\ x_2\ y_2\ z_2\ x_a\ y_a\ z_a\ x_b\ y_b\ z_b$

Color Number   Endpoint 2   Control Point b